Stem Cells in Clinical Applications

Series Editor
Phuc Van Pham

More information about this series at http://www.springer.com/series/14002

Phuc Van Pham
Editor

Pancreas, Kidney and Skin Regeneration

Springer

Editor
Phuc Van Pham
Laboratory of Stem Cell Research and Application
University of Science
Vietnam National University
Ho Chi Minh City, Vietnam

ISSN 2365-4198 ISSN 2365-4201 (electronic)
Stem Cells in Clinical Applications
ISBN 978-3-319-55686-4 ISBN 978-3-319-55687-1 (eBook)
DOI 10.1007/978-3-319-55687-1

Library of Congress Control Number: 2017934047

Printed on acid-free paper

This Springer imprint is published by Springer Nature
The registered company is Springer International Publishing AG
The registered company address is: Gewerbestrasse 11, 6330 Cham, Switzerland

Preface

Diabetes mellitus (DM), commonly referred to as diabetes, is a group of metabolic diseases in which high blood sugar levels are present over a prolonged period. According to the World Health Organization, the number of diabetic patients rapidly increased for 10 years and will continuously increase in the coming years. DM is divided into two main groups: type 1 DM and type 2 DM. Type 1 DM is considered as an autoimmune disease in which beta cells in the pancreas are destroyed, while type 2 DM is considered as the insulin resistance of body cells. In recent years, both type 1 and type 2 DMs are demonstrated as results of abnormal immune systems in patients. DM can cause many complications including diabetic ketoacidosis, nonketotic hyperosmolar coma, or death. Kidney failure and foot ulcers are two serious complications of DM. Generally, pancreas, kidney, or skin injuries are resulting from injuries of the immune system and tissue aging.

In the current years, stem cell therapies for pancreas, kidney, and skin regeneration are moved to the clinic with exciting results. A majority of clinical applications used hematopoietic stem cells and mesenchymal stem cells. Hematopoietic stem cells can replace the abnormal immune system with a new normal immune system, while mesenchymal stem cells are used with two strategies including immune modulation and cell replacement.

This volume of *Stem Cells in Clinical Applications* book series with the title *Pancreas, Kidney, and Skin Regeneration* aims to provide an updated invaluable resource for advanced undergraduate students, graduate students, researchers, and clinicians in stem cell applications for pancreas, kidney, and skin regeneration. The book with *13* chapters covers almost the present applications of stem cells in DM treatment and kidney and skin regeneration. Chapters one through six discuss pancreas regeneration – diabetes mellitus treatment. Chapters seven and eight discuss stem cell therapies for kidney regeneration. And chapters nine through thirteen discuss wound/ulcer healing by stem cell transplantation as well as skin regeneration using stem cells.

We are indebted to our authors who graciously accepted their assignments and who have infused the text with their energetic contributions. We are incredibly thankful to the staff of Springer Science + Business Media that published this book.

Ho Chi Minh City, Vietnam Phuc Van Pham

Contents

Part III Skin Regeneration

Contributors

Işil Albeniz Department of Biophysics, Istanbul University, Istanbul Faculty of Medicine, Capa-Istanbul, Turkey

Division of Plastic Surgery, Royal Free Hospital, NHS Foundation Trust Hospital, London, UK

Aurora Almadori UCL Division of Surgery & Interventional Science, London, UK

Department of Plastic Surgery, Royal Free Hospital NHS Foundation Trust Hospital, London, UK

Saeid Amini-Nik Sunnybrook Research Institute, Toronto, Canada

Laboratory Medicine and Pathobiology, University of Toronto, Toronto, Canada

Department of Surgery, Division of Plastic Surgery, University of Toronto, Toronto, Canada

Alexander E. Berezin Cardiology Unit, Internal Medicine Department, State Medical University, Zaporozhye, Ukraine

Oltjon Cobani ProEstetica Hospital, Research Center in Regenerative Medicine and Surgery of Romanian Academy of Medical Sciences, Bucharest, Romania

Hakan Darici Department of Histology and Embryology, Istinye University Hospital, Faculty of Medicine, Topkapı-Istanbul, Turkey

Arefeh Jafarian Immunology, Asthma and Allergy Research Institute (IAARI), Tehran University of Medical Sciences, Tehran, Iran

Marc G. Jeschke Institute of Medical Science, University of Toronto, Toronto, Canada

Sunnybrook Research Institute, Toronto, Canada

Department of Surgery, Division of Plastic Surgery, University of Toronto, Toronto, Canada

Department of Immunology, University of Toronto, Toronto, Canada

Ross-Tilley Burn Centre, Sunnybrook Health Sciences Centre, Toronto, Canada

Dana Jianu ProEstetica Hospital, Research Center in Regenerative Medicine and Surgery of Romanian Academy of Medical Sciences, Bucharest, Romania

Stefan Jianu ProEstetica Hospital, Research Center in Regenerative Medicine and Surgery of Romanian Academy of Medical Sciences, Bucharest, Romania

Deepak M. Kalaskar UCL Division of Surgery & Interventional Science, London, UK

Erdal Karaöz Department of Histology and Embryology, Istinye University Hospital, Faculty of Medicine, Topkapı-Istanbul, Turkey

Lei Lei Department of Endocrinology, Aerospace Center Hospital, Peking University Affiliate, Beijing, China

Andrew M.L. Lever Department of Medicine, Addenbrooke's Hospital, University of Cambridge, Cambridge, UK

Yuanjie Mao Department of Medicine, Cleveland Clinic, Akron General Medical Center, Akron, OH, USA

Leyla Türker Şener Department of Biophysics, Istanbul University, Istanbul Faculty of Medicine, Capa-Istanbul, Turkey

Masoud Soleimani Department of Hematology, Faculty of Medical Sciences, Tarbiat Modares University, Tehran, Iran

Umang G. Thakkar Department of Regenerative Medicine and Stem Cell Therapy and Department of Pediatrics
G.R. Doshi and K.M. Mehta Institute of Kidney Diseases & Research Centre (IKDRC)-
Dr. H.L. Trivedi Institute of Transplantation Sciences (ITS), Ahmedabad, Gujarat, India

Hargovind L. Trivedi Department of Regenerative Medicine and Stem Cell Therapy and Department of Pediatrics
G.R. Doshi and K.M. Mehta Institute of Kidney Diseases & Research Centre (IKDRC)-
Dr. H.L. Trivedi Institute of Transplantation Sciences (ITS), Ahmedabad, Gujarat, India

Department of Nephrology and Transplantation Medicine
G.R. Doshi and K.M. Mehta Institute of Kidney Diseases & Research Centre (IKDRC)-
Dr. H.L. Trivedi Institute of Transplantation Sciences (ITS), Ahmedabad, Gujarat, India

Jian Tu Prince of Wales Clinical School, University of New South Wales, Sydney, NSW, Australia

Sicong Tu Prince of Wales Clinical School, University of New South Wales, Sydney, NSW, Australia

Phuc Van Pham Laboratory of Stem Cell Research and Application, University of Science, Vietnam National University, Ho Chi Minh City, Vietnam

Aruna V. Vanikar Department of Regenerative Medicine and Stem Cell Therapy and Department of Pediatrics
G.R. Doshi and K.M. Mehta Institute of Kidney Diseases & Research Centre (IKDRC)-Dr. H.L. Trivedi Institute of Transplantation Sciences (ITS), Ahmedabad, Gujarat, India

Department of Pathology, Laboratory Medicine, Transfusion Services and Immunohematology
G.R. Doshi and K.M. Mehta Institute of Kidney Diseases & Research Centre (IKDRC)-Dr. H.L. Trivedi Institute of Transplantation Sciences (ITS), Ahmedabad, Gujarat, India

Yusef Yousuf Institute of Medical Science, University of Toronto, Toronto, Canada

Sunnybrook Research Institute, Toronto, Canada

Paolo Zamboni Department of Morphology, Surgery and Experimental Medicine, Section of Translational of Medicine and Surgery, University of Ferrara, Ferrara, Italy

Sant'Anna University Hospital, Vascular Diseases Center and Institute of Translational Medicine and Surgery, University of Ferrara, Ferrara, Italy

Jing Zhao Department of Medicine, Addenbrooke's Hospital, University of Cambridge, Cambridge, UK

Ilaria Zollino Department of Morphology, Surgery and Experimental Medicine, Section of Translational of Medicine and Surgery, University of Ferrara, Ferrara, Italy

Part I
Pancreas Regeneration

Chapter 1
Stem Cell Transplantation in Diabetes Mellitus Type I and Type II

Sicong Tu and Jian Tu

1.1 Introduction

Diabetes mellitus is a chronic disease with a worldwide prevalence of 346 million patients in 2011 (WHO 2015). It is estimated that death related to diabetes will be doubled from 2005 to 2030 and diabetes is one of the leading causes of blindness, renal failure, and leg amputation (WHO 2015). Diabetes has been mainly classified into type I and type II according to its pathogenesis and manifestations; however, other less common types of diabetes have been identified, including monogenic maturity-onset diabetes of the young (Stride and Hattersley 2002) and mitochondrial diabetes (Mazzaccara et al. 2012) or late-onset autoimmune diabetes. In type I diabetes, genetic susceptibility and environmental factors are the initial triggers of an autoimmune process, which eventually results in destruction of all β-cells, insulin depletion, and lifelong insulin-dependent diabetes (Cnop et al. 2005). Type II diabetes generally occurs later in life, a sequence of insulin resistance, glucose intolerance, β-cell exhaustion, and subsequently ever-increasing insufficient insulin production that leads to insulin depletion and development of hyperglycaemia and diabetes (Stumvoll et al. 2010; Kahn et al. 2001).

In type I diabetes, patients are thoroughly dependent on exogenous insulin from the very early stage; however, treatment of type II diabetes could initially be diet and exercise and then go on medications that increase insulin sensitivity or insulin secretion (Stumvoll et al. 2010). However, most of the type II diabetic patients finally become completely dependent on exogenous insulin. Therefore, insulin therapy constitutes the main therapeutic approach for type I diabetes and advanced type II diabetes.

S. Tu • J. Tu (✉)
Prince of Wales Clinical School, University of New South Wales,
Sydney, NSW 2052, Australia
e-mail: s.tu@neura.edu.au; j.tu@unsw.edu.au

P.V. Pham (ed.), *Pancreas, Kidney and Skin Regeneration*, Stem Cells in Clinical Applications, DOI 10.1007/978-3-319-55687-1_1

Current insulin or any other pharmacologic therapy is neither curative of the disease nor preventive of its complications (Krentz and Bailey 2005). The reason for this is the lack of normal glucose homeostasis (tight regulation of plasma glucose concentration between 70 and 99 mg/dl). Normally, glucose homeostasis is achieved through various nutritional (Gannon and Nuttall 2010), neuronal, and hormonal (insulin, glucagon, glucagon-like peptide-1 (GLP-1), and glucose-dependent insulinotropic polypeptide (GIP) (Yabe and Seino 2011; Tolhurst et al. 2009; Lee et al. 2012)), nutrients (leucine alone or in a combination with glutamine), vasoactive intestinal polypeptide (VIP), and 5-hydroxytryptamine (Takahara et al. 2001) and pituitary adenylate cyclise-activating polypeptide (PCAP) (Ahrén 2008) mechanisms.

Insulin secretion from β-cells is by far the most important glucose regulator. During postprandial periods, increase in the blood glucose levels normally leads to increments in the secretion of insulin in a biphasic pattern, starting within few minutes of stimulus and continuing as long as high level of blood glucose is sensed (Tolhurst et al. 2009; Lee et al. 2012). Therefore, glucose-induced insulin secretion (GIIS) from the β-cells is the main mechanism for glucose homeostasis. Insulin increases glycogenesis, lipogenesis, and glucose transportation into insulin-sensitive cells (muscle and adipose cells) and most importantly reduces hepatic glucose output. All these effects result in the reduction of blood glucose level.

Ideally, treatment of diabetes must mimic natural blood glucose haemostasis to prevent diabetic complications. As depicted in Fig. 1.1, several therapeutic modalities with the potential of glucose sensing have been proposed for this purpose (Levine and Leibowitz 1999). Insulin pumps were designed to release insulin according to its sensing of glucose. However, none of these mechanical devices has been completely successful (Weinzimer et al. 2008).

Another therapeutic option is the implantation of live insulin-producing cells as they are capable of sensing glucose levels and secrete insulin accordingly. Edmonton protocol was devised and offered temporary insulin independence after transplantation of cadaveric human pancreatic islet cells (Vrochides et al. 2009; Ryan et al. 2005). In another study, this therapeutic modality could eliminate the need for insulin administration after transplantation of insulin-producing cells and result in better glycaemic control (Robertson 2004). Also, Shapiro et al. reported seven cases of type I diabetes becoming completely insulin independent after islet transplantation (Shapiro et al. 2000). However, islet transplantation has at least two noticeable obstacles: short-term insulin independence (Ryan et al. 2005) and shortage of donor islets (Hakim 2002).

Based on our understanding of how islet transplant can correct hyperglycaemia, many researchers are focusing on developing the new potentially curative cell-based therapies for diabetes. Stem cells from various sources (Table 1.1) have been differentiated into insulin-producing cells (IPCs). Therefore, an excellent source for replacing donor insulin-producing cells has been revealed. Furthermore, stem cell therapy is expected to produce insulin-producing cells that are functionally similar to normal pancreatic beta cells. Therefore, IPCs should be able to produce and store a sufficient amount of insulin and secrete it appropriately in response to the stimuli. If completely accomplished, diabetes can be cured, and its complications can be prevented.

Fig. 1.1 Therapeutic approaches for diabetes

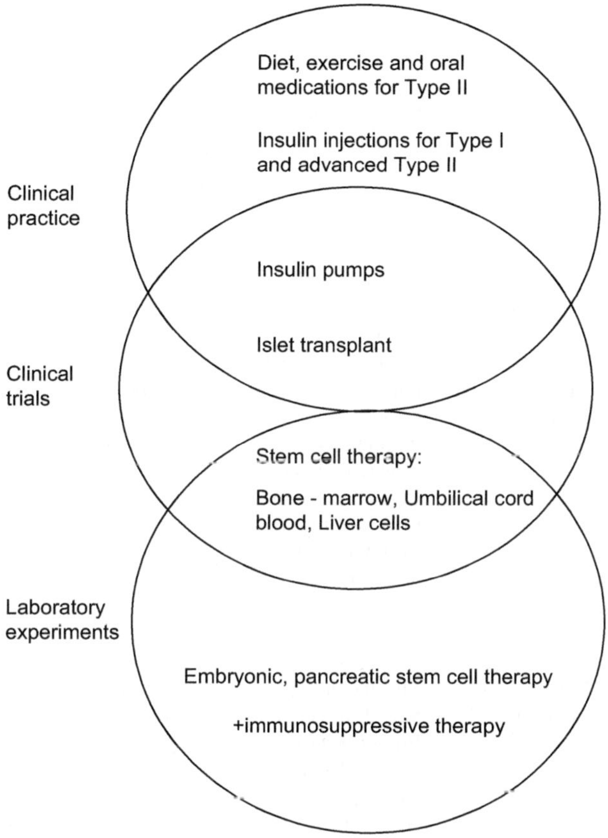

To improve this potential treatment from an idea to a routinely administered clinical practice, an increasing number of investigations on various types of stem cell have been performed. Insulin-producing cells have been derived from embryonic stem cells (ESCs), mesenchymal stem cells (MSCs), umbilical cord blood stem cells (UCBSCs), bone marrow stem cells (BMSCs), adipose tissue-derived stem cells (ATDSCs), pancreatic progenitor cells, and various adult cell lines after genetic manipulation, such as pancreatic or hepatic cell lines. In this chapter, we systematically review stem cells that have been studied for potential cell therapy of diabetes and the current limitations of stem cell transplantation in diabetes.

1.2 Stem Cells

Stem cells have unique self-renewal and multilineage differentiation capacities that could be harnessed for therapeutic purposes. "Self-renewal" is the ability to undergo cycles of mitotic division while maintaining the same undifferentiated state as the parent cell (Weissman 2000). A potential therapeutic strategy could be to aim for the

Table 1.1 Stem cells differentiable to IPCs and their characteristics

Stem cells	Advantages	Disadvantages	References
Embryonic stem cells	Pluripotentiality Self-renewal capacity	Ethical issues Limited resource Immaturity Immune rejection Tumorigenicity	Soria (2000), Kahan (2003) Houard (2003), Shiroi (2002) Lumelsky (2001), Shi (2005) Miyazaki 2004 Blyszczuk (2003) Segev 2004 Rezania 2012, Bose (2012)
Mesenchymal stem cells	Pluripotentiality Self-renewal capacity Immunomodulatory	Immaturity Tumorigenicity	Santamaria (2011) Duffy (2011) Rackham (2011) Ito (2010)
Umbilical cord blood stem cells	Large donor resource Availability Low rejection risk Low ethical issues Low infection risk Complication reversal	Immaturity Tumorigenicity Technical issues	Reddi (2010) Sun (2007) Yoshida (2005) Ende (2004) Tsai (2012) He (2011) Wang (2004) Broxmeyer (1989) Koblas (2005) Parekh (2009) Haller (2008)
Bone marrow stem cells	Multipotentiality, Self-differentiation Immunomodulatory Standardized technique Availability	Immaturity Accessibility Tumorigenicity Safety issues	Ianus (2003) Karnieli (2007) Hess (2003) Wu (2007) Iskovich (2007)
Adipose stem cells	Multipotentiality, Accessibility Availability	Immaturity Tumorigenicity	Mohamad Buang (2012) Li (2012), Kim (2010) Bassi (2012), Lee (2008)
Adult pancreatic stem cells Pancreatic ductal cells Pancreatic acinar cells	Partial differentiation Low ethical issues Low immune rejection	Difficult cell isolation and differentiation	Li (2003), Kim (2004) Xia (2009), Inada (2008) Ramiya (2000) Bonner-Weir (2000) Noguchi (2003), Yatoh (2007), Bulotta (2002), Zhou (2008)
Hepatic cells	Similar to β-cells Low immune rejection Low tumorigenicity Low ethical issues	Difficult cell propagation and expansion	Simpson (1997), Sapir (2005) Zalzman (2005), Liu (2007) Zheng (2007)

replacement of these stem cells; when an injury occurs, they can reconstitute the tissue. However, not all tissues in the body contain a resident pool of stem cells, a notable example being the endocrine pancreas (Grompe 2012). Damage to pancreatic islets can only be treated by islet transplantation. The another characteristic of stem cells is the cell potency that refers to the capacity of stem cells being able to differentiate into different cell types. As such, their therapeutic potentials for diabetes are clear to see, and efforts are ongoing to produce insulin-producing cells for transplantation in diabetes.

1.2.1 Embryonic Stem Cells for Transplantation in Diabetes

Embryonic stem cells (ESCs) are the cells derived from the preimplantation embryo at the blastocystic stage. They have the capability to grow indefinitely in cultures and always keep their potential to differentiate into adult cells of various lineages (Evans and Kaufman 1981). Twenty years after introduction of ESCs from mice embryos (Evans and Kaufman 1981; Martin 1981) and 2 years after isolation of human ESCs from early human embryo by Thompson et al. (Thomson 1998), Soria et al. published the first report of producing IPCs from mouse embryonic stem cells in 2000 (Soria et al. 2000). Using a cell-trapping system, they induced the mouse embryonic stem cells into insulin-expressing cells that could normalize glucose levels after transplantation under the kidney capsule (Soria et al. 2000). Some investigators later showed that ESCs can spontaneously differentiate into IPCs after forming embryoid bodies (Kahan et al. 2003; Houard et al. 2003; Shiroi et al. 2002). Although this differentiation is highly important showing ESCs' differentiation potential, because of low concentration of IPCs – 1% of the cells expressed insulin – these IPCs have yet provided practical benefit.

Since that time, several protocols have been designed to improve IPC yield. These protocols are mainly based on the sequential activation and inhibition of signalling pathways through adding respective ligands or biologically active small chemical compounds to the cell culture media. In fact, the aim is to make culture condition as similar as possible to what happens in the embryonic development of beta cells (Fig. 1.2). Several studies have reported successful differentiation of embryonic stem cells into insulin-producing cells by improving culture condition and/or genetic manipulation (Soria et al. 2000; Kaczorowski et al. 2002; Lumelsky et al. 2001). Lumelsky et al. demonstrated a five-step protocol: production of a highly enriched population of nestin-positive cells from embryoid bodies (EBs) by placing the EBs into a serum-free medium (ITSFn) eliminating many other cell types from the original EBs and then expanding nestin-positive cells in the presence of a mitogen "basic fibroblast growth factor (bFGF)" in N2 serum-free medium, followed by mitogen withdrawal to promote cessation of cell division and differentiation and adding a B27 supplement and nicotinamide to improve the yield of pancreatic endocrine cells (Lumelsky et al. 2001). While they showed that the cells expressed insulin, later Rajagopal et al. and Sipione et al. questioned their results and claimed it was related to the uptake of insulin from the cell culture media rather than de novo insulin biosynthesis (Rajagopal et al. 2003; Sipione et al. 2004). A simpler three-step protocol was then introduced by Shi et al., which used a combined induction using activin A, all-trans retinoic acid, basic fibroblast growth

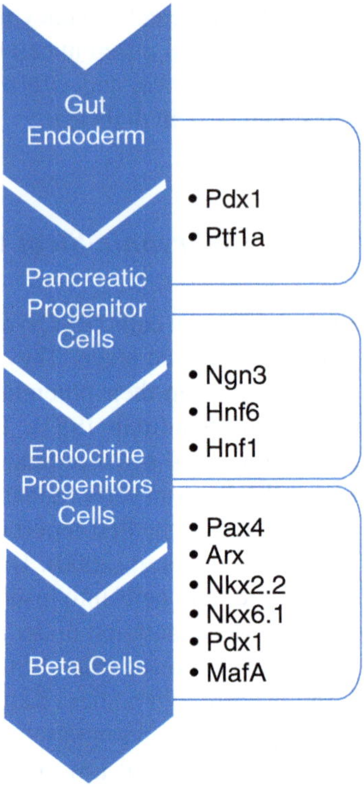

Fig. 1.2 Stages and differentiation factors during embryonic development of β-cells

factor, N2 and B27 supplements, laminin, and nicotinamide for differentiating murine ESCs into insulin-producing cells in 2 weeks (Shi et al. 2005).

The currently accepted protocol for differentiating ESCs into insulin-producing cells includes three successive steps: differentiation of ESCs via embryoid bodies into definitive endoderm, formation of committed pancreatic progenitors, and directed differentiation into the pancreatic lineage (Schroeder et al. 2006). The first step is to develop definitive endoderm, expressing Pdx1 (Chen et al. 2012), Sox 17 (Fu et al. 2011), FoxA2 (Fu et al. 2011), brachyury protein, CXCR4, and Cerberus. In vitro spontaneous differentiation of ESCs can be facilitated by adding growth factors, such as activin A, tumour growth factor beta, fibroblast growth factor 10, and retinoic acid (Assady et al. 2001; Brolén et al. 2005). For the first step of differentiation, TGF-beta/activin signalling family is necessary to develop endoderm progenitors, as they positively regulate Wnt signalling that is essential for the pancreas formation (Sulzbacher et al. 2009). For the next step, developing committed pancreatic progenitors, protocols basically use growth and extracellular matrix factors, such as laminin, nicotinamide (Vaca et al. 2003), and insulin (Schroeder et al. 2006). Moreover, addition of cyclopamine, retinoic acid, and TGF 10 and inhibition of Hedgehog signalling are shown to assist pancreatic formation from endoderm (Hebrok 2003; DiIorio et al. 2002). The final step, which is differentiation into endocrine cells, can be promoted by inactivation of Notch signalling pathways

(Apelqvist et al. 1999), up-regulation of Ngn3 (Rukstalis and Habener 2009), and pancreatic polypeptide expression (Gu et al. 2003).

In addition to the above-mentioned protocols and other two-dimensional (2D) culture techniques, some investigators introduced three-dimensional (3D) protocols. In these methods, basically embryoid bodies produce all three germinal layers, and pancreas eventually differentiates from the endodermal layer (Ku et al. 2004). By this approach, 3D protocols can optimize stem cell differentiation into insulin-producing cells by improving the cell-cell interactions and extracellular interactions of the differentiated cells. The improved extracellular interactions can optimize insulin production and secretion. Although, there is no experiment comparing the efficiency of 2D and 3D cell cultures, it is necessary to optimize both 2D and 3D techniques to achieve the best outcomes of stem cell differentiation into insulin-producing cells.

To improve the results of differentiation, some investigators have combined cell culture modulation protocols with genetic manipulation. The genetic manipulation is focused on the delivery of several growth and transcription factors involved in beta cell differentiation. The most important of transcription factors in beta cell differentiation are Pdx1 and Pax4. Pdx1 is expressed throughout the epithelium of early pancreatic buds. Pdx1 is one of the most important intrinsic factors that signal the region of the gut with pancreatic fate (Ohlsson et al. 1993). In adult beta cells, it is only expressed in beta cells and plays an important role in regulation of important genes such as insulin (Chen et al. 2012), glucose transporter 2 (Waeber et al. 1996), and glucokinase (Watada et al. 1996) which are significant in pancreatic beta cell identity and function. Like Pdx1, mice lacking Pax4 do not have beta cells and become diabetic (Sosa-Pineda et al. 1997). Genetic manipulation of Pdx1 and Pax4 has resulted in improvement in the development of IPCs (Miyazaki et al. 2004; Blyszczuk et al. 2003). Blyszczuk et al. demonstrated that constitutive expression of Pax4 and to a lesser extent Pdx1 significantly facilitates the development of IPCs (Blyszczuk et al. 2003). Miyazaki et al. illustrated that Pdx1 induction in mouse ES cells could enhance the transcription of insulin and other beta cell markers (Miyazaki et al. 2004).

Differentiation methods for human ESCs (hESCs) are based on similar methods used in animal ESCs (Segev et al. 2004). For instance, Segev et al. used a multistep differentiation method for hESCs, which is basically similar to the previously described protocols for the differentiation of animal ESCs (Segev et al. 2004). They initially cultured and plated embryoid bodies in the insulin-transferrin-selenium-fibronectin medium, followed by culturing EBs in the medium supplemented with N2, B27, and basic fibroblast growth factor (bFGF) and then reduced the glucose concentration in the medium, withdrew bFGF, and added nicotinamide, resulting in high yield of IPCs. Although these IPCs respond to glucose stimulation in insulin secretion, they were considered immature as they co-expressed glucagon or somatostatin (Segev et al. 2004). This protocol could result in 70% IPC production that is considerably higher than spontaneous differentiation of hESCs into IPCs as previously reported by Assady et al. (2001). A limitation of Segev's study is that the function of differentiated IPCs had yet assessed in vivo. In addition, IPCs underwent apoptosis. Therefore, it is necessary to determine the survival of IPCs in vivo (Segev et al. 2004). Rezania et al. initially differentiated hESCs into a highly enriched Pdx1 population of IPCs in vitro and allowed them further maturation towards pancreatic beta cells in vivo (Rezania et al. 2012). Transplantation of the graft resulted in an increase in insulin secretion over time and improved glucose tolerance in the immune-deficient diabetic mice (Rezania et al. 2012). In another study, Bose et al. applied a five-step protocol and 3D culture in amino acid-rich media supplemented with anti-hyperglycaemic hormone glucagon-like peptide-1 (Glp1), analog liraglutide, and exendin 4 (Bose et al. 2012). They harvested 3D clusters that constituted 65% of IPCs and transplanted them into diabetic NOD/SCID mice, achieving reduced blood glucose level 96 days after transplantation (Bose et al. 2012). Advances in protocols developed for differentiating hESCs have paved the way for what is widely expected to be the first human embryonic stem cell-based clinical trials (Fig. 1.3). The US Food and Drug Administration (FDA) has accepted ViaCyte, Inc.'s (*San Diego, California*, USA) Investigational New Drug Application (IND) for its VC-01™ candidate cell replacement therapy to treat type I diabetes on the *August 19, 2014* (Laikind 2014). ViaCyte's VC-01 product consists of pancreatic progenitor cells, called PEC-01™ cells, which are derived from a proprietary human embryonic stem cell line, encapsulated in ViaCyte's proprietary Encaptra® device. When implanted under the skin, the PEC-01 cells are specifically designed to mature and further differentiate into insulin-producing beta and other endocrine cells that regulate blood glucose in a manner similar or identical to the islets. The company plans to promptly initiate a phase 1/2 clinical trial which it believes to be the first ever clinical evaluation of a stem cell-derived islet replacement therapy for the treatment of patients with type I diabetes. The study is designed to demonstrate the effectiveness of the VC-01 product in replacing the lost endocrine function. The trial would also evaluate clinically relevant secondary endpoints related to the need for administration of pharmaceutical insulin to control type I diabetes and the incidence of hypoglycaemia, a common side effect associated with pharmaceutical insulin usage (Laikind 2014).

All these studies confirmed that ESCs offer the potential to be a source of insulin-producing cells for the transplantation treatment of diabetes mellitus type I and the late-stage insulin-dependent diabetes type II. Pluripotentiality and capability to unlimitedly grow in vitro are the two most important advantages of ESC transplantation for the treatment of diabetes. Application of various protocols, ESCs can be differentiated into many different types of cell lines. Therefore, it seems that differentiation of ESCs into IPCs might be easier than using other types of stem cells. In addition, they can infinitely grow in culture media. In contrast to other sources of IPCs, such as adult hepatic and pancreatic cell lines, there is no need to induce cell growth in vitro.

The limitations in the application of ESCs are ethical controversies, lack of long-term studies, and technical issues. In some countries, clinical application of ESCs is not ethically accepted. The most important technical problems include difficulties with in vitro work, immaturity of stem cells, tumorigenesis in vivo, and immune rejection post-transplantation. Culturing techniques of embryonic stem cells are

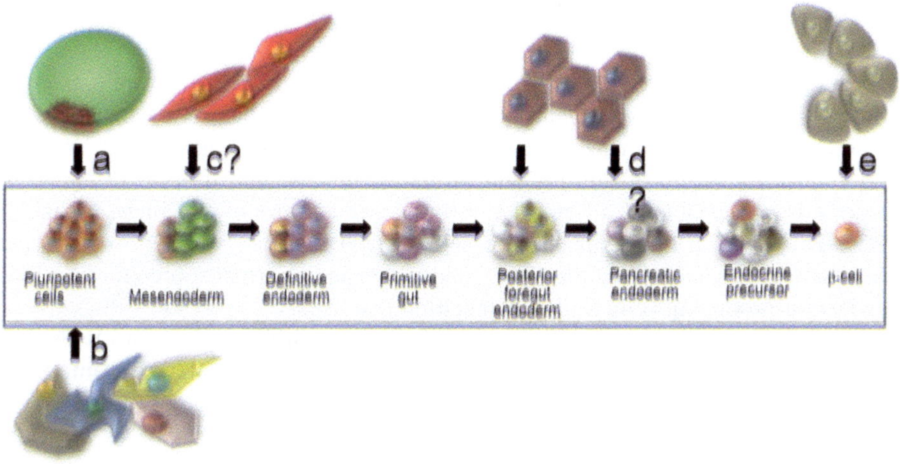

Fig. 1.3 Insulin-producing β-cell differentiation approaches. Conventional means of pluripotent stem cell procurement entail the extraction and culture of the inner cell mass of preimplantation blastocysts. These are known as embryonic stem (ES) cells (**a**). An alternative source derived from terminally differentiated somatic cells makes use of reprogramming technology. These reprogrammed cells are called iPS cells (**b**) and are functionally equivalent to ES cells. Pluripotent cells of both origins can be differentiated along the pancreatic β-cell lineage following multistep protocols that recapitulate the key biological events of pancreatic development. The sequential activation of genes such as Sox17 (definitive endoderm), Pdx1 (pancreatic progenitors), and Ngn3 (pancreatic endocrine progenitors) is commonly observed in these methods. The robust expression of terminal differentiation markers, such as MafA, insulin, glucokinase, or GLUT2, is observed only sporadically in most reported methods, as the efficiency of the last differentiation step is limited thus far. This is why the state of the art calls for the transplantation of pancreatic progenitors rather than terminally differentiated cells. Although some types of adult MSCs could hypothetically be compared with the mesendodermal precursors (which undergo an epithelial-to-mesenchymal transition before becoming definitive endoderm), the ability of such cells to seamlessly join the β-cell differentiation cascade at that juncture is debatable (**c**). Recently published evidence suggests that endodermal progenitor cells residing in the adult liver and extrahepatic biliary tree may also retain the ability to become either hepatocytes or pancreas, depending on the nature of the ECM onto which they are seeded. If isolated and expanded, these could be potentially used for β-cell differentiation either at the foregut endoderm-equivalent or the pancreatic endoderm-equivalent level (**d**). Last, reprogramming techniques previously developed for iPS cell generation have also been successfully used to transdifferentiate (or laterally reprogram) either liver or pancreatic acinar tissue (**e**) directly into insulin-producing β-cells (Orlando et al. 2013, 2014; reuse licence number 3620791332648 including electronic rights obtained from BMJ Publishing Group Ltd.)

more complex than other types of stem cells. The current cell culture technique has yet produced functionally mature insulin-producing cells suitable for transplantation. As mentioned above, the final stage of ESCs differentiation occurs in vivo after transplantation (Bose et al. 2012). There is a risk of development of teratoma under post-transplant immunosuppressed condition (Cells et al. 2005). ESCs or the IPCs differentiated from ESCs express human leukocyte antigens; therefore, they might be rejected by the host immune system. All these issues and the possible solutions are discussed in details below in Sect. 1.3 "Discussion".

1.2.2 Mesenchymal Stem Cells for Transplantation in Diabetes

In contrast to embryonic stem cells, mesenchymal stem cells (MSCs) are present in adult tissues after birth. These pluripotent stromal cells can be derived from various tissues, such as bone marrow and peripheral blood, skin and adipose tissue, muscle and connective tissue of uterine, and probably all postnatal tissues. In addition, these cells are capable of differentiating into various types of cell lines, including IPCs. They have a high capacity for self-renewal and have immunosuppressive characteristics. All these features have made them an important source for developing insulin-producing cells for transplantation in diabetes.

MSCs have been isolated from various tissues and differentiated into IPCs. For instance, Santamaria et al. differentiated human endometrial stromal cells into IPCs using a non-transfection protocol. Five weeks after transplantation of differentiated cells under the kidney capsules of diabetic mice, normoglycaemia was achieved (Santamaria et al. 2011). For the purpose of deriving MSCs, umbilical blood cord (Bieback et al. 2004), bone marrow (Karnieli et al. 2007), and adipose tissue (Timper et al. 2006) are probably the most commonly used tissues. Thus, they are discussed in more details below.

The application of MSCs in treatment of diabetes type I has demonstrated immunomodulation capacity (Duffy et al. 2011) which can both prevent further autoimmune destruction of the beta cells – the main mechanism for development of diabetes type I – and also improve IPCs' function after transplantation. Furthermore, co-transplantation of mesenchymal stem cells with islet cells into kidney capsule can improve both morphology and function of transplanted islets (Rackham et al. 2011). These cells can surround islet cells and increase vascularization into the graft (Ito et al. 2010).

When MSCs are cocultured with IPCs, they can serve as a gene delivery vehicle for the greater benefit of glucose-responsive insulin secretion. For example, islets were cocultured with MSCs expressing human HGF and human IL-1Ra resulting in protection of insulin-producing cells from apoptotic death, improvement in maintaining 3D structures, and better insulin secretion (Wu et al. 2011).

1.2.3 Umbilical Cord Blood Stem Cells for Transplantation in Diabetes

Cord blood after birth is another important source of stem cells that are able to be differentiated into IPCs (Van Pham et al. 2014). In comparison to other sources, they have yet adequately studied and can be assumed as a new and emerging field of investigation (Reddi et al. 2010; Sun et al. 2007; Yoshida et al. 2005; Ende et al. 2004; Tsai et al. 2012; He et al. 2011). It has been confirmed that umbilical cord blood contains mesenchymal stem cells (MSCs) (Bieback et al. 2004; Wang et al. 2004; Van Pham et al. 2014), endothelial stem cells (ESCs) (Murohara et al. 2000), haematopoietic stem cells (Broxmeyer et al. 1989), and a not yet fully characterized population of multipotent cells (Koblas et al. 2005) that possess the potential to differentiate into IPCs in vivo (Parekh et al. 2009). Since the expression of key pancreatic transcription factors, such as pdx1, ngn3, isl1, brn4, and pax6 on freshly derived cells from umbilical cord blood, it has been hypothesized that a subset of "pancreas-committed" cells can be obtained from the umbilical cord blood (Parekh et al. 2009; Van Pham et al. 2014). Transfection of pdx1 mRNA into umbilical cord blood-derived mesenchymal stem cells successfully differentiated them into insulin-producing cells (Van Pham et al. 2014).

Although no pancreatic progenitors were directly identified in umbilical cord blood, it is still possible that umbilical cord blood stem cells differentiate into pancreatic cells. Several studies demonstrated that infusion of umbilical cord blood can reverse diabetes type I in human and animal (Reddi et al. 2010; Ende et al. 2004; Haller et al. 2008). For instance, transplantation of human UCBSCs into type I diabetic mice reduced their blood glucose levels (Ende et al. 2004). In another study, transfusion of human umbilical cord blood into type I diabetic children reduced their daily requirement of insulin dose (Reddi et al. 2010). However, not all evidence is supportive. In Bleich's study, infusion of umbilical cord blood did not improve hyperglycaemia in type I diabetic children (Bleich 2009).

Once a specific stem cell population has been isolated; there are three approaches for obtaining mature IPCs from umbilical cord blood. The first approach is the direct transplantation of not manipulated umbilical cord blood stem cells for the purpose of subsequent differentiation within recipient's pancreas. The second method is the differentiation of umbilical cord blood stem cells in vitro before transplantation of them into recipient's liver. The third way is the expansion of umbilical cord blood stem cells prior to their differentiation and transplantation (Koblas et al. 2005). All these approaches have been investigated, and various degrees of success have been achieved. For instance, Sun et al. derived colonies from the umbilical cord blood that had the potential for differentiation into pancreatic beta cells (Sun et al. 2007). These colonies could secrete both insulin and C-peptide. However, the glucose-responsive insulin production and secretion has yet assessed (Sun et al. 2007). Tsai et al. applied a three-stage protocol to differentiate mesenchymal stem cells derived from human umbilical cord blood to IPCs. They achieved an increasing C-peptide release after glucose challenge of these IPCs in vitro and reduction of blood glucose level of diabetic rats after transplantation of differentiated cells into their livers (Tsai et al. 2012).

UCBSCs have several important advantages over other types of insulin-producing stem cells. Probably the most important of which is that they are easily collectable and freely available. Having rich number of haematopoietic stem cells, they can be considered as a valuable alternative to bone marrow (BM)-derived stem cells. The collection technique of UCBSCs is easier and less invasive than bone marrow-derived stem cells, and the harvested UCBSCs can be cryopreserved and kept in stem cell banks for future application. Another advantage of other types of insulin-producing stem cells is that because of harvesting of UCBSCs at the beginning of life, they are less imposed to viral and other biological factors. Therefore, they have a lower risk of viral transmission from the donor to the recipient (Behzad-Behbahani et al. 2005). UCBSCs share common features with hESCs (Sun et al. 2007; Zhao et al. 2006; Chao et al. 2008). In general, they have less ethical issues, larger potential donor resources, and lower risk of immune rejection (Chao et al. 2008). Another important feature of umbilical cord blood-derived stem cells for the treatment of diabetes is that they contain other types of stem cells, such as endothelial progenitor cells, that can be used for the treatment of diabetic neuropathy (Naruse et al. 2005). Application of umbilical cord blood-derived stem cells has also been reported to reverse other diabetic complications, but it has yet cleared whether this important benefit is a direct result of improvement in glucose metabolism or from the trophic influence of UCBSCs.

While the application of UCBSCs has been declared no side effects and can be safely given to children suffering from type I diabetes, the graft-versus-host disease (GVHD) related to the contamination of immune cells in the grafts and transplanted to the recipients is still a considerable complication. In comparison to bone marrow stem cells, umbilical cord blood-derived stem cells have lower number of T cells, higher immaturity of B cells, and lower function of dendritic cells, therefore, much lower risk of GVHD possibly caused by UCBSCs than that of BMSCs. The risk of GVHD related to UCBSCs can be reduced to minimal by depleting immune cells from the grafts using advanced purification techniques. Large-scale clinical application of umbilical cord blood-derived stem cells to treat diabetes requires us to improve the efficacy of graft preparations. It is highly necessary to improve current methods of expansion of stem cells, selection of the appropriate subpopulations of cells, and administration of the proper dose of umbilical cord blood-derived stem cells. Therefore, further investigation must be performed to bring this promising source of stem cell therapy to the bedsides of diabetic patients.

1.2.4 Bone Marrow Stem Cells for Transplantation in Diabetes

Adult bone marrow contains pluripotent cells with the potential to be differentiated into various mesenchymal lineage cells, including insulin-producing cells. Ianus et al. first declared that bone marrow-derived cells can produce insulin-producing cells that have the characteristic of glucose-dependent and incretin-enhanced insulin secretion (Ianus et al. 2003). Karnieli et al. differentiated bone marrow mesenchymal stem cell into insulin-producing cells in vitro. These insulin-producing cells were further functionally matured in vivo and successfully reduced hyperglycaemia in diabetic immunodeficient mice post-transplantation (Karnieli et al. 2007). Although others suggested limited possibility of differentiation of bone marrow-derived cells into pancreatic beta cells (Choi et al. 2003; Lechner et al. 2004), there are considerable number of publications of transdifferentiation of bone marrow-derived cells in vivo (Hess et al. 2003; Gao et al. 2008; Hasegawa et al. 2007; Huang et al. 2010; Rosengren et al. 2009). Hess et al. transplanted bone marrow-derived stem cells into the diabetic mice and regenerated pancreatic tissue and reduced hyperglycaemia (Hess et al. 2003). Others also reported successful repair of pancreatic damage after transplantation of bone marrow-derived cells (Gao et al. 2008; Hasegawa et al. 2007; Huang et al. 2010; Rosengren et al. 2009). Wu et al. transplanted bone marrow mesenchymal stem cells through the portal vein and alleviated the hyperglycaemia of diabetic rats (Wu et al. 2007). Although it is unclear whether pancreatic regeneration following the transplantation of bone marrow-derived cells is through neogenesis or endogenesis (Iskovich et al. 2007; Rosenberg 1995), the reduction in blood glucose promises an effective treatment of diabetes. Several clinical trials have confirmed the efficacy of BMSC transplantation to patients with type I diabetes (Fotino et al. 2010; Couri et al. 2009; Voltarelli et al. 2007). For instance, Couri et al. transplanted autologous non-myeloablative haematopoietic stem cells into newly diagnosed type I diabetic patients, resulting in insulin independence (Couri et al. 2009). In most of the clinical trials, however, the clinical outcome is possibly due to the immunomodulatory effects of the BMSCs rather insulin secretion from IPCs.

The application of BMSCs for the treatment of diabetes has several significant advantages, including multipotentiality and self-differentiation potential of these cells in vivo, their immunomodulatory effects, available techniques for extraction of BMSCs, and the possibility of autologous transplantation offering a relatively adequate supply of BMSCs. On the other hand, there are two major disadvantages in the clinical application of BMSCs for the treatment of diabetes, which are tumorigenicity and accessibility. Tumorigenicity has remained unsolved because immunosuppression after transplantation makes this a significant risk to the recipients. Although successful clinical trials demonstrate that the application of BMSC transplantation for the treatment of diabetes is currently safe, longer-term follow-ups are necessary. In addition, using the traditional procedure of bone marrow aspiration to obtain bone marrow-derived stem cells is painful and requires general or spinal anaesthesia and imposes the risk of severe infection. The number of stem cells obtained using the traditional procedure of bone marrow aspiration might not be sufficient for the clinical application and requires an expansion step ex vivo in order to obtain a clinically significant number of BMSCs. Such a step is time-consuming, is expensive, and carries the risks of contamination and loss of graft volume.

1.2.5 Adipose Stem Cells for Transplantation in Diabetes

Adipose stem cells (ASCs) were first introduced in 2001 by Zuk et al. (2001). ASCs have been studied for their potential in the treatment of heart disease, osteoarthritis, and diabetes mellitus (Minteer et al. 2012). ASCs could be derived from adipose tissue using liposuction procedure, many multiple lineage-committed progenitor cells or multipotent cells from other sources, such as pericytes, and marrow-derived MSCs from peripheral blood (Zuk et al. 2001). Although much less work has been reported about the differentiation of ASCs into IPCs in comparison to other types of stem cells, ASCs are a freely available source and offer the potential to differentiate into IPCs. Mohamad et al. differentiated human ASCs into IPCs using a two-step induction protocol (Mohamad Buang et al. 2012). After 3 weeks, IPCs secreted a significant amount of insulin from the differentiated cells in response to glucose (Mohamad Buang et al. 2012). The injection of ASCs into diabetic rats could effectively suppress pancreatic islet damage, increase the expression of insulin in pancreatic beta cells, and efficiently decrease the fasting blood glucose of the rats (Li et al. 2012).

The application of ASCs for the treatment of diabetes offers significant advantages, such as multipotentiality, accessibility and availability, and immunomodulatory features similar to other types of MSCs (Kim et al. 2010; Bassi et al. 2012; Puissant et al. 2005). In addition to other features of MSCs mentioned earlier, ASCs can be easily obtained using less invasive procedures, such as liposuction or lipoaspiration of abdominal fat tissue. There are large quantities of human adipose tissue available because of the popular cosmetic liposuction procedure used for the weight loss purpose, which could be a valuable source for adipose-derived mesenchymal stem cells. The availability and immunomodulatory features of hASCs can be considered as an alternative source of MSCs to replace bone marrow or embryonic stem cells for the clinical application in the treatment of diabetes (Lee et al. 2008; MacIsaac et al. 2012).

1.2.6 Pancreatic Stem Cells for Transplantation in Diabetes

Isolation of pancreatic stem cells or differentiation of other pancreatic adult cells through transdifferentiation offers another source of insulin-producing cells (Noguchi et al. 2009; Yamamoto et al. 2006). Since pancreatic cells have acquired some important phenotypes of normal beta cells during their initial differentiation, they can be theoretically considered as one of the most reasonable and practical sources to produce IPCs. The idea to apply these cells came from the advanced knowledge in the regenerative capacity of pancreatic beta cells (Hayashi et al. 2003). Hayashi et al. found that endocrine cells regenerated after 90% pancreatectomy (Hayashi et al. 2003). It has been believed that regeneration following pancreatectomy or other physiologic conditions, such as pregnancy, is a result of proliferation of pre-existing islet cells (Furuyama et al. 2011; Solar et al. 2009; Dor et al. 2004). It has been suggested that various types of pancreatic cells, including acinar cells, duct epithelial cells, and other types of endocrine cells, especially, alpha cells of pancreatic islets that share many important specialized genes and transcriptional factors, such as glucokinase with beta cells (Tu et al. 1999), are able to differentiate into beta cells.

As depicted in Fig. 1.2, pancreatic progenitors originate from endoderm progenitors following activation of Pdx1 and Ptf1a/p48. These cells can further differentiate into three types of cells that are endocrine progenitors, acinar cells, and pancreatic ducts. Activation of Ngn3, Hnf6, and Hnf1 is necessary for the differentiation of endocrine progenitors. At the next phase, beta cells are produced following activation of Pax4, Arx, Nkx2.2, Nkx6.1, Pdx1, and MafA. Theoretically, all these cells including pancreatic progenitor cells, acinar cells, ductal cells, endocrine progenitors, and other pancreatic endocrine cells can be differentiated into IPCs. The advantages of differentiating these cells into IPCs are that they have been partially differentiated because of their pancreatic origin, little ethical concerns regarding manipulation of them, and no immune rejection after autologous transplantation. The main disadvantage of their application could be limited to cell isolation and differentiation techniques.

1.2.7 Adult Pancreatic Stem Cells for Transplantation in Diabetes

Adult pancreatic stem cells (APSCs) are poorly identified and a controversial type of cells. Because of regenerative characteristic of the pancreas, it is assumed that adult pancreas contains APSCs (Gong et al. 2012; Montanya 2004; Fernandes et al. 1997) and they are possibly located among ductal (Bonner-Weir 2000; Bouwens and Klöppel 1996) or islet cells (Zulewski et al. 2001; Li et al. 2003). Nestin, an intermediate filament protein, was initially introduced as a marker for endocrine progenitor cells (Abraham et al. 2002). Later, however, the expression of nestin was identified in various types of cells, such as neuroepithelial stem cells and endothelial (Klein et al. 2003) and other pancreatic cells (Kim et al. 2004). Therefore, nestin is no longer a specific marker for the identification of endocrine precursor cells (Street et al. 2004); in fact, nestin is a common marker of precursor multilineage stem cells. Gong et al. proposed c-kit protein to be the possible marker for pancreatic progenitor stem cells (Gong et al. 2012). There is still controversy regarding the nature of pancreatic stem/progenitor cells because lineage-tracing experiments indicate that pancreatic duct cells are the progenitor cells and able to differentiate into beta cells (Xia et al. 2009). On the other hand, Li et al. reported that progenitor cells are somatostatin-positive islet cells that can be differentiated into IPCs and improve hyperglycaemia in diabetic mice post-transplantation (Li et al. 2003).

1.2.8 Duct Epithelial Cells for Transplantation in Diabetes

At the early stage of embryogenesis, the primary pancreatic ducts can be differentiated into multipotential endocrine cells (Yang et al. 2006). Even after birth or following injury, duct epithelial cells demonstrate their capacity as pancreatic progenitors. For instance, Inada et al. created transgenic mice that express human carbonic anhydrase II (CAII) promoter (Inada et al. 2008). CAII-expressing cells within the pancreas are progenitors that can produce new islets and acinar cells after injury. In their model, duct epithelial cells first replicate, then regress to a less differentiated state, and then produce new endocrine and exocrine pancreas (Inada et al. 2008). Ramiya et al. isolated pancreatic duct epithelial cells from prediabetic adult mice, grow them in long-term cultures in vitro, and induce these cells becoming functional islets (Ramiya et al. 2000). After transplantation of these islets into diabetic mice, they could reverse hyperglycaemia.

Various culturing strategies and in vitro genetic manipulation techniques (through protein transduction and/or gene transfer technology) of duct epithelial cells have been devised for producing IPCs (Yamamoto et al. 2006; Bonner-Weir et al. 2000; Noguchi et al. 2003, 2005, 2010; Yatoh et al. 2007). Culturing duct epithelial cells with growth factors, such as GLP-1 (Bulotta et al. 2002), exendin-4, nicotinamide, keratinocyte growth factor (Noguchi et al. 2010), and fibroblast growth factor (Hardikar et al. 2003), could differentiate them into IPCs. For instance, Bulotta et al. cultured rat pancreatic duct in the presence of GLP-1 and demonstrated a time-dependent expression of GLUT-2, insulin, and glucokinase and a glucose-dependent insulin secretion (Bulotta et al. 2002). Expression of ngn3 in clonal pancreatic precursor duct cells could induce duct cell differentiation, resulting in endocrine cells and expression of insulin (Yamamoto et al. 2006). After 2 weeks culture of human pancreatic duct cells with exendin-4, nicotinamide, keratinocyte growth factor, Pdx-1 protein, or BETA2/NeuroD protein in vitro, human pancreatic duct cells differentiated into IPCs (Noguchi et al. 2010). Culturing islet-like clusters in a 3D condition resulted in pancreatic endocrine cell budding and produced glucose-responsive IPCs with similar ultrastructure of normal beta cells (Bonner-Weir et al. 2000). Although the design of some lineage-tracing studies is not perfect, the contribution of these studies to the transdifferentiation of duct cells into IPCs is significant (Xia et al. 2009; Venkatesan et al. 2011).

1.2.9 Acinar Cells for Transplantation in Diabetes

Reprogramming procedure has been applied to produce IPCs from acinar cells, which consists of dedifferentiation of acinar cells into duct progenitor cells, proliferation, and redifferentiation. Dedifferentiation of acinar cells can be mediated by Notch and stromelysin-3 signalling (Mukhi and Brown 2011). The proliferation and differentiation are similar to the process of differentiation in pancreatic duct cells (Bonner-Weir et al. 2000). Other methods have also applied to produce IPCs from acinar cells; including coculture manipulations and/or transdifferentiation. For instance, Zhou et al. reprogrammed differentiated exocrine cells into IPCs using adenoviral delivery of three transcription factors (Ngn3, Pdx1, and Mafa) and resulted in successful correction of hyperglycaemia in diabetic mice (Zhou et al. 2008). Transdifferentiation might be associated with lower risk of tumorigenesis compared to the process that applies dedifferentiation. However, it still relied on a viral vector to deliver transcription factor genes; application of the current viral vector-free gene delivery technique would improve its potential for clinical application.

1.2.10 Hepatic Cells for Transplantation in Diabetes

Hepatic cells have attracted much attention in their application as host cells of insulin biosynthesis and secretion due to their similar embryonic origin to pancreatic beta cells, capacity to synthesize and secrete proteins, and sharing important transcriptomic features, such as GLUT-2 and glucokinase. In our previous studies, the full-length human insulin cDNA was transfected into a human hepatoma cell line (HEP G2) that was able to synthesize insulin, secrete insulin in response to glucose, and reverse diabetic state in mice (Simpson et al. 1995, 1997). It has been reported that single gene transfection of Pdx-1 alone can result in insulin production in adult human hepatic cells (Sapir et al. 2005); however, these cells have low level of insulin content and abnormal glucose responsiveness since hepatic cells are rich in hexokinase and poor in glucokinase and prohormone convertase (Zalzman et al. 2005). Several pancreatic transcription factors have been applied to promote transdifferentiation of hepatic cells and improve their glucose responsiveness (Simpson et al. 1997; Liu et al. 2007). Co-transfection of hepatoma cells with human insulin genes and the GLUT2 improved the glucose responsiveness of hepatoma cells (Simpson et al. 1997; Zheng et al. 2007), which is possibly via pathways dependent on K_{ATP} (Liu et al. 2007). There is significant difference between manipulated hepatic cells and pancreatic insulin-producing cells in glucose-responsive insulin secretion, which requires additional regulators other than glucokinase to achieve a normal insulin secretion in response to glucose (Matschinsky et al. 1993; Tu and Tuch 1997; German 1993).

Hepatic cells can be considered as one of the best autologous sources of tissue for transplantation since they are accessible and have a great regenerative capacity. Diabetic patients have often maintained normal liver function, and their livers could be a source of hepatic cells for producing IPCs. Post-transplantation tumorigenicity and ethical issues are not a major concern in the application of hepatic cells for the treatment of diabetes. The main difficulty with hepatic cells is about propagation and expansion of hepatic cells. Culturing hepatic cells with betacellulin (a beta cell-stimulating hormone) (Kojima et al. 2003), hepatocyte growth factor, interleukin-6 (Hayek et al. 1995), and activin A (Zalzman et al. 2005) benefits propagation and expansion of hepatic cells. Transduction of human telomerase gene into hepatic cells is another method to immortalize hepatic cells (Zalzman et al. 2003) in order to an unlimited expansion of hepatic cells for transplantation.

1.3 Discussion

One in four diabetic patients including both type I and II require a daily insulin injection (2011 National Diabetes Fact Sheet, http://www.cdc.gov/diabetes/pubs/factsheet11.htm). The current therapeutic options for diabetes are neither curative nor preventive for its complication. There is an urgent need for a better diabetic management approach. Considerable achievements in the field of stem cell therapy have attracted investigations on this novel potentially curative therapeutic modality. Theoretically, diabetes especially type I diabetes is curable because it is caused by monocellular damage, insulin-producing cell and only requires monofunctional cell replacement. The required function of the cell is to produce only one hormone, insulin, and secrete it into the bloodstream in response to blood glucose fluctuation. Therefore, stem cell replacement of damaged insulin-producing cell is the simplest type of stem cell therapy for a chronic disease. In contrast to stem cell therapy for other diseases, such as heart failure and neuronal diseases, stem cells can be transplanted to more convenient and easily accessible sites, such as under the kidney capsule for diabetes, and not exactly into the pancreas. Theoretically, any sites where have sufficient blood circulation could serve as graft location. To achieve the optimal function of cells, it is ideal for the graft to have the natural interactions with other endocrine and mesenchymal cells of the pancreas. In some clinical trials, bone marrow- or umbilical cord blood-derived stem cells were directly transplanted into the pancreas; however, the current census for graft location is under the kidney capsule. However, long-term studies and clinical trials are required to verify this.

The efficacy of IPCs reflects on de novo insulin synthesis and reversal of hyperglycaemia post-transplantation. Detection of insulin mRNA or other products of de novo insulin synthesis, e.g. C-peptide, in the differentiated IPCs and correction of hyperglycaemia in human and animal models have been achieved after successful transplantation of stem cells. After graft removal, reversion of hyperglycaemia and diabetic state was maintained in animal studies confirming the efficacy of IPCs. There are several obstacles preventing the clinical application of stem cell replacement for diabetes, including functional immaturity in glucose-responsive insulin secretion, immune rejection of stem cells after transplantation, and the risk of development of tumours.

1.3.1 Functional Immaturity of Stem Cell-Derived Insulin-Producing Cells

Significant progress has been achieved in inducing stem cells to differentiate into insulin-producing cells (Hosoya et al. 2012). However, there is no evidence of insulin secretion from the differentiated IPCs prior to transplantation. To the best of our knowledge, the differentiated IPCs available for the current clinical application are immature since they co-express other endocrine hormones, such as glucagon (Segev et al. 2004; Kroon et al. 2008), and cannot appropriately secrete insulin in response to glucose. In spite of a high percentage of IPCs identified in the differentiated cell clusters and their capacity for the expression and storage of insulin, their functional maturation has partially achieved in vitro and fully functional maturation requires further differentiation in vivo (Kroon et al. 2008). For instance, the final maturation of human ESCs occurred in vivo and corrected hyperglycaemia 96 days post-transplantation (Bose et al. 2012).

Figure 1.4 simplifies insulin synthesis and secretion in a mature β-cell. The reason for the functional immaturity of IPCs is that the current differentiation protocols do not completely mimic the in vivo microenvironment where rich in transcription factors that beta cells need for their develop and lack of the interactions between beta cells and their surrounding other five different types of islet cells (alpha, delta,

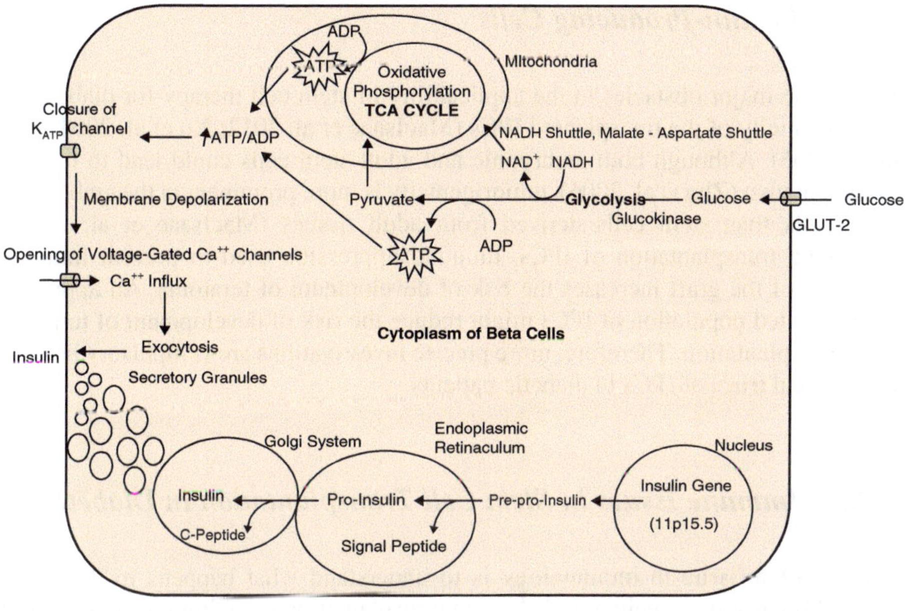

Fig. 1.4 Insulin biosynthesis and secretion from a mature β-cell. In a mature pancreatic β-cell, human insulin gene, which is located on chromosome 11p15.5, encodes preproinsulin in the cytoplasm. It is translocated into the endoplasmic reticulum (ER) through the interaction between its signalling peptide and the signalling recognition particle on the ER membrane. Within the lumen of ER, the signalling peptide is split, and proinsulin is made. The formation of three disulphide bonds leads to its folding; it is delivered to Golgi apparatus and is stored in the secretory granules in the form of insulin. Secretion of insulin through exocytosis is a highly dynamic process, regulated by various nutrients and hormonal and neural stimuli. Glucose-induced insulin secretion (GIIS) is the principal mechanism of insulin secretion, and glucose is the most important physiologic regulator of insulin secretion. After glucose is transported into β-cells through glucose transporter-2 (GLUT-2), it is metabolized leading to an increase in ATP/ADP ratio. This, in turn, results in the closure of ATP-sensitive K^+ channels (K_{ATP} channels) and subsequently depolarization of the cell membrane, which leads to opening of voltage-dependent Ca^{2+} channels and influx of Ca^{2+}. The increase in the intracellular Ca^{2+} leads to fusion of insulin granules to the plasma membrane and secretion of insulin

pancreatic polypeptide, and epsilon cells), endothelial cells, and fibroblasts. The current protocols can be improved by coculturing IPCs with endothelial cells as the endothelial cell signalling promotes the functional maturation of IPCs (Jaramillo and Banerjee 2012). Recognition of currently unknown glucose-responsive transcription factors that are involved in beta cell differentiation might help for optimizing the current differentiation protocols (Hang and Stein 2011; Nishimura et al. 2006; Artner et al. 2007). Further advancement of our knowledge of the glucose sensing of beta cells and the mechanism of insulin secretion will improve the current techniques for differentiating stem cells into IPCs.

1.3.2 Tumorigenicity of Stem Cell-Derived Insulin-Producing Cells

One of the major obstacles in the applicability of stem cell therapy for diabetes is tumorigenicity of the transplanted IPCs (MacIsaac et al. 2012; Xu et al. 2004; Zhu et al. 2006). Although both embryonic and adult stem cells could lead to tumour growth in vivo (Zhu et al. 2006), tumorigenicity is more prominent in the embryonic stem cells than stem cells derived from adult tissues (MacIsaac et al. 2012). Following transplantation of IPCs, immunosuppression used to prevent immune rejection of the graft increases the risk of development of teratoma. An appropriately purified population of IPCs might reduce the risk of development of tumours post-transplantation. Therefore, more precise investigations are compulsory prior to any clinical trials of IPCs in diabetic patients.

1.3.3 Immune Issues in Stem Cell Transplantation in Diabetes

An important issue in immunology is to understand what happens to the transplanted IPCs or their recipients following transplantation since immune rejection of stem cells is often observed after transplantation of ESCs and the graft-versus-host disease is common after transplantation of UCBSCs and BMSCs. After transplantation of stem cells, the immune reaction of the recipients results in the destruction of IPCs. For this reason, post-transplantation immunosuppression is mandatory to accomplish reversal of diabetes in animal models or human (Couri et al. 2009; Voltarelli et al. 2007). After transplantation of autologous non-myeloablative haematopoietic stem cell, high-dose immunosuppression improved beta cell function and prolonged insulin independence in the majority of the patients (Voltarelli et al. 2007). However, immunosuppression is associated with other risks, such as infection, sterility, and the toxicity of immunosuppressive drugs. There is considerable number of regulatory T cells in the stem cell grafts derived from bone marrow or umbilical cord blood cells. These regulatory T cells have immunosuppressive and anti-inflammatory characteristics (Ende et al. 2004; Haller et al. 2008); therefore, immune rejection is a less important issue in these grafts comparing with other sources for stem cells. In Haller's report, transplantation of hUCBSCs improved hyperglycaemia without the need for immunosuppression (Haller et al. 2008).

An alternative approach to overcoming immune issues is to apply gene therapy, engineering immune cells, or beta cells for preventing immune destruction or apoptotic destruction of transplanted IPCs. This can be achieved by direct delivery of protective genes, such as caspase 3 (Wang et al. 2011), IL-10 (Dénes et al. 2006; Zhang et al. 2003), bcl2 (Gene et al. 2001), and other genes (Plesner et al. 2010; Bertera et al. 2003; Emamaullee and Shapiro 2006) into the grafts. The most practical solution is to encapsulate the IPCs to protect the grafts from the recipients' immune system (Taniguchi et al. 1997). This technique can prevent inflammatory and immune cells and immunoglobulin entering capsule to attack/bind to the IPCs while allowing insulin outflow into recipients' blood circulation. Alginate has been the most commonly used polymer to encapsulate IPCs. Alginate-based capsules have been shown to remain stable for several years in animals and humans (Orlando et al. 2013, 2014; Soon-Shiong et al. 1994; Duvivier-Kali et al. 2001).

1.4 Conclusion

We have reviewed significant advances related to the clinical application of stem cell transplantation in diabetes mellitus type I and type II. The successful outcome has been achieved in the standardization of stem cell identification, isolation, purification, differentiation, expansion, and storage procedures. Transplantation of stem cell-derived insulin-producing cells effectively reduced hyperglycaemia, although normalization was not achieved. This is partially due to the complex architecture and physiology of the native pancreas, a poor understanding of aetiology of the disease, and the interactions between β-cells and other endocrine cells, growth factors, and pancreatic extracellular matrix. While current research endeavours are promising, the transition to safe and effective clinical implementation faces significant obstacles. Collaborative efforts are required to overcome tumorigenicity, immune rejection, and functional immaturity of stem cell-derived insulin-producing cells, speed up this novel therapy for restoring β-cell function in diabetic patients to prevent diabetic complications.

References

Abraham EJ, Leech CA, Lin JC, Zulewski H, Habener JF (2002) Insulinotropic hormone glucagon-like peptide-1 differentiation of human pancreatic islet-derived progenitor cells into insulin-producing cells. Endocrinology 143(8):3152–3161

Ahrén B (2008) Role of pituitary adenylate cyclase-activating polypeptide in the pancreatic endocrine system. Ann N Y Acad Sci 1144:28–35

Apelqvist A, Li H, Sommer L, Beatus P, Anderson DJ, Honjo T, Hrabe de Angelis M, Lendahl U, Edlund H (1999) Notch signalling controls pancreatic cell differentiation. Nature 400(6747):877–881. doi:10.1038/23716

Artner I, Blanchi B, Raum JC, Guo M, Kaneko T, Cordes S, Sieweke M, Stein R (2007) MafB is required for islet beta cell maturation. Proc Natl Acad Sci U S A 104(10):3853–3858. doi:10.1073/pnas.0700013104

Assady S, Maor G, Amit M, Itskovitz-Eldor J, Skorecki KL, Tzukerman M (2001) Insulin production by human embryonic stem cells. Diabetes 50(8):1691–1697

Bassi EJ, Moraes-Vieira PMM, Moreira Sá CSR, Almeida DC, Vieira LM, Cunha CS, Hiyane MI, Basso AS, Pacheco-Silva A, Câmara NOS (2012) Immune regulatory properties of allogeneic adipose-derived mesenchymal stem cells in the treatment of experimental autoimmune diabetes. Diabetes 61:2534

Behzad-Behbahani A, Pouransari R, Tabei SZ, Rahiminejad MS, Robati M, Yaghobi R, Nourani H, Ramzi MM, Farhadi-Andarabi A, Mojiri A, Rahsaz M, Banihashemi M, Zare N (2005) Risk of viral transmission via bone marrow progenitor cells versus umbilical cord blood hematopoietic stem cells in bone marrow transplantation. Transplant Proc 37(7):3211–3212. doi:10.1016/j.transproceed.2005.07.007

Bertera S, Crawford ML, Alexander AM, Papworth GD, Watkins SC, Robbins PD, Trucco M (2003) Gene transfer of manganese superoxide dismutase extends islet graft function in a mouse model of autoimmune diabetes. Diabetes 52(2):387–393

Bieback K, Kern S, Klüter H, Eichler H (2004) Critical parameters for the isolation of mesenchymal stem cells from umbilical cord blood. Stem Cells (Dayt Ohio) 22(4):625–634. doi:10.1634/stemcells.22-4-625

Bleich D (2009) Umbilical cord blood and type 1 diabetes: a road ahead or dead end? Diabetes Care 32(11):2138–2139. doi:10.2337/dc09-1456

Blyszczuk P, Czyz J, Kania G, Wagner M, Roll U, St-Onge L, Wobus AM (2003) Expression of Pax4 in embryonic stem cells promotes differentiation of nestin-positive progenitor and insulin-producing cells. Proc Natl Acad Sci U S A 100(3):998–1003

Bonner-Weir S (2000) Life and death of the pancreatic beta cells. Trends Endocrinol Metab: TEM 11(9):375–378

Bonner-Weir S, Taneja M, Weir GC, Tatarkiewicz K, Song KH, Sharma A, O'Neil JJ (2000) In vitro cultivation of human islets from expanded ductal tissue. Proc Natl Acad Sci U S A 97(14):7999–8004

Bose B, Shenoy PS, Konda S, Wangikar P (2012) Human embryonic stem cell differentiation into insulin secreting beta cells for diabetes. Cell Biol Int 36:1013. doi:10.1042/CBI20120210

Bouwens L, Klöppel G (1996) Islet cell neogenesis in the pancreas. Virchows Arch: Int J Pathol 427(6):553–560

Brolén GKC, Heins N, Edsbagge J, Semb H (2005) Signals from the embryonic mouse pancreas induce differentiation of human embryonic stem cells into insulin-producing beta-cell-like cells. Diabetes 54(10):2867–2874

Broxmeyer HE, Douglas GW, Hangoc G, Cooper S, Bard J, English D, Arny M, Thomas L, Boyse EA (1989) Human umbilical cord blood as a potential source of transplantable hematopoietic stem/progenitor cells. Proc Natl Acad Sci U S A 86(10):3828–3832

Bulotta A, Hui H, Anastasi E, Bertolotto C, Boros LG, Di Mario U, Perfetti R (2002) Cultured pancreatic ductal cells undergo cell cycle re-distribution and beta-cell-like differentiation in response to glucagon-like peptide-1. J Mol Endocrinol 29(3):347–360

Chao KC, Chao KF, Fu YS, Liu SH (2008) Islet-like clusters derived from mesenchymal stem cells in Wharton's jelly of the human umbilical cord for transplantation to control type 1 diabetes. PLoS One 3(1):e1451. doi:10.1371/journal.pone.0001451

Chen C, Leavitt T, Sibley E (2012) Intestinal Pdx1 mediates nutrient metabolism gene networks and maternal expression is essential for perinatal growth in mice. Biochem Biophys Res Commun:2–6. doi:10.1016/j.bbrc.2012.06.153

Choi JB, Uchino H, Azuma K, Iwashita N, Tanaka Y, Mochizuki H, Migita M, Shimada T, Kawamori R, Watada H (2003) Little evidence of transdifferentiation of bone marrow-derived cells into pancreatic beta cells. Diabetologia 46(10):1366–1374

Cnop M, Welsh N, Jonas JC, Jorns A, Lenzen S, Eizirik DL (2005) Mechanisms of pancreatic – cell death in type 1 and type 2 diabetes: many differences, few similarities. Diabetes 54(suppl_2):S97–S107. doi:10.2337/diabetes.54.suppl_2.S97

Couri CEB, Oliveira MCB, Stracieri ABPL, Moraes DA, Pieroni F, Barros GMN, Madeira MIA, Malmegrim KCR, Foss-Freitas MC, Simões BP, Martinez EZ, Foss MC, Burt RK, Voltarelli JC (2009) C-peptide levels and insulin independence following autologous nonmyeloablative hematopoietic stem cell transplantation in newly diagnosed type 1 diabetes mellitus. JAMA 301(15):1573–1579. doi:10.1001/jama.2009.470

Dénes B, Yu J, Fodor N, Takátsy Z, Fodor I, Langridge WHR (2006) Suppression of hyperglycemia in NOD mice after inoculation with recombinant vaccinia viruses. Mol Biotechnol 34(3):317–327. doi:10.1385/MB:34:3:317

Dilorio PJ, Moss JB, Sbrogna JL, Karlstrom RO, Moss LG (2002) Sonic hedgehog is required early in pancreatic islet development. Dev Biol 244(1):75–84. doi:10.1006/dbio.2002.0573

Dor Y, Brown J, Martinez OI, Melton DA (2004) Adult pancreatic beta-cells are formed by self-duplication rather than stem-cell differentiation. Nature 429(6987):41–46. doi:10.1038/nature02520

Duffy MM, Ritter T, Ceredig R, Griffin MD (2011) Mesenchymal stem cell effects on T-cell effector pathways. Stem Cell Res Ther 2(4):34–34. doi:10.1186/scrt75

Duvivier-Kali VF, Omer A, Parent RJ, O'Neil JJ, Weir GC (2001) Complete protection of islets against allorejection and autoimmunity by a simple barium-alginate membrane. Diabetes 50(8):1698–1705

Emamaullee JA, Shapiro AMJ (2006) Interventional strategies to prevent beta-cell apoptosis in islet transplantation. Diabetes 55(7):1907–1914. doi:10.2337/db05-1254

Ende N, Chen R, Reddi AS (2004) Effect of human umbilical cord blood cells on glycemia and insulitis in type 1 diabetic mice. Biochem Biophys Res Commun 325(3):665–669

Evans MJ, Kaufman MH (1981) Establishment in culture of pluripotential cells from mouse embryos. Nature 292(5819):154–156

Fernandes A, King LC, Guz Y, Stein R, Wright CV, Teitelman G (1997) Differentiation of new insulin-producing cells is induced by injury in adult pancreatic islets. Endocrinology 138(4):1750–1762

Fotino C, Ricordi C, Lauriola V, Alejandro R, Pileggi A (2010) Bone marrow-derived stem cell transplantation for the treatment of insulin-dependent diabetes. Rev Diabet Stud: RDS 7(2):144–157. doi:10.1900/RDS.2010.7.144

Fu S, Fei Q, Jiang H, Chuai S, Shi S, Xiong W, Jiang L, Lu C, Atadja P, Li E, Shou J (2011) Involvement of histone acetylation of Sox17 and Foxa2 promoters during mouse definitive endoderm differentiation revealed by microRNA profiling. PLoS One 6(11):e27965

Fujikawa T, Oh S-H, Pi L, Hatch HM, Shupe T, Petersen BE (2005) Teratoma formation leads to failure of treatment for type I diabetes using embryonic stem. Am J Pathol 166(6):1781–1791. Cells C-dI-p

Furuyama K, Kawaguchi Y, Akiyama H, Horiguchi M, Kodama S, Kuhara T, Hosokawa S, Elbahrawy A, Soeda T, Koizumi M, Masui T, Kawaguchi M, Takaori K, Doi R, Nishi E, Kakinoki R, Deng JM, Behringer RR, Nakamura T, Uemoto S (2011) Continuous cell supply from a Sox9-expressing progenitor zone in adult liver, exocrine pancreas and intestine. Nat Genet 43(1):34–41. doi:10.1038/ng.722

Gannon MC, Nuttall FQ (2010) Amino acid ingestion and glucose metabolism – a review. IUBMB Life 62(9):660–668

Gao X, Song L, Shen K, Wang H, Niu W, Qin X (2008) Transplantation of bone marrow derived cells promotes pancreatic islet repair in diabetic mice. Biochem Biophys Res Commun 371(1):132–137

Gene AB, Contreras Juan L, Bilbao GU, CHA S, Jiang Xiao L, Eckhoff DEE, Jenkins STM, Thomas FRT, Curiel DAT, Thomas JUM (2001) Cytoprotection of pancreatic islets before and soon after transplantation by gene transfer of the anti-apoptotic bcl-2 gene1. Transplantation 71(8):1015–1023

German MS (1993) Glucose sensing in pancreatic islet beta cells: the key role of glucokinase and the glycolytic intermediates. Proc Natl Acad Sci U S A 90(5):1781–1785

Gong J, Zhang G, Tian F, Wang Y (2012) Islet-derived stem cells from adult rats participate in the repair of islet damage. J Mol Histol 43:745. doi:10.1007/s10735-012-9447-6

Grompe M (2012) Tissue stem cells: new tools and functional diversity. Cell Stem Cell 10(6):685–689

Gu G, Brown JR, Melton DA (2003) Direct lineage tracing reveals the ontogeny of pancreatic cell fates during mouse embryogenesis. Mech Dev 120(1):35–43

Hakim NS (2002) Pancreatic transplantation for patients with type I diabetes. HPB: Off J Int Hepatol Pancreatol Biliary Assoc 4(2):59–61. doi:10.1080/136518202760378407

Haller MJ, Viener H-L, Wasserfall C, Brusko T, Atkinson MA, Schatz DA (2008) Autologous umbilical cord blood infusion for type 1 diabetes. Exp Hematol 36(6):710–715. doi:10.1016/j.exphem.2008.01.009

Hang Y, Stein R (2011) MafA and MafB activity in pancreatic β cells. Trends Endocrinol Metab: TEM 22(9):364–373

Hardikar AA, Marcus-Samuels B, Geras-Raaka E, Raaka BM, Gershengorn MC (2003) Human pancreatic precursor cells secrete FGF2 to stimulate clustering into hormone-expressing islet-like cell aggregates. Proc Natl Acad Sci U S A 100(12):7117–7122. doi:10.1073/pnas.1232230100

Hasegawa Y, Ogihara T, Yamada T, Ishigaki Y, Imai J, Uno K, Gao J, Kaneko K, Ishihara H, Sasano H, Nakauchi H, Oka Y, Katagiri H (2007) Bone marrow (BM) transplantation promotes beta-cell regeneration after acute injury through BM cell mobilization. Endocrinology 148(5):2006–2015

Hayashi KY, Tamaki H, Handa K, Takahashi T, Kakita A, Yamashina S (2003) Differentiation and proliferation of endocrine cells in the regenerating rat pancreas after 90% pancreatectomy. Arch Histol Cytol 66(2):163–174

Hayek A, Beattie GM, Cirulli V, Lopez AD, Ricordi C, Rubin JS (1995) Growth factor/matrix-induced proliferation of human adult beta-cells. Diabetes 44(12):1458–1460

He D, Wang J, Gao Y, Zhang Y (2011) Differentiation of PDX1 gene-modified human umbilical cord mesenchymal stem cells into insulin-producing cells in vitro. Int J Mol Med 28(6):1019–1024

Hebrok M (2003) Hedgehog signaling in pancreas development. Mech Dev 120(1):45–57

Hess D, Li L, Martin M, Sakano S, Hill D, Strutt B, Thyssen S, Gray DA, Bhatia M (2003) Bone marrow-derived stem cells initiate pancreatic regeneration. Nat Biotechnol 21(7):763–770

Hosoya M, Kunisada Y, Kurisaki A, Asashima M (2012) Induction of differentiation of undifferentiated cells into pancreatic beta cells in vertebrates. Int J Dev Biol 56(5):313–323

Houard N, Rousseau GG, Lemaigre FP (2003) HNF-6-independent differentiation of mouse embryonic stem cells into insulin-producing cells. Diabetologia 46(3):378–385. doi:10.1007/s00125-003-1041-8

Huang Y, Kucia M, Hussain L-R, Wen Y, Xu H, Yan J, Ratajczak MZ, Ildstad ST (2010) Bone marrow transplantation temporarily improves pancreatic function in streptozotocin-induced diabetes: potential involvement of very small embryonic-like cells. Transplantation 89(6):677–685

Ianus A, Holz GG, Theise ND, Hussain MA (2003) In vivo derivation of glucose-competent pancreatic endocrine cells from bone marrow without evidence of cell fusion. J Clin Invest 111(6):843–850

Inada A, Nienaber C, Katsuta H, Fujitani Y, Levine J, Morita R, Sharma A, Bonner-Weir S (2008) Carbonic anhydrase II-positive pancreatic cells are progenitors for both endocrine and exocrine pancreas after birth. Proc Natl Acad Sci U S A 105(50):19915–19919. doi:10.1073/pnas.0805803105

Iskovich S, Kaminitz A, Yafe MP, Mizrahi K, Stein J, Yaniv I, Askenasy N (2007) Participation of adult bone marrow-derived stem cells in pancreatic regeneration: neogenesis versus endogenesis. Curr Stem Cell Res Ther 2(4):272–279

Ito T, Itakura S, Todorov I, Rawson J, Asari S, Shintaku J, Nair I, Ferreri K, Kandeel F, Mullen Y (2010) Mesenchymal stem cell and islet co-transplantation promotes graft revascularization and function. Transplantation 89(12):1438–1445

Jaramillo M, Banerjee I (2012) Endothelial cell co-culture mediates maturation of human embryonic stem cell to pancreatic insulin producing cells in a directed differentiation approach. J Visual Exp: JoVE 61:3759

Kaczorowski DJ, Patterson ES, Jastromb WE, Shamblott MJ (2002) Glucose-responsive insulin-producing cells from stem cells. Diabetes/Metab Res Rev 18(6):442–450

Kahan BWBW, Jacobson LMLM, Hullett DADA, Ochoada JM, Oberley TD, Lang KM, Odorico JS (2003) Pancreatic precursors and differentiated islet cell types from murine embryonic stem cells: an in vitro model to study islet differentiation. Diabetes 52(8):2016–2024

Kahn SE, Prigeon RL, Schwartz RS, Fujimoto WY, Knopp RH, Brunzell JD, Porte D (2001) Obesity, body fat distribution, insulin sensitivity and islet β-cell function as explanations for metabolic diversity. J Nutrition 131(2):354–360

Karnieli O, Izhar-Prato Y, Bulvik S, Efrat S (2007) Generation of insulin-producing cells from human bone marrow mesenchymal stem cells by genetic manipulation. Stem Cells (Dayt Ohio) 25(11):2837–2844

Kim S-Y, Lee S-H, Kim B-M, Kim E-H, Min B-H, Bendayan M, Park I-S (2004) Activation of nestin-positive duct stem (NPDS) cells in pancreas upon neogenic motivation and possible cytodifferentiation into insulin-secreting cells from NPDS cells. Dev Dyn: Off Publ Am Assoc Anat 230(1):1–11. doi:10.1002/dvdy.20012

Kim SC, Han DJ, Lee JY (2010) Adipose tissue derived stem cells for regeneration and differentiation into insulin-producing cells. Curr Stem Cell Res Ther 5(2):190–194

Klein T, Ling Z, Heimberg H, Madsen OD, Heller RS, Serup P (2003) Nestin is expressed in vascular endothelial cells in the adult human pancreas. J Histochem Cytochem: Off J Histochem Soc 51(6):697–706

Koblas T, Harman SM, Saudek F (2005) The application of umbilical cord blood cells in the treatment of diabetes mellitus. Rev Diabet Stud: RDS 2(4):228–234. doi:10.1900/RDS.2005.2.228

Kojima H, Fujimiya M, Matsumura K, Younan P, Imaeda H, Maeda M, Chan L (2003) NeuroD-betacellulin gene therapy induces islet neogenesis in the liver and reverses diabetes in mice. Nat Med 9(5):596–603. doi:10.1038/nm867

Krentz AJ, Bailey CJ (2005) Oral antidiabetic agents: current role in type 2 diabetes mellitus. Drugs 65(3):385–411

Kroon E, Martinson LA, Kadoya K, Bang AG, Kelly OG, Eliazer S, Young H, Richardson M, Smart NG, Cunningham J, Agulnick AD, D'Amour KA, Carpenter MK, Baetge EE (2008) Pancreatic endoderm derived from human embryonic stem cells generates glucose-responsive insulin-secreting cells in vivo. Nat Biotechnol 26(4):443–452

Ku HT, Zhang N, Kubo A, O'Connor R, Mao M, Keller G, Bromberg JS (2004) Committing embryonic stem cells to early endocrine pancreas in vitro. Stem Cells (Dayt Ohio) 22(7):1205–1217

Laikind P (2014) ViaCyte, Inc. Announces FDA acceptance of IND to commence clinical trial of VC-01™ candidate cell replacement therapy for type 1 diabetes. http://viacyte.com/press-releases/viacyte-inc-announces-fda-acceptance-of-ind-to-commence-clinical-trial-of-vc-01-candidate-cell-replacement-therapy-for-type-1-diabetes/. Accessed 13 May 2015

Lechner A, Yang Y-G, Blacken RA, Wang L, Nolan AL, Habener JF (2004) No evidence for significant transdifferentiation of bone marrow into pancreatic beta-cells in vivo. Diabetes 53(3):616–623

Lee J, Han D-J, Kim S-C (2008) In vitro differentiation of human adipose tissue-derived stem cells into cells with pancreatic phenotype by regenerating pancreas extract. Biochem Biophys Res Commun 375(4):547–551. doi:10.1016/j.bbrc.2008.08.064

Lee J, Cummings BP, Martin E, Sharp JW, Graham JL, Stanhope KL, Havel PJ, Raybould HE (2012) Glucose sensing by gut endocrine cells and activation of the vagal afferent pathway is impaired in a rodent model of type 2 diabetes mellitus. Am J Physiol Regul Integr Comp Physiol 302(6):R657–R666

Levine F, Leibowitz G (1999) Towards gene therapy of diabetes mellitus. Mol Med Today 5(4):165–171

Li L, Seno M, Yamada H, Kojima I (2003) Betacellulin improves glucose metabolism by promoting conversion of intraislet precursor cells to beta-cells in streptozotocin-treated mice. Am J Phys Endocrinol Metab 285(3):E577–E583. doi:10.1152/ajpendo.00120.2003

Li Y-Y, Liu H-H, Chen H-L, Li Y-P (2012) Adipose-derived mesenchymal stem cells ameliorate STZ-induced pancreas damage in type 1 diabetes. Biomed Mater Eng 22(1):97–103

Liu GJ, Simpson AM, Swan MA, Tao C, Tuch BE, Russell M, Jovanovic A, Martin DK (2007) UKPMC Funders Group Author manuscript ATP-sensitive potassium channels induced in liver cells after transfection with insulin cDNA and the GLUT2 transporter regulate glucose-stimulated insulin secretion. FASEB J 17(12):1682–1684

Lumelsky N, Blondel O, Laeng P, Velasco I, Ravin R, McKay R (2001) Differentiation of embryonic stem cells to insulin-secreting structures similar to pancreatic islets. Science (New York NY) 292(5520):1389–1394

MacIsaac ZM, Shang H, Agrawal H, Yang N, Parker A, Katz AJ (2012) Long-term in-vivo tumorigenic assessment of human culture-expanded adipose stromal/stem cells. Exp Cell Res 318(4):416–423. doi:10.1016/j.yexcr.2011.12.002

Martin GR (1981) Isolation of a pluripotent cell line from early mouse embryos cultured in medium conditioned by teratocarcinoma stem cells. Proc Natl Acad Sci U S A 78(12):7634–7638

Matschinsky F, Liang Y, Kesavan P, Wang L, Froguel P, Velho G, Cohen D, Permutt MA, Tanizawa Y, Jetton TL (1993) Glucokinase as pancreatic beta cell glucose sensor and diabetes gene. J Clin Invest 92(5):2092–2098

Mazzaccara C, Iafusco D, Liguori R, Ferrigno M, Galderisi A, Vitale D, Simonelli F, Landolfo P, Prisco F, Masullo M, Sacchetti L (2012) Mitochondrial diabetes in children: seek and you will find it. PLoS One 7(4):e34956. doi:10.1371/journal.pone.0034956

Minteer D, Marra KG, Rubin JP (2012) Adipose-derived mesenchymal stem cells: biology and potential applications. Advances in biochemical engineering/biotechnology. Springer, Berlin

Miyazaki S, Yamato E, Miyazaki J-i (2004) Regulated expression of pdx-1 promotes in vitro differentiation of insulin-producing cells from embryonic stem cells. Diabetes 53(4):1030–1037

Mohamad Buang ML, Seng HK, Chung LH, Saim AB, Idrus RBH (2012) In vitro generation of functional insulin-producing cells from lipoaspirated human adipose tissue-derived stem cells. Arch Med Res 43(1):83–88

Montanya E (2004) Islet- and stem-cell-based tissue engineering in diabetes. Curr Opin Biotechnol 15(5):435–440. doi:10.1016/j.copbio.2004.08.011

Mukhi S, Brown DD (2011) Transdifferentiation of tadpole pancreatic acinar cells to duct cells mediated by notch and stromelysin-3. Dev Biol 351(2):311–317. doi:10.1016/j.ydbio.2010.12.020

Murohara T, Ikeda H, Duan J, Shintani S, Ki S, Eguchi H, Onitsuka I, Matsui K, Imaizumi T (2000) Transplanted cord blood-derived endothelial precursor cells augment postnatal neovascularization. J Clin Invest 105(11):1527–1536. doi:10.1172/JCI8296

Naruse K, Hamada Y, Nakashima E, Kato K, Mizubayashi R, Kamiya H, Yuzawa Y, Matsuo S, Murohara T, Matsubara T, Oiso Y, Nakamura J (2005) Therapeutic neovascularization using cord blood-derived endothelial progenitor cells for diabetic neuropathy. Diabetes 54(6):1823–1828

Nishimura W, Kondo T, Salameh T, El Khattabi I, Dodge R, Bonner-Weir S, Sharma A (2006) A switch from MafB to MafA expression accompanies differentiation to pancreatic beta-cells. Dev Biol 293(2):526–539

Noguchi H, Kaneto H, Weir GC, Bonner-Weir S (2003) PDX-1 protein containing its own antennapedia-like protein transduction domain can transduce pancreatic duct and islet cells. Diabetes 52(7):1732–1737

Noguchi H, Bonner-Weir S, Wei F-Y, Matsushita M, Matsumoto S (2005) BETA2/NeuroD protein can be transduced into cells due to an arginine- and lysine-rich sequence. Diabetes 54(10):2859–2866

Noguchi H, Oishi K, Ueda M, Yukawa H, Hayashi S, Kobayashi N, Levy MF, Matusmoto S (2009) Establishment of mouse pancreatic stem cell line. Cell Transplant 18(5):563–571

Noguchi H, Naziruddin B, Shimoda M, Fujita Y, Chujo D, Takita M, Peng H, Sugimoto K, Itoh T, Tamura Y, Olsen GS, Kobayashi N, Onaca N, Hayashi S, Levy MF, Matsumoto S (2010) Induction of insulin-producing cells from human pancreatic progenitor cells. Transplant Proc 42(6):2081–2083

Ohlsson H, Karlsson K, Edlund T (1993) IPF1, a homeodomain-containing transactivator of the insulin gene. EMBO J 12(11):4251–4259

Orlando G, Dominguez-Bendala J, Shupe T, Bergman C, Bitar KN, Booth C, Carbone M, Koch KL, Lerut JP, Neuberger JM, Petersen B, Ricordi C, Atala A, Stratta RJ, Soker S (2013) Cell and organ bioengineering technology as applied to gastrointestinal diseases. Gut 62(5):774–786

Orlando G, Gianello P, Salvatori M, Stratta RJ, Soker S, Ricordi C, Dominguez-Bendala J (2014) Cell replacement strategies aimed at reconstitution of the beta-cell compartment in type 1 diabetes. Diabetes 63(5):1433–1444

Parekh VS, Joglekar MV, Hardikar AA (2009) Differentiation of human umbilical cord blood-derived mononuclear cells to endocrine pancreatic lineage. Differ: Res Biol Divers 78(4):232–240

Plesner A, Soukhatcheva G, Korneluk RG, Verchere CB (2010) XIAP inhibition of β-cell apoptosis reduces the number of islets required to restore euglycemia in a syngeneic islet transplantation model. Islets 2(1):18–23. doi:10.4161/isl.2.1.9997

Puissant B, Barreau C, Bourin P, Clavel C, Corre J, Bousquet C, Taureau C, Cousin B, Abbal M, Laharrague P, Penicaud L, Casteilla L, Blancher A (2005) Immunomodulatory effect of human adipose tissue-derived adult stem cells: comparison with bone marrow mesenchymal stem cells. Br J Haematol 129(1):118–129

Rackham CL, Chagastelles PC, Nardi NB, Hauge-Evans AC, Jones PM, King AJF (2011) Co-transplantation of mesenchymal stem cells maintains islet organisation and morphology in mice. Diabetologia 54(5):1127–1135. doi:10.1007/s00125-011-2053-4

Rajagopal J, Anderson WJ, Kume S, Martinez OI, Da M (2003) Insulin staining of ES cell progeny from insulin uptake. Science (New York NY) 299(5605):363–363. doi:10.1126/science.1077838

Ramiya VK, Maraist M, Arfors KE, Schatz DA, Peck AB, Cornelius JG (2000) Reversal of insulin-dependent diabetes using islets generated in vitro from pancreatic stem cells. Nat Med 6(3):278–282. doi:10.1038/73128

Reddi AS, Kuppasani K, Ende N (2010) Human umbilical cord blood as an emerging stem cell therapy for diabetes mellitus. Curr Stem Cell Res Ther 5(4):356–361

Rezania A, Bruin JE, Riedel MJ, Mojibian M, Asadi A, Xu J, Gauvin R, Narayan K, Karanu F, O'Neil JJ, Ao Z, Warnock GL, Kieffer TJ (2012) Maturation of human embryonic stem cell-derived pancreatic progenitors into functional islets capable of treating pre-existing diabetes in mice. Diabetes 61(8):2016–2029. doi:10.2337/db11-1711

Robertson RP (2004) Islet transplantation as a treatment for diabetes – a work in progress. N Engl J Med 350(7):694–705. doi:10.1056/NEJMra032425

Rosenberg L (1995) In vivo cell transformation: neogenesis of beta cells from pancreatic ductal cells. Cell Transplant 4(4):371–383

Rosengren AH, Taneera J, Rymo S, Renström E (2009) Bone marrow transplantation stimulates pancreatic β-cell replication after tissue damage. Islets 1(1):10–18

Rukstalis JM, Habener JF (2009) Neurogenin3: a master regulator of pancreatic islet differentiation and regeneration. Islets 1(3):177–184. doi:10.4161/isl.1.3.9877

Ryan EA, Paty BW, Senior PA, Bigam D, Alfadhli E, Kneteman NM, Lakey JRT, Shapiro AMJ (2005) Five-year follow-up after clinical islet transplantation. Diabetes 54(7):2060–2069

Santamaria X, Massasa EE, Feng Y, Wolff E, Taylor HS (2011) Derivation of insulin producing cells from human endometrial stromal stem cells and use in the treatment of murine diabetes. Mol Ther: J Am Soc Gene Ther 19(11):2065–2071. doi:10.1038/mt.2011.173

Sapir T, Shternhall K, Meivar-Levy I, Blumenfeld T, Cohen H, Skutelsky E, Eventov-Friedman S, Barshack I, Goldberg I, Pri-Chen S, Ben-Dor L, Polak-Charcon S, Karasik A, Shimon I, Mor E, Ferber S (2005) Cell-replacement therapy for diabetes: generating functional insulin-producing tissue from adult human liver cells. Proc Natl Acad Sci U S A 102(22):7964–7969. doi:10.1073/pnas.0405277102

Schroeder IS, Kania G, Blyszczuk P, Wobus AM (2006) Insulin-producing cells. Methods Enzymol 418:315–333

Segev H, Fishman B, Ziskind A, Shulman M, Itskovitz-Eldor J (2004) Differentiation of human embryonic stem cells into insulin-producing clusters. Stem Cells (Dayt Ohio) 22(3):265–274. doi:10.1634/stemcells.22-3-265

Shapiro AM, Lakey JR, Ryan EA, Korbutt GS, Toth E, Warnock GL, Kneteman NM, Rajotte RV (2000) Islet transplantation in seven patients with type 1 diabetes mellitus using a glucocorticoid-free immunosuppressive regimen. N Engl J Med 343(4):230–238. doi:10.1056/NEJM200007273430401

Shi Y, Hou L, Tang F, Jiang W, Wang P, Ding M, Deng H (2005) Inducing embryonic stem cells to differentiate into pancreatic beta cells by a novel three-step approach with activin A and all-trans retinoic acid. Stem Cells (Dayt Ohio) 23(5):656–662

Shiroi A, Yoshikawa M, Yokota H, Fukui H, Ishizaka S, Tatsumi K, Takahashi Y (2002) Identification of insulin-producing cells derived from embryonic stem cells by zinc-chelating dithizone. Stem Cells (Dayt Ohio) 20(4):284–292. doi:10.1634/stemcells.20-4-284

Simpson AM, Tuch BE, Swan MA, Tu J, Marshall GM (1995) Functional expression of the human insulin gene in a human hepatoma cell line (HEP G2). Gene Ther 2(3):223–231

Simpson AM, Marshall GM, Tuch BE, Maxwell L, Szymanska B, Tu J, Beynon S, Ma S, Camacho M (1997) Gene therapy of diabetes: glucose-stimulated insulin secretion in a human hepatoma cell line (HEP G2ins/g). Gene Ther 4(11):1202–1215. doi:10.1038/sj.gt.3300527

Sipione S, Eshpeter A, Lyon JG, Korbutt GS, Bleackley RC (2004) Insulin expressing cells from differentiated embryonic stem cells are not beta cells. Diabetologia 47(3):499–508. doi:10.1007/s00125-004-1349-z

Solar M, Cardalda C, Houbracken I, Martín M, Maestro MA, De Medts N, Xu X, Grau V, Heimberg H, Bouwens L, Ferrer J (2009) Pancreatic exocrine duct cells give rise to insulin-producing beta cells during embryogenesis but not after birth. Dev Cell 17(6):849–860. doi:10.1016/j.devcel.2009.11.003

Soon-Shiong P, Heintz RE, Merideth N, Yao QX, Yao Z, Zheng T, Murphy M, Moloney MK, Schmehl M, Harris M et al (1994) Insulin independence in a type 1 diabetic patient after encapsulated islet transplantation. Lancet 343(8903):950–951

Soria B, Roche E, Berná G, León-Quinto T, Reig JA, Martín F (2000) Insulin-secreting cells derived from embryonic stem cells normalize glycemia in streptozotocin-induced diabetic mice. Diabetes 49(2):157–162

Sosa-Pineda B, Chowdhury K, Torres M, Oliver G, Gruss P (1997) The Pax4 gene is essential for differentiation of insulin-producing beta cells in the mammalian pancreas. Nature 386(6623):399–402. doi:10.1038/386399a0

Street CN, Lakey JRT, Seeberger K, Helms L, Rajotte RV, Shapiro AMJ, Korbutt GS (2004) Heterogenous expression of nestin in human pancreatic tissue precludes its use as an islet precursor marker. J Endocrinol 180(2):213–225

Stride A, Hattersley AT (2002) Different genes, different diabetes: lessons from maturity-onset diabetes of the young. Ann Med 34(3):207–216

Stumvoll M, Goldstein BJ, van Haeften TW (2010) Type 2 diabetes: principles of pathogenesis and therapy. Lancet 365(9467):1333–1346. doi:10.1016/S0140-6736(05)61032-X

Sulzbacher S, Schroeder IS, Truong TT, Wobus AM (2009) Activin A-induced differentiation of embryonic stem cells into endoderm and pancreatic progenitors-the influence of differentiation factors and culture conditions. Stem Cell Rev 5(2):159–173

Sun B, Roh K-H, Lee S-R, Lee Y-S, Kang K-S (2007) Induction of human umbilical cord blood-derived stem cells with embryonic stem cell phenotypes into insulin producing islet-like structure. Biochem Biophys Res Commun 354(4):919–923. doi:10.1016/j.bbrc.2007.01.069

Takahara H, Fujimura M, Taniguchi S, Hayashi N, Nakamura T, Fujimiya M (2001) Changes in serotonin levels and 5-HT receptor activity in duodenum of streptozotocin-diabetic rats. Am J Physiol Gastrointest Liver Physiol 281(3):G798–G808

Taniguchi H, Fukao K, Nakauchi H (1997) Constant delivery of proinsulin by encapsulation of transfected cells. J Surg Res 70(1):41–45

Thomson JA (1998) Embryonic stem cell lines derived from human blastocysts. Science 282(5391):1145–1147. doi:10.1126/science.282.5391.1145

Timper K, Seboek D, Eberhardt M, Linscheid P, Christ-Crain M, Keller U, Müller B, Zulewski H (2006) Human adipose tissue-derived mesenchymal stem cells differentiate into insulin, somatostatin, and glucagon expressing cells. Biochem Biophys Res Commun 341(4):1135–1140. doi:10.1016/j.bbrc.2006.01.072

Tolhurst G, Reimann F, Gribble FM (2009) Nutritional regulation of glucagon-like peptide-1 secretion. J Physiol 587(Pt 1):27–32

Tsai P-J, Wang H-S, Shyr Y-M, Weng Z-C, Tai L-C, Shyu J-F, Chen T-H (2012) Transplantation of insulin-producing cells from umbilical cord mesenchymal stem cells for the treatment of streptozotocin-induced diabetic rats. J Biomed Sci 19(1):47–47

Tu J, Tuch BE (1997) Expression of glucokinase in glucose-unresponsive human fetal pancreatic islet-like cell clusters. J Clin Endocrinol Metab 82(3):943–948

Tu J, Tuch BE, Si Z (1999) Expression and regulation of glucokinase in rat islet beta- and alpha-cells during development. Endocrinology 140(8):3762–3766

Vaca P, Berná G, Martín F, Soria B (2003) Nicotinamide induces both proliferation and differentiation of embryonic stem cells into insulin-producing cells. Transplant Proc 35(5):2021–2023

Van Pham P, Thi-My Nguyen P, Thai-Quynh Nguyen A, Minh Pham V, Nguyen-Tu Bui A, Thi-Tung Dang L, Gia Nguyen K, Kim Phan N (2014) Improved differentiation of umbilical cord blood-derived mesenchymal stem cells into insulin-producing cells by PDX-1 mRNA transfection. Differentiation 87(5):200–208. doi:http://dx.doi.org/10.1016/j.diff.2014.08.001

Venkatesan V, Gopurappilly R, Goteti SK, Dorisetty RK, Bhonde RR (2011) Pancreatic progenitors: the shortest route to restore islet cell mass. Islets 3(6):295–301. doi:10.4161/isl.3.6.17704

Voltarelli JC, Couri CEB, Stracieri ABPL, Oliveira MC, Moraes DA, Pieroni F, Coutinho M, Malmegrim KCR, Foss-Freitas MC, Simões BP, Foss MC, Squiers E, Burt RK (2007) Autologous nonmyeloablative hematopoietic stem cell transplantation in newly diagnosed type 1 diabetes mellitus. JAMA 297(14):1568–1576. doi:10.1001/jama.297.14.1568

Vrochides D, Paraskevas S, Papanikolaou V (2009) Transplantation for type 1 diabetes mellitus. Whole organ or islets? Hippokratia 13(1):6–8

Waeber G, Thompson N, Nicod P, Bonny C (1996) Transcriptional activation of the GLUT2 gene by the IPF-1/STF-1/IDX-1 homeobox factor. Mol Endocrinol (Baltimore Md) 10(11):1327–1334

Wang H-S, Hung S-C, Peng S-T, Huang C-C, Wei H-M, Guo Y-J, Fu Y-S, Lai M-C, Chen C-C (2004) Mesenchymal stem cells in the Wharton's jelly of the human umbilical cord. Stem Cells (Dayt Ohio) 22(7):1330–1337

Wang P, Yigit MV, Medarova Z, Wei L, Dai G, Schuetz C, Moore A (2011) Combined small interfering RNA therapy and in vivo magnetic resonance imaging in islet transplantation. Diabetes 60(2):565–571. doi:10.2337/db10-1400

Watada H, Kajimoto Y, Umayahara Y, Matsuoka T, Kaneto H, Fujitani Y, Kamada T, Kawamori R, Yamasaki Y (1996) The human glucokinase gene beta-cell-type promoter: an essential role of insulin promoter factor 1/PDX-1 in its activation in HIT-T15 cells. Diabetes 45(11):1478–1488

Weinzimer SA, Steil GM, Swan KL, Dziura J, Kurtz N, Tamborlane WV (2008) Fully automated closed-loop insulin delivery versus Semiautomated hybrid control in pediatric patients with

type 1 diabetes using an artificial pancreas. Diabetes Care 31:934. doi:10.2337/dc07-1967. S.A.W.
Weissman IL (2000) Stem cells: units of development, units of regeneration, and units in evolution. Cell 100(1):157–168
WHO (2015) Diabetes. http://www.who.int/mediacentre/factsheets/fs312/en/. Accessed 10 May 2015
Wu X-H, Liu C-P, Xu K-F, Mao X-D, Zhu J, Jiang J-J, Cui D, Zhang M, Xu Y, Liu C (2007) Reversal of hyperglycemia in diabetic rats by portal vein transplantation of islet-like cells generated from bone marrow mesenchymal stem cells. World J Gastroenterol: WJG 13(24):3342–3349
Wu H, Lu W, Mahato RI (2011) Mesenchymal stem cells as a gene delivery vehicle for successful islet transplantation. Pharm Res 28(9):2098–2109. doi:10.1007/s11095-011-0434-5
Xia B, Zhan X-R, Yi R, Yang B (2009) Can pancreatic duct-derived progenitors be a source of islet regeneration? Biochem Biophys Res Commun 383(4):383–385. doi:10.1016/j.bbrc.2009.03.114
Xu W, Qian H, Zhu W, Chen Y, Shao Q, Sun X, Hu J, Han C, Zhang X (2004) A novel tumor cell line cloned from mutated human embryonic bone marrow mesenchymal stem cells. Oncol Rep 12(3):501–508
Yabe D, Seino Y (2011) Two incretin hormones GLP-1 and GIP: comparison of their actions in insulin secretion and β cell preservation. Prog Biophys Mol Biol 107(2):248–256
Yamamoto T, Yamato E, Taniguchi H, Shimoda M, Tashiro F, Hosoi M, Sato T, Fujii S, Miyazaki JI (2006) Stimulation of cAMP signalling allows isolation of clonal pancreatic precursor cells from adult mouse pancreas. Diabetologia 49(10):2359–2367. doi:10.1007/s00125-006-0372-7
Yang K-m, A-d L, Mei Y, H-y Z, Li H, Yang H-j (2006) Islet formation and regeneration. Chin Med Sci J Chung-kuo i hsüeh k'o hsüeh tsa chih/Chin Acad Med Sci 21(1):27–32
Yatoh S, Dodge R, Akashi T, Omer A, Sharma A, Weir GC, Bonner-Weir S (2007) Differentiation of affinity-purified human pancreatic duct cells to beta-cells. Diabetes 56(7):1802–1809. doi:10.2337/db06-1670
Yoshida S, Ishikawa F, Kawano N, Shimoda K, Nagafuchi S, Shimoda S, Yasukawa M, Kanemaru T, Ishibashi H, Shultz LD, Harada M (2005) Human cord blood – derived cells generate insulin-producing cells in vivo. Stem Cells (Dayt Ohio) 23(9):1409–1416. doi:10.1634/stemcells.2005-0079
Zalzman M, Gupta S, Giri RK, Berkovich I, Sappal BS, Karnieli O, Zern MA, Fleischer N, Efrat S (2003) Reversal of hyperglycemia in mice by using human expandable insulin-producing cells differentiated from fetal liver progenitor cells. Proc Natl Acad Sci U S A 100(12):7253–7258. doi:10.1073/pnas.1136854100
Zalzman M, Anker-Kitai L, Efrat S (2005) Differentiation of human liver-derived, insulin-producing cells toward the beta-cell phenotype. Diabetes 54(9):2568–2575
Zhang YC, Pileggi A, Agarwal A, Molano RD, Powers M, Brusko T, Wasserfall C, Goudy K, Zahr E, Poggioli R, Scott-jorgensen M, Campbell-thompson M, Crawford JM, Nick H, Flotte T, Ellis TM, Ricordi C, Inverardi L, Atkinson MA (2003) Inhibits diabetes recurrence in syngeneic islet cell transplantation of NOD mice. Diabetes 52:708
Zhao Y, Wang H, Mazzone T (2006) Identification of stem cells from human umbilical cord blood with embryonic and hematopoietic characteristics. Exp Cell Res 312(13):2454–2464. doi:10.1016/j.yexcr.2006.04.008
Zheng HT, Deng HC, Huang CJ, Lan NZ, Fang F, Jian R (2007) Co-transfection of GK and mhPINS genes into HepG2 cells confers glucose-stimulated insulin secretion. Cytotherapy 9(6):580–586. doi:10.1080/14653240701411350
Zhou Q, Brown J, Kanarek A, Rajagopal J, Melton DA (2008) In vivo reprogramming of adult pancreatic exocrine cells to beta-cells. Nature 455(7213):627–632. doi:10.1038/nature07314
Zhu W, Xu W, Jiang R, Qian H, Chen M, Hu J, Cao W, Han C, Chen Y (2006) Mesenchymal stem cells derived from bone marrow favor tumor cell growth in vivo. Exp Mol Pathol 80(3):267–274

Zuk PA, Zhu M, Mizuno H, Huang J, Futrell JW, Katz AJ, Benhaim P, Lorenz HP, Hedrick MH (2001) Multilineage cells from human adipose tissue: implications for cell-based therapies. Tissue Eng 7(2):211–228. doi:10.1089/107632701300062859

Zulewski H, Abraham EJ, Gerlach MJ, Daniel PB, Moritz W, Müller B, Vallejo M, Thomas MK, Habener JF (2001) Multipotential nestin-positive stem cells isolated from adult pancreatic islets differentiate ex vivo into pancreatic endocrine, exocrine, and hepatic phenotypes. Diabetes 50(3):521–533

Chapter 2
Stem Cell Therapy for Type-1 Diabetes Mellitus

Umang G. Thakkar, Aruna V. Vanikar, and Hargovind L. Trivedi

Abbreviations

AD-MSC	Adipose tissue-derived mesenchymal stem cell
BM	Bone marrow
DC	Dendritic cells
DCCTR	Diabetes control and complication trial research
DCDM	Diagnostic criteria for DM

U.G. Thakkar, MBBS, DCH (✉)
Department of Regenerative Medicine and Stem Cell Therapy and Department of Pediatrics, G.R. Doshi and K.M. Mehta Institute of Kidney Diseases & Research Centre (IKDRC)- Dr. H.L. Trivedi Institute of Transplantation Sciences (ITS), Civil Hospital Campus, Asarwa, Ahmedabad, Gujarat 380016, India
e-mail: umangpaedia@yahoo.co.in

A.V. Vanikar
Department of Regenerative Medicine and Stem Cell Therapy and Department of Pediatrics, G.R. Doshi and K.M. Mehta Institute of Kidney Diseases & Research Centre (IKDRC)- Dr. H.L. Trivedi Institute of Transplantation Sciences (ITS), Civil Hospital Campus, Asarwa, Ahmedabad, Gujarat 380016, India

Department of Pathology, Laboratory Medicine, Transfusion Services and Immunohematology, G.R. Doshi and K.M. Mehta Institute of Kidney Diseases & Research Centre (IKDRC)- Dr. H.L. Trivedi Institute of Transplantation Sciences (ITS), Gujarat 380016, India

H.L. Trivedi
Department of Regenerative Medicine and Stem Cell Therapy and Department of Pediatrics, G.R. Doshi and K.M. Mehta Institute of Kidney Diseases & Research Centre (IKDRC)- Dr. H.L. Trivedi Institute of Transplantation Sciences (ITS), Civil Hospital Campus, Asarwa, Ahmedabad, Gujarat 380016, India

Department of Nephrology and Transplantation Medicine, G.R. Doshi and K.M. Mehta Institute of Kidney Diseases & Research Centre (IKDRC)- Dr. H.L. Trivedi Institute of Transplantation Sciences (ITS), Gujarat 380016, India

P.V. Pham (ed.), *Pancreas, Kidney and Skin Regeneration*, Stem Cells in Clinical Applications, DOI 10.1007/978-3-319-55687-1_2

DM	Diabetes mellitus
ESC	Embryonic stem cell
EV	Extracellular vesicles
HbA1c	Glycosylated hemoglobin
HSC	Hematopoietic stem cell
IPC	Insulin-producing cell
iPSC	Induced pluripotent stem cells
IPSC	Insulin-producing stem cell
ISC	Insulin-secreting cell
MSC	Mesenchymal stem cell
MODY	Maturity onset diabetes of the young
SCT	Stem cell therapy
T1DM	Type-1 diabetes mellitus
T2DM	Type-2 diabetes mellitus
UCB	Umbilical cord blood
WHO	World Health Organization

2.1 Review

2.1.1 Introduction of Disease

Diabetes mellitus (DM) is a metabolic disease characterized by chronic hyperglycemia resulting from defects in insulin secretion, insulin action, or both. The abnormalities are found in carbohydrate, lipid, and protein metabolism in diabetics, because of deficient action of insulin on target organs. If ketones are present in blood/urine, treatment is required urgently, because ketoacidosis can evolve rapidly.

2.1.2 Historical Witnesses of Diabetes Mellitus and Discovery of Insulin

DM was described for the first time as "Madhumeha" (sweet-tasting urine) by noted Indian physician Charaka and surgeon Sushruta in the fifth and sixth century AD who observed that urine of diabetics attracted ants and was sticky to touch. The Egyptian physicians also described it about 3500 years ago. The word "diabetes" ("to go through") was coined by Apollonius of Memphis around 250 BC. These patients drained more fluid than the amount consumed; hence, this term was coined (Mac Cracken et al. 1997). Then, in the second century AD, the term "diabetes" was once again adopted by Aretaeus of Cappadocia who stated that it was the fault of the kidney by which diabetes developed. This theory was also accepted by a Roman physician named Galen (AD 131–201). He proposed an alternative name to diabetes, "diarrhea urinosa" for excessive urinary output and "dipsakos" for excessive thirst and drinking (Pickup and Willium 2003). However, the Indian description of diabetes was more scholastic, differentiating two types of diabetes: one affecting older and obese persons and the second affecting younger and thin persons. They also had observed that young diabetics had shorter life span. This description is followed as of today also: as type-1 (T1) and type-2 (T2) DM (Kahn 1994). In 1809, John Rollo, an apothecary and chemist, was the first to use the term "mellitus," the Latin and Greek root for honey. He documented high level of sugar in blood and in urine (MacCracken and Hoel 1997). It took about 60 years for further research in diabetes when Paul Langerhans noticed small clusters of cells in the pancreas in the year 1869 and described these structures without speculating about their possible function. In 1893, Edouard Laguesse suggested that these clusters of cells in the pancreas might constitute endocrine tissue of the pancreas, which he named as "islets of Langerhans" (after Paul Langerhans) (Pickup and Willium 2003). In 1909, a Belgian physician, Jean de Meyer isolated glucose-lowering hormone from pancreatic islet cells and named it as insulin (Latin, insula=Island) (De Meyer 1904). In 1921, Frederick G. Banting and Charles Herbert Best (his student under Macleod's patronization) proved that insulin is an active substance of the pancreas associated with hypoglycemia in diabetic dogs. On 1st January 1922, the insulin extract made by Banting and Best was injected for the first time in history, into Leonard Thompson, a 14-year-old boy with DM. This was repeated by Collip. Thompson's blood sugar returned to normal in about 24 h along with disappearance of his glycosuria and ketonuria (MacCracken and Hoel 1997)! In 1922, Banting treated Elizabeth Evans Hughes. Her recovery from hyperglycemia due to DM was hailed around the world as a true medical miracle. Banting and Macleod won the Nobel Prize for the discovery of insulin in 1923.The first commercial product of human insulin was developed by recombinant DNA technology in 1979 by Goeddel et al. (Ullrich et al. 1977), and human insulin was first prepared by Graham Bell et al. in 1980. In July 1996, the Food and Drug Administration of the USA approved the first recombinant DNA human insulin analogue, Lispro (Humalog). Currently more than 300 insulin analogues have been identified, including 70 from animals, 80 chemically modified, and 150 biosynthetic insulin preparations (Drjer 1992). This historical review shows that most of the advancement in discovery, treatment, and management of diabetes occurred in the twentieth century and that has given the greatest benefit to mankind in the form of longer life with better quality.

2.1.3 Classification of Diabetes Mellitus

The etiological classification was recommended by the "American Diabetes Association" (Diagnosis and Classification of Diabetes Mellitus 2009) and "WHO" expert committee on the classification and diagnosis of diabetes (World Health Organization WHO/NCD/NCS/99.2. 1999). Classification with minor modification is as below.

I. Type-1 diabetes (β-cell destruction, usually leading to absolute insulin deficiency)
   1. Immune mediated
   2. Idiopathic
II. Type-2 diabetes (may range from predominantly insulin resistance with relative insulin deficiency to a predominantly secretory defect with insulin resistance)
III. Other specific types
   1. Genetic defects of β-cell function
   2. Genetic defects in insulin action
   3. Diseases of the exocrine pancreas
   4. Endocrinopathies
   5. Drug or chemical induced
   6. Infections
   7. Uncommon forms of immune-mediated diabetes
   8. Other genetic syndromes sometimes associated with diabetes

      Genetic defects of β-cell function or insulin action, formerly termed "maturity onset diabetes of the young" (MODY), was originally described as a disorder with the following characteristics: onset before 25 years of age, autosomal dominant inheritance, and nonketotic diabetes mellitus (Fajans et al. 2001; Murphy et al. 2008).
IV. Gestational diabetes mellitus

(Patients with any form of DM may require insulin treatment at some stage of their disease. Such use of insulin does not, of itself, classify the patient.)

2.1.4 Criteria for Diagnosis of Diabetes

Diagnostic criteria for DM are based on blood glucose measurements and the presence or absence of symptoms (DCDM 2009; WHO/NCD/NCS/99.2. 1999). Three ways to diagnose DM are possible and each, in the absence of unequivocal hyperglycemia, must be confirmed, on a subsequent day, by any one of the three methods given in Table 2.1.

2.1.5 Spectrum of DM

T1DM, T2DM, and T1.5DM are differentiated from each other in Table 2.2 (Islets of Hope 2006; Zimmet et al. 2001; Leroith et al. 2003; Yoon and Jun 2005; Daneman 2006; Wilkin 2008).

Table 2.1 Criteria for the diagnosis of diabetes mellitus

1. Symptoms of diabetes plus casual plasma glucose concentration ≥11.1 mmol/L (200 mg/dL) [a]Casual is defined as any time of day without regard to time since last meal. **Or**
2. Fasting plasma glucose ≥7.0 mmol/L (≥126 mg/dL). Fasting is defined as no caloric intake for at least 8 h. **Or**
3. 2-h post-load glucose ≥11.1 mmol/L (≥200 mg/dL) during an OGTT. The test should be performed as described by WHO (86), using a glucose load containing the equivalent of 75 g anhydrous glucose dissolved in water or 1.75 g/kg of body weight to a maximum of 75 g (65)

[a]Corresponding values (mmol/L) are ≥10.0 for venous whole blood and ≥11.1 for capillary whole blood and ≥6.3 for both venous and capillary whole blood
Or
HbA1C ≥6.5%. The test should be performed in a laboratory using a method that is NGSP certified and standardized to the DCCT assay (In the absence of unequivocal hyperglycemia, criteria 1–3 should be confirmed by repeat testing)

Table 2.2 Spectrum of diabetes

Feature	T1DM (IDDM)	T2DM (NIDDM)	T1.5DM (LADA)
Nature of onset Age of diagnosis	Usually rapid onset Occurs in usually childhood age	Onset is months/years. Occur mainly in older adults	Onset is slow. Occurs in 35–40 years/earliest at 25 years
Genes, triggers factors	Autoimmune, idiopathic, genetic	Hereditary, sedentary lifestyle, obesity, etc.	Autoimmune
ICA(islet cell antibodies) IAA(insulin autoantibodies) IA2 (islet antigen 2) GAD(GAD65-AAGAD) HLA C-peptide	ICA = 80% IAA = often detected IA2 = 50–70% GAD = positive HLA = yes C-peptides – always low	ICA – no IAA – no IA2 – no GAD – negative HLA – no C-peptides – normal to high	ICA – positive IAA – yes, often IA2 – often GAD – positive HLA – yes, often C-peptides – low
Treatment	Insulin	Oral hypoglycemic agents ± insulin	Insulin ± OHA
Prognosis and Complications	No complete cure	No complete cure	No complete cure

2.1.6 Epidemiology of T1DM

Occurrence of T1DM is increasing internationally, and it accounts for an approximate of 5–10% of all cases of DM (Daneman 2006) or 11–22 million globally (http://ghr.nlm.nih.gov/condition/type-1-diabetes 2013). According to Diabetes Control and Complications Trial (DCCT), the prevalence of T1DM in 2010 was 0.1–0.5% worldwide among general population, more than six million patients (1 out of 100–300 newborns), and its incidence was 30–50 new patients per every 100,000 individuals, with a 3% increase yearly, mainly in developing nations acquiring a Western lifestyle and diet (DCCTR 1993; Zimmet et al. 2001; Leroith et al. 2003, Lippincott and Williams, Yoon and Jun 2005, Daneman 2006; Wilkin

2008). T1DM is one of the most common chronic diseases of childhood; however, it can be diagnosed at any age (Gale 2005). Peak in presentation of T1DM usually noticed in between 5 and 7 years of age and at or near puberty (Harjutsalo et al. 2008). Whereas most autoimmune disorders disproportionately affect female, T1DM is slightly more common in male (Ostman et al. 2008). The incidence of T1DM varies with seasonal changes and birth month. More cases are diagnosed in autumn and winter (Moltchanova et al. 2009), and being born in the spring is associated with a higher chance of having T1DM (Kahn et al. 2009). Development of T1DM-associated autoimmunity (i.e., formation of islet autoantibodies) in the months or years before onset of symptoms also shows some seasonal synchronization (Kukko et al. 2005). The incidence of T1DM has been increasing globally for several decades (Dabelea 2009); still its incidence and prevalence vary substantially (Fig. 2.1) (Maahs et al. 2010). T1DM is most common in Finland (>60 cases/100,000 people each year) and Sardinia (around 40 cases/100,000 people each year) (Patterson et al. 2009). By contrast, the disorder is uncommon in China, India, and Venezuela (around 0.1 cases per 100,000 people each year) (Thunander et al. 2008). A plethora of environmental influences have been purported to affect the epidemiology of T1DM (Maclaren and Atkinson 1992), with infant and adolescent diets (Knip et al. 2010), vitamin-D and its pathway constituents (Svoren et al. 2009; Blanton et al. 2011; Cooper et al. 2011), and viruses receiving the most focus (Yeung et al. 2011; Stene and Rewers 2012).

2.1.7 Management of T1DM

The discovery of insulin in 1921–1922 was the most significant therapeutic event in the history of T1DM; however, exogenous insulin administration is not always necessary to provide the metabolic regulation to avoid disease related-complications. Because of selective destruction of the insulin-producing β cells within the pancreatic islets by triggers resulting in complete insulin deficiency, which causes hyperglycemia eventually leading to acute (ketoacidosis) and chronic (retinopathy, nephropathy, neuropathy) complications, hypercoagulability, dyslipidemia, and accelerated atherosclerosis, increased cardiovascular diseases, and reduced life expectancy (DCCTR 1993; Zimmet et al. 2001; Leroith et al. 2003). The impact of DM involves ≈10% of total health-care budget in developed nations with over US $100 billion spent every year in the USA alone and over US $200 billion worldwide. The treatment of choice in association with tailored diet and physical exercise programs is daily exogenous insulin administration. Novel insulin formulations (e.g., glargine and lispro analogues) together with infusion-pump and glucose-sensor technologies have improved metabolic control with limited benefits (DCCTR 1993; Zimmet et al. 2001; Leroith et al. 2003; Daneman 2006). Treatment targets to achieve better glycemic control with T1DM in children and adolescents by plasma glucose and glycosylated hemoglobin (HbA1c) established in the

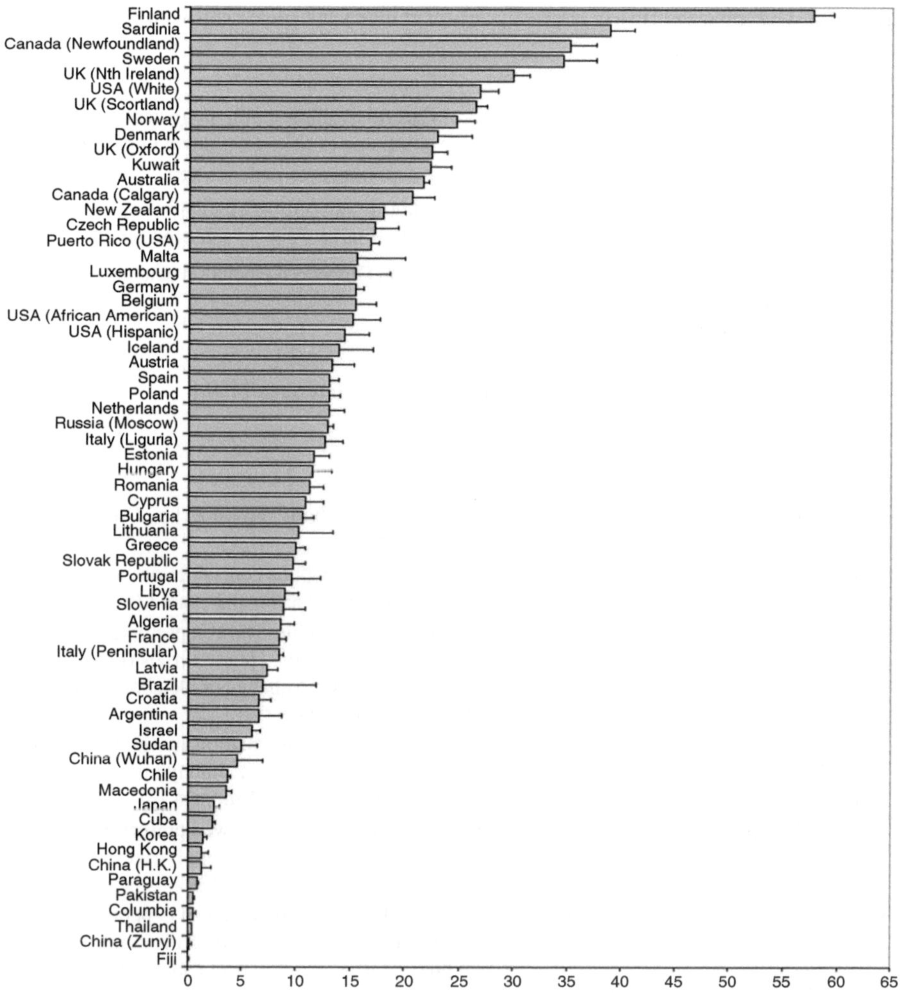

Fig. 2.1 Global incidence and prevalence of T1DM

evidence-based 2003 clinical practice guidelines of the Canadian Diabetes Association (2003) (Table 2.3).

Currently no approved agents are available to arrest the autoimmune destruction of β cells. The interest in reversing T1DM has grown in the last 5 years (Greenbaum and Atkinson 2011). In addition to preserving C-peptide production, the key goal is to induce immune tolerance against β cells. Majority of the approaches involve provision of self-antigens like vaccination with specific islet-cell proteins like insulin or glutamic acid decarboxylase (GAD) or immune suppression (Table 2.4).

Disappointingly, after promising phase 1–2 trials in patients with recent-onset T1DM and detectable endogenous insulin production, phase 3 trials of anti-CD3 antibodies (otelixizumab and teplizumab) and Diamyd vaccine (GAD-alum

Table 2.3 Glycemic and HbA1c targets by age for children and adolescents with T1DM

Age (years)	Plasma glucose (mmol/L)	HbA1c (%)	Considerations
<5	6.0–12.0	≤9.0	Careful avoidance of hypoglycemia in this age group due to risk of cognitive impairment
5–12	4.0–10.0	8.0	Adapt targets to patient's age
13–18	4.0–7.0	7.0	Appropriate for most patients
>18	4.0–6.0	6.0	Only if targets can be achieved safely

Source: 2003 clinical practice guidelines of the Canadian Diabetes Association (Canadian Diabetes Association 2003)
HbA1c hemoglobin A1c

Table 2.4 Agents assessed as immunomodulatory therapy to reverse T1DM

Study	Study phase and year	Main finding
Insulin APL (NBI-6042)	Phase 2; 2009	No change in metabolic response (i.e., C-peptide preservation) 135
Anti-CD20 (rituximab)	Phase 2; 2011	Preservation of C-peptide concentrations at 1 year, but no difference from placebo at 2 years 136
Anti-CD3 (teplizumab)	Phase 3; 2011	Although phase 2 studies showed preservation of C-peptide concentrations, phase I trials (Protégé study) 137 showed no change in metabolic response and the study stopped early
CTLA4, (abatacept) immunoglobulin fusion protein	Phase 2; 2011	T-cell co-stimulatory modulation slowed reduction in β-cell function over 2 years, although preservation of C-peptide was seen for 9–6 months 138
Anti-CD3 (otelixizumab)	Phase 3; 2011	Although phase 2 studies showed preservation of C-peptide concentrations, a phase 3 trial showed no change in metabolic response 139
GAD65 protein (Diamyd)	Phase 3; 2012	Phase 2 studies reported preserved C-peptide concentration, with no improvements in insulin needs. Two phase 3 trials did not meet end points140,141
HSP60 (DiaPep277)	Phase 3; 2012	Phase 2 trials suggested increased C-peptide concentrations; a phase 3 trial noted C-peptide preservation at 1 year but only in adults (age 16–45 years) with type-1 diabetes 142

immunotherapy) did not meet with primary end points (Walter et al. 2009; Pescovitz et al. 2009; Sherry et al. 2011; Orban et al. 2011; Bach 2011; Ludvigsson et al. 2012; Wherrett et al. 2011). Administration of DiaPep277, a synthetic immunomodulator at 3-month intervals, resulted in less decline in stimulated C-peptide concentrations at 1 year in adults with T1DM than in the cohort that received placebo (Buzzetti et al. 2011; Schloot et al. 2007). Other phase 2 studies of immunomodulators showed evidence of therapeutic efficacy in settings of recent-onset T1DM; however, even with continued use, majority did not show long-lasting effects. For example, the fusion protein CTLA4-Ig (abatacept) preserved stimulated C-peptide concentration for only 9 months despite continuous intravenous administration for 2 years (Orban et al. 2011). These results imply that single-agent immunosuppression alone might be insufficient to completely control the autoimmune destruction of β cells, or more specific and targeted therapies are required.

Glucose homeostasis requires finely regulated insulin secretion by pancreatic β cells present in islets of Langerhans (Mering and Minkowski 1889). In healthy person, insulin is secreted at a rate of ~2 pmol/kg/min, under fasting basal conditions (Eaton et al. 1980; Polonsky et al. 1984), and it increases by rate of ~5 to tenfold, after meal ingestion (Meier and Butler 2005). To accomplish this requirement, normally functioning β cells, and adequate number of β cells, means β-cell mass should be present. The human pancreas contains ≈ 1 million islets, each containing approximately 2000 β cells in healthy individual (Langerhans 1869; Stefan et al. 1982; Rahier et al. 1983; Bonner-Weir 1991). Thus, the β cells constitute ~1.5% of the total pancreas (1–2 g in total) (Bonner-Weir 1991).

The pancreatic islets of Langerhans containing β cells are functionally complex endocrine structures which produce insulin, that detect minimal changes in blood glucose levels and other metabolites, and also maintain metabolic homeostasis by a fine real-time secretion of specific hormones. A single β cell with a size of 15 μm can store about 10,000 insulin granules, and a single insulin granule of the size of 300 nm contains approximately 200,000 molecules of the crystallized insulin (Halban 2004). This is a well-orchestrated process, which is initially triggered by the glucose intake at the cell membrane and then eventually ends up with the glucose-responding insulin secretion (Ball and Barber 2003). Therefore, the generation of reliable pancreatic β cells is quite a difficult task.

These studies imply that a combination of therapies targeting multiple pathogenic pathways and improving β-cell viability is needed to preserve endogenous insulin production in this group of patients.

2.1.8 *Pancreatic Transplantation*

After ≈40 years of unsuccessful attempts by various researchers to control DM using partial pancreas transplantation, an English surgeon Charles Pybus (1882–1975) made a statement in 1924 that resonates even today: "Not much can be said about the principles of grafting, but it seems that until we are able to understand them (and I feel we do not understand them at present, especially the chemical factors), then we must continue to fail in such operations, although they may appear the most rational treatment for the diseases for which they are attempted" (Benedum 1999; Pybus 1924). The very first pancreatic transplant was surgically operated in 1966 by Kelly and colleagues (1967), and in the same year, pancreatic transplantation for T1DM was achieved at the University of Minnesota by Lillehei and coworkers (Jahansouz et al. 2011). Afterward, >25,000 pancreatic transplants have been performed worldwide (Gruessner and Sutherland 2005). The apparent easiest solution was deceased donors' pancreatic transplantation to replace diseased/

destroyed β cells of islet of Langerhans, in T1DM. The pancreas can be grafted heterotopically under strict immunosuppressive regimen to avoid immune rejection. Data from the International Pancreas Transplant Registry shows encouraging results of metabolic control following whole pancreas transplantation with consequential discontinuation of the exogenous insulin administration. However, there are two major drawbacks with this approach, significant morbidity related to major surgery, and life-long immunosuppression requirement (Cefalu 2012).

As compared to kidney transplantation, pancreas transplantation has higher risk of surgical complications; also non-immunological complications of pancreas transplantation (including thrombosis and graft pancreatitis) account for graft loss in 5–10% of cases. These usually occur within 6 months of transplantation and are as important as etiology of pancreas graft loss in simultaneous pancreas and kidney transplantation (Ciancio et al. 1996b; Gruessner et al. 1996; Gruessner and Sutherland 2001, 2005).

All these problems related to pancreatic surgery led to introduce alternative approach of islet transplantation (Ballinger and Lacy 1972; Scharp et al. 1990). First clinical practice for pancreatic islet transplant was reported in 1977 to treat diabetic patients (Lakey et al. 2003). With further developments of the Ricordi method in 1989 for islet extraction (Ricordi et al. 1989), Edmonton Protocol in 2000 (Shapiro et al. 2000), and recent refinements, islet transplantation is the feasible option to treat T1DM, but potential immunosuppression is needed to alleviate the graft survival.

2.1.9 Islet Cell Transplantation

Islets of Langerhans are clusters of different endocrine cells scattered throughout the pancreas, each type secreting a specific hormone: α cells (glucagon), β cells (insulin and amylin), δ cells (somatostatin), and PP cells (pancreatic polypeptide). It is estimated that a normal human pancreas hosts about one million islet cells; however, the number varies with age, sex, weight of the donor, organ size, and with functional integrity (Ricordi 1992; Leroith et al. 2003; Cabrera et al. 2006; Leibiger and Berggren 2008). In 2000, a breakthrough protocol was developed for islet transplantation without the use of glucocorticoids for immune suppression; (Shapiro et al. 2000) the initially promising results deteriorated so that at 5 years, only 10% of patients remained independent of exogenous insulin (Ryan et al. 2005). Therefore, islet transplantation still remains an experimental procedure with ongoing research focusing on new methods using biomaterials (e.g., encapsulation), immunomodulation, site of delivery, and improved vascularization (Gibly et al. 2011). There are many reasons for poor outcome of clinical trials of islet transplantation, and only some of them are identified. In particular, during islet infusion, an intravascular instant blood-mediated inflammatory reaction (IBMIR) is believed to be responsible for destroying 50–70% of the infused β cells. An upregulation of tissue factor and other molecules on islet cell surface after isolation process is capable of triggering innate immunity via activation of coagulation, complements, inflammation, and natural antibodies thereby destroying the islets. Peri-transplant anticoagulant prophylaxis with heparin can counteract this reaction (Moberg et al. 2002; Johansson et al. 2005; Eich et al. 2007).

Progressive decline in functioning of active β-cell mass from retrieval to the grafting procedure and immediate posttransplant time period is due to immune and nonimmune factors including activation of coagulation cascade and hostile microenvironment (Balamurugan et al. 2014; Nilsson et al. 2011). Calcineurin inhibitor/mTOR inhibitor and the process of islet isolation itself may contribute to low islet viability (Posselt et al. 2010; Balamurugan et al. 2010; Webb et al. 2012). The main hindrance to achieve consistent positive results is diabetogenic effect of corticosteroids and CNIs on β-cell function and survival, as well as on the development of peripheral insulin resistance. Posttransplant DM occurs in more than 50% solid organ transplant recipients, and the incidence increases with dose and duration of immunosuppressive therapy. Moreover, drug-dependent increment of lipids is associated with increased allograft loss and toxicity. Glucolipotoxicity may cause β-cell dysfunction and loss (Subramanian and Trence 2007; Vantyghem et al. 2007; Poitout and Robertson 2008). These immunosuppressants result in both local and systemic toxicity that invariably impairs survival and functional competence of the grafted islet cells. Thus both pancreas and islet transplantation require immunosuppressive therapy, which carry the threat of recurrence of autoimmunity. Experiments were carried out in Japan for using living donors, but these may lead to clinical complications jeopardizing the donor to a risk of developing DM (Matsumoto et al. 2005, 2006).

Recipient- and graft-related complications post-islet transplantation include intra-abdominal bleeding, pleural/abdominal effusion, peripheral portal vein branch thrombosis, transient transaminitis, intrahepatic focal steatosis and amyloid deposits, common or opportunistic infections, profound neutropenia, pneumonia, ovarian cysts, viral reactivation, dyslipidemia, and renal toxicity. Tacrolimus may cause acute vasomotor vasculopathy with tubular necrosis and/or chronic fibrotic vasculopathy with glomerulosclerosis and interstitial fibrosis. Moreover, sirolimus may induce acute renal dysfunction and/or chronic proteinuria by increasing glomerular permeability and injury or by suppressing the compensatory renal cell proliferation and repair capacity. Papillary thyroid carcinomas, squamous and basal-cell skin carcinomas, ovarian and breast cancer, and pulmonary nodule have also been reported (Mineo et al. 2010).

Thus surgeons/researchers must consider the chance of infections (zoonotic diseases) into the hosts from the donating animal species by xenotransplantation, and also the limited availability of islet for transplantation from donated organs have driven efforts to introduce other potential sources of glucose-responsive insulin-producing tissue, as well as the use of stem cells (SC) (human embryonic SCs, non-pancreatic SCs, pancreatic SCs, and induced pluripotent SCs) as insulin-producing surrogates for β cells will provide a therapeutically meaningful advance.

Thus, therapeutic interventions to cure T1DM mainly focus on preservation of residual β cells, restoration of glucose-responsive, insulin-producing β cells using replacement or regeneration strategies, protection of replaced β cells from allo-/

autoimmune destruction and/or restoration of β-cell-specific unresponsiveness in the absence of chronic immunosuppression (Chhabra and Brayman 2013).

The need for unlimited supply, of a substitute for β cells of primary human islet of Langerhans, has led to a research on the suitability of stem or progenitor cells to generate insulin-secreting cells (ISC), in replacement therapies for DM. Other than downregulating the immune system for subsequently preserving residual β cells, another way is to offer a cell-based therapy, with differentiation of SCs into functional insulin-secreting β cell or a β-like cell, as the use of SCs to treat T1DM has been proposed for many years.

The main goal of stem cell therapy (SCT) is to achieve stable and normalized glycemic control with the absence of severe hypoglycemic attacks and improving quality of life, preventing long-term complications related to T1DM, and reducing procedure ± immunosuppression-related adverse effects. Insulin independence is not necessarily the primary goal of SCT, although desirable; however, a reduction in insulin requirement and restoration of C-peptide secretion should be desired and with beneficial effects.

2.1.10 Alternative Approach to Islet/Pancreas Transplantation

In 1925, Nobel Prize winner Banting described in his lecture that, "Insulin is not a cure for diabetes; it is a treatment" (Banting 1965). Advances in clinical transplantation of pancreas/pancreatic islets of Langerhans have some limitations to generate insulin-producing cells from renewable SCs to treat DM. A recently reported work in Nature Biotechnology strengthens the evidence that SCs can give rise to cells that secrete insulin in a glucose-responsive manner, which is the characteristic of pancreatic β cells. This encourages hope of curing T1DM patients with cell therapy.

Cell therapy in actual sense involves "immunological resetting" by SC rescue through its reproducibility under strict proliferating control to generate sufficient quantity of tissue by differentiation into desired cell type(s) with surrounding tissue integrity and survival even after transplantation via proper function throughout the life of recipients without any untoward effects. The replaced cells must have the ability to synthesize, store, and release insulin in response to ambient glycemia and to avoid development of hyper-insulinemic hypoglycemia from induction of insulin-producing cells (IPCs) (pancreatic β cells) either by differentiation of SCs in vivo or transplantation of ex vivo differentiated cells in the pancreas in T1DM.

2.1.11 Stem Cells Transplantation

Canadian scientists Ernest A. McCulloch and James Till from the Ontario Cancer Institute in Toronto, with their colleagues Andy Becker and Lou Siminovitch, reported the presence of self-renewing cells within the bone marrow (BM) of mice and postulated that these cells were regenerative SCs. In 1924, cell morphologist Alexander A. Maximow has been the first to discover a type of cell within the mesenchyme that develops into various types of blood cells. However, McCulloch and Till were the first to reveal the clonal nature of marrow cells (Becker et al. 1963; Siminovitch et al. 1963; Zhang et al. 1999), now identified as the first described adult SCs, the hematopoietic SCs (HSC).

Then, SCs were reported in 1963 by Becker and colleagues (1963) and are defined by two characteristics; (I) they can differentiate into many cell types in response to appropriate signals, (II) in undifferentiated state, and SCs have the ability to regenerate themselves by cell division. SCs can be divided into two subtypes: embryonic SCs and adult SCs. Embryonic SCs are pluripotent cells derived from inner cell mass of blastocysts and have the ability to differentiate into any of the three germ cell types (Thomson et al. 1998). These were initially derived from embryos of mouse in 1981 (Evans and Kaufman 1981; Martin 1981), and in 1998 James Thomson and colleagues successfully cultivated and continuously cultured embryonic SCs from human blastocysts (Thomson et al. 1998). On the other hand, adult SCs are self-renewed and undifferentiated cells found in adult organ niches and also known as somatic SCs. Adult SCs function as repair cells to regenerate damaged tissues. For example, mesenchymal stem cells (MSC) (capable of generating bones, fats, and cartilage), HSCs (derived from mesoderm, give rise to adult blood lineages), mammary SCs, intestinal SCs, endothelial SCs, neural SCs, and testicular cells.

2.1.12 Strategies for β-Cell Repair by Stem Cells

SC-based strategies represent significant therapeutic potential owing to both the intrinsic regenerative capacity and the immunomodulatory potential of SCs to restore glycol-metabolic and immune homeostasis (Fig. 2.2). This regenerative capacity can be harnessed to make available a self-replenishing supply of glucose-responsive insulin-producing cells, and the immunomodulatory properties help in arresting β-cell destruction, preserving residual β-cell mass, facilitating endogenous β-cell regeneration, ameliorating innate/alloimmune graft rejection, and preventing the recurrence of autoimmunity (Fiorina et al. 2011; Barcala Tabarrozzi et al. 2013; Fandrich and Ungefroren 2010; Sims and Evans-Molina 2012). Thus, SCs with immunomodulatory properties can potentially be used to reverse hyperglycemia, alone or in combination with β-cell replacement strategies (Madec et al. 2009; Jurewicz et al. 2010; Rackham et al. 2011).

SCs obtained from a different sources, have been tested for their β-cell regenerative capacity and ability to restore immune homeostasis or promote longitudinal islet graft survival. These include embryonic SCs (ESCs), induced pluripotent SCs (iPSCs), BM-HSCs and umbilical cord blood-derived MSCs (UCB-MSCs), adipose tissue-derived MSCs, and pancreas-derived multipotent precursor cells, as well as

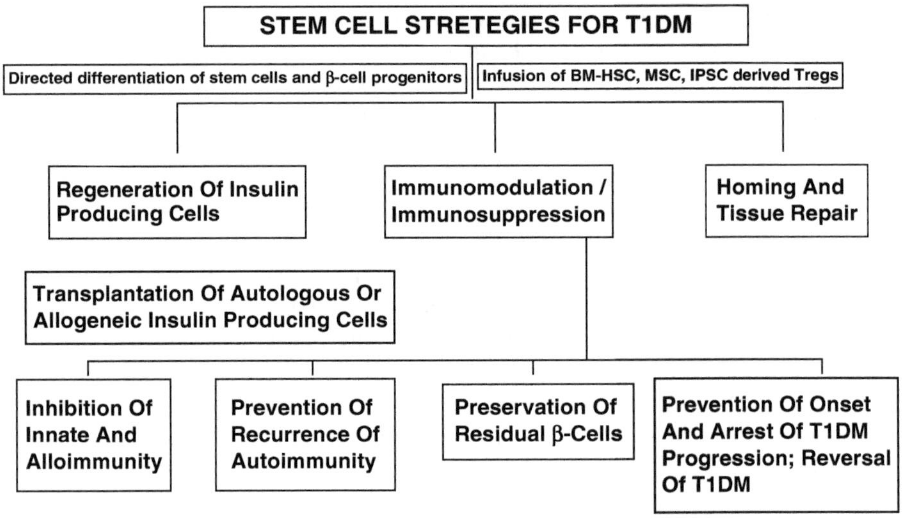

Fig. 2.2 Stem cell-based strategies for β-cell regeneration and immunomodulation

pancreatic β-cell progenitors that reside in the ductal epithelium, exocrine tissue, and within islet proper, neural progenitor cells, and facultative β-cell progenitors from spleen, liver, and the endometrium.

2.2 Role of Stem Cells in Treatment of Diabetes

2.2.1 Embryonic Stem Cells

The continued need for an alternate of replacing β cells in T1DM patients has fostered scientific and public interest in ESC as a potential therapy. ESCs have characteristic of pluripotency and self-renewal ability, which allowed researchers to explore the application of SCs in a number of medical conditions contributing to destructive etiology. T1DM fits in category of such diseases. Islet transplantation has revealed significant potential to achieve insulin independence (Shapiro et al. 2006). Theoretically, ESCs can be differentiated into any cell line, including pancreatic β cell over an appropriate time span and when exposed to appropriate signal in correct sequence (Martin 1981). It is also hypothesized that if pancreatic islet cells have been developed from ESCs differentiation, the shortage of β cell of the pancreas would be overcome in diabetics. The discovery of methods of isolation and growth of human ESCs in 1998 renewed the hopes of researchers, clinicians, and diabetic patients and their families to cure the T1DM, and perhaps T2DM as well may be within striking distance.

First in vitro attempt to produce islet cells from mouse ESCs was reported in 2000; Soria and team were able to achieve a degree of control in hyperglycemia which lasted for few months (Soria et al. 2000). Bernat Soria and his colleagues added DNA containing part of the insulin gene to ESC from mice. The insulin gene was linked to another gene which rendered the mice resistant to an antibiotic drug. In the presence of antibiotics, only activating insulin promoter cells were able to survive. These cells were cloned and then cultured under different conditions. Cells were cultured in low concentrated glucose medium and responded accordingly with changes in glucose concentration by increasing insulin secretion nearly sevenfold. Unfortunately, production of insulin-positive cells was low due to contamination of non-islet insulin-producing cells and selection of cells before full differentiation. Other results have also been reported in different experiments over the next few years with variable degrees of successes. These experiments were carried out on human (Assady et al. 2001; Segev et al. 2004) and mouse (Lumelsky et al. 2001; Blyszczuk et al. 2004) ESCs, but factors like differentiated cells, immaturity (Segev et al. 2004), low glucose-insulin response (Assady et al. 2001), low number of insulin-producing cells (Hori et al. 2005), and by low cell homogeneity (Lumelsky et al. 2001) limit the success of the strategy. Ron McKay and his colleagues described many experiments, in which they induced mouse ESC to differentiate into insulin-secreting structures, which resembled pancreatic islets (Lumelsky et al. 2001). Several research groups are trying to apply McKay's results with mice, to induce human ESCs to differentiate into insulin-producing islets. All these studies made the researchers to rethink the existing strategies for cell differentiation techniques. Kubo's method (Kubo et al. 2004) to retain culture conditions to convert mouse ESCs into definitive endoderm was updated and refined by D'Amour et al. (2005) to achieve to nearly 100% output to produce pure definitive endoderm cell population. This research work was continued until a five stage in vitro process was introduced after in vivo development of the pancreas (D'Amour et al. 2006).

According to Jon Odorico from the University of Wisconsin in Madison, ESCs can differentiate and express the insulin gene. Itskovitz-Eldor and colleague further characterized insulin-producing β cells ~1–3% from embryoid bodies (Assady et al. 2001). However as compared to previous experiments, there was no insulin secretion found in response to glucose even when C-peptide was released in response to stimuli like KCI and cAMP. The cells resembled to 6–9 weeks of embryo; in vitro differentiation was stopped at that time to achieve more specific and successful results (Kroon et al. 2008). These differentiated cells were transplanted into epididymal fat pad in immune-deficient mice. Insulin secretion was measured in glucose-dependent manner. Posttransplantation C-peptide secretion levels were low at 1 month, and at the 90th day, it was reached at those levels where they could be seen with 3000–5000 human islets transplantation.

2.2.2 Obstacles in Application of Embryonic Stem Cells in Clinics

ESCs-based β-cell replacement therapy for T1DM has several major obstacles that must be overcame to before this approach can be considered as a therapeutic option.

Ethical and religious sensitivities: There are ethical and religious sensitivities concerning with the use of human embryos which still hamper its use for research purposes, and many governments ban or at least highly restrict association with this research (Watson 2003; Gruss 2003). The Roman Catholic Church has repeatedly demanded an international ban on the use of human ESC for research purpose, because it requires the destruction of embryos who has all the moral rights and protection as any other human being (Copland 2004; Oakley 2002). Owing to these moral concerns, the United States Congress has enacted a broad ban on federal funding for human ESC research. Later, this ban was loosened to allow research on human SC lines that already existed. A highly debated question is whether SCs from human embryos should be given to use for generation of SC lines, which were unsuitable for fertilization programs and discarded. The US President George W. Bush stated that, "There is no such thing as a spare embryo" to the religious authorities on this matter (New York Times, May 26, 2005). Landry and Zucker (2004) pointed out that death before the onset of neural development, a significant fraction of these human embryos will be found to be "organismically" dead. So, using such embryos to generate human SC lines would not contradict the ethics of the Catholic Church and other religious authorities. Recently, novel techniques to derive mouse SCs without affecting the subsequent development of the embryo have been described (Chung et al. 2006; Meissner and Jaenisch 2006). These types of techniques may resolve some of the ethical concerns regarding the generation of ESC line.

There were some potential risks imposed by continued replication of ESCs-derived transplants including loss of cell cycle control and the induction of tumor cell growth. Teratomas have been observed in IPSC lines derived from ESCs consistent with it (Fujikawa et al. 2005). To provide an ultimate cure for diabetes, human ESCs-derived islets or β cells would either need to maintain their ability to proliferate or the transplantation procedure would have to be repeated at regular intervals. Therefore, maintaining of a physiologic balance between replication and cell death appears to be a major challenge for β-cell replacement therapies based on ESCs.

Functional β-cell mass requires more than 18 months to establish in developing humans (Bouwens et al. 1997; Kassem et al. 2000), which is not yet clear if it will be possible to drive ESCs to a useful mass of β cells or β and other cell-type aggregates, ex vivo tend to undergo senescence and differentiation within days in culture thereby losing their pluripotency (Halvorsen et al. 2000).

2.3 Adult Stem Cells

2.3.1 Adult Pancreatic Stem Cells

Islet comprises functional cells of four types: glucagon-producing α cells, insulin-producing β cells, somatostatin-producing δ cells, and pancreatic polypeptide-producing cells (Liu et al. 2013). Adult pancreatic SCs can also be another source for pancreatic β cells as they carry the characteristics of multipotency and clonogenic potential. By applying pancreatic duct ligation model of injury, differentiation and proliferation characteristics of ductal cells of the pancreas have been proposed as the major source for β cells for regeneration of tissues (Wang et al. 1995). Experiments on rats have revealed that β cells and pancreatic tissues would be regenerated expeditiously if 90% of the pancreas undergo resection (Bonner-Weir et al. 1993). β-cells mass was reported after the activation of duct lining NGN3+ endocrine precursors in adult mice (Xu et al. 2008). Advanced strategies are needed to be developed to isolate and grow adult pancreatic cells and to differentiate into β cells. Epithelial cells of pancreatic duct were harvested and induced in vitro to become functional islets (Ramiya et al. 2000). Clonal characteristics of multiple progenitor cells from the pancreas of adult mouse were reported; at differentiation stage, endocrine, exocrine, neuronal, and glial cell populations were produced by clonal colonies. Produced β cells express the insulin secretion and glucose-dependent reactivity (Seaberg et al. 2004). To identify pancreatic SCs, clonal analysis was performed which is capable of differentiating the pancreatic SCs into pancreatic endocrine and exocrine cells (Suzuki et al. 2004). Future targets will be to tackle all challenges of harvesting, purifying, and growing various populations of pancreatic progenitor cells and also for inducing the β-cell differentiation without genetic mutations.

2.3.2 Adult Non-pancreatic Stem Cells

MSCs and HSCs have the capability to proliferate and refill the damaged or dead tissues and cells as these possess multipotency.

2.3.3 Mesenchymal Stem Cells

MSCs were first identified by Friedenstein and his colleagues (Friedenstein et al. 1966). They described it as bone-forming progenitor cells from rat BM. MSCs can be harvested from adipose tissues, BM, and other organs, but the richest source remains BM. They are reported as pericytes which were localized in the blood vessels' wall (Meirelles Lda and Nardi 2009; Masoud et al. 2012). MSCs have high potential to self-renew and to differentiate in vitro and in vivo. Immunomodulatory properties of MSCs inhibit several components of immune systems in vitro (Hoogduijn et al. 2010). Hence, severe refractory diseases have been treated using MSCs in human (Dazzi and Marelli-Berg 2008). Several lines of evidence have shown that under appropriate environments, MSCs are able to differentiate into mesodermal, endodermal, and even ectodermal cells. MSCs are known as hypoimmunogenic cells because of its properties like escaping immune recognition and inhibiting immune responses. Therefore, MSCs appear to be a very promising tool for regenerative and immunoregulatory cell therapy. The Mesenchymal and Tissue Stem Cell Committee of the International Society for Cellular Therapy has proposed a minimal set of four criteria to define human MSC (Dominici et al. 2006):

1. MSCs have to be plastic adherent when maintained under standard culture conditions.
2. MSC must have the ability for osteogenic, adipogenic, and chondrogenic differentiation.
3. MSC must express CD73, CD90, and CD105.
4. MSC must lack expression of the hematopoietic lineage markers c-kit, CD14, CD11b, CD34, CD45, CD19, CD79α, and human leukocyte antigen (HLA)-DR.

MSCs obtained from BM, adipose tissue, and umbilical cord were same for its morphology, immune phenotype, success rate of isolating MSC, colony frequency, and differentiation capacity. (Kern et al. 2006; Izadpanah et al. 2006). Adipose tissue is the most attractive source for generation of MSC for researchers and clinicians of nearly all medical subspecialties as it requires simple surgical procedure and uncomplicated enzyme-based isolation procedures, and easy and repeatable access to the subcutaneous adipose tissue is possible (Casteilla et al. 2005; Oedayrajsingh-Varma et al. 2006). Therefore, ADSC do represent an alternative source of autologous adult SCs which can be obtained repeatedly in large quantities under local anesthesia with a minimum patient discomfort.

These advantage and properties of MSCs triggered the researchers to find out the effects of MSCs in autoimmune diseases like T1DM. Differentiation of MSCs gives rise to mesodermal tissues, but other tissues including IPSCs have also been reported (Davani et al. 2009). Many protocols have been devised to differentiate MSCs into IPSCs. Some researchers used pancreatectomy model to release unknown pancreas regenerative factors. MSCs were grown in this extract of regenerative factors transformed to insulin-producing islet-like clusters which were responsive to glucose concentration (Jahr and Bretzel 2003). Another research group used defined culture conditions and successfully differentiated the BM-derived stromal cells to insulin-producing islet-like aggregates which resemble to β cells in gene expression pattern as well as active production of insulin (Oh et al. 2004). Serial differentiation of BM-derived MSCs in a three step differentiation mechanism using endocrine differentiation inducers and primed with stromal-derived factor-1-α to enhance its therapeutic potential and survival has been reported, and the resultant IPSCs were found to respond to glucose tolerance test as well as lowered blood glucose levels when transplanted in streptozotocin-induced diabetic rats (Tariq et al. 2013). In animal models of T1DM alone (Lee et al. 2006; Fiorina et al. 2009) and in combination with HSCs (Urban et al. 2008), MSCs exhibit beneficiary effects in glycemic control. The HSCs were used along with AD-MSC because HSC transplantation with immunosuppression conditioning is believed to create active and passive tolerance by clonal deletion/T-cell suppression and helps in angiogenesis. Researchers also used them to revive nonfunctioning pancreatic β cells. Co-infusion of HSCs with Ad-MSC-ISCs that was examined on five patients with T1DM showed a decrease in exogenous insulin requirement and an increase in C-peptide level (Trivedi et al. 2008). In another trial on mice, researchers injected a mixture of MSCs and HSCs into the BM of mice with STZ-/radiation-induced injury, and the normal blood glucose level was regained successfully. It is reported that MSCs have the capability to inhibit the proliferation of pancreatic β-cell-specific T-cells, ultimately reducing the damage induced by T-cells on new β cells (Liu et al. 2013). Although it is impossible to elaborate all mechanisms in detail, BM SCs can stimulate the regeneration of damaged pancreatic β cells. Without any doubt, BM SCs have therapeutic effects on DM, and these are ideal adjuvants for cell treatment and therapy in future.

2.3.4 *MSC-Derived Extracellular Vesicles as Novel Immunomodulator in T1DM*

Depending on the cellular sources, extracellular vesicles (EV) may induce immune cell activation or inhibition because of their different immunomodulatory functions (Théry et al. 2009). The EVs have emerged as paracrine factors of MSC actions. In fact, MSC-derived EVs released proteins and nucleic acids, capable to mimic the effect of originating cells. Antigen-presenting cells/B lymphocytes-released vesicles activate the T cells by direct peptide-MHC complex presentation to transfer antigen and/or peptide-MHC complex, from dendritic cells (DCs), natural killer (NK) cells, macrophages, and B cells and to induce DC maturation (Théry et al. 2009; Robbins and Morelli 2014). Conversely, EV released from tumor cells or from MSC exhibited inhibitory functions on T cells, NK cells, and DCs (Robbins and Morelli 2014). Moreover, EVs of MSCs promoted the regulatory T-cell activity and induced the monocyte differentiation into myeloid-derived suppressor cells (MDSCs). In an autoimmune encephalomyelitis experiment, MSC-derived EVs inhibited autoreactive lymphocyte proliferation and induced tolerogenic signaling, via PD-L1, TGF-β, IL-10, and CD4+ CD25+ Foxp3+ Treg cells (Mokarizadeh et al. 2012).

EV released from heterologous human BM-MSC mimic the immunomodulatory characteristic of the cells in T1DM by inducing a shift toward an anti-inflammatory and regulatory T-cell. After an integrin-mediated EV internalization in patient's peripheral blood mononuclear cells (PBMCs), a downregulation of Th1 responses, observed as IFN-γ production, number of Th17 cells, and levels of pro-inflammatory IL-17 was observed (Favaro et al. 2014). These Th17 effecter cells participate in T1DM pathways paralleling Th1 cells, and their secreted IL-17 contributes to β-cell death.

T-cells have shown to produce PGE2 and TGF-β, involved in immunomodulation which is the property of MSC (Spaggiari et al. 2009). TGF-β conveyed as protein and as mRNA within and on the surface of EV (Pap et al. 2011). It also inhibits the lymphocyte proliferation and promotes the Treg generation (Bruno et al. 2015). Thus, EV-associated functional miRNAs can be transferred to target cells (Ratajczak et al. 2006; Deregibus et al. 2007; Valadi et al. 2007). MSC-derived EV expressed miRNAs, miR-21, known to enhance TGF-β signaling. RNA depletion of EV reduced the TGF-β transcript in PBMCs that suggested increasing the TGF-β production by transfer of TGF-β mRNA or miRNA. Hence, EV may increase the TGF-β activation pathway and its release, in a paracrine/autocrine way in T-lymphocytes. MSC-derived EVs may restore Th1/Th2 balance and preserve Treg cells in T1DM. In fact, EV) enhanced the production of IL-10 and induced higher frequencies of Foxp3+ Treg cells (Favaro et al. 2014). The production of IL-6 was increased in the presence of EV by PBMCs, which is known to suppress maturation of inflammatory DC and mediate β-cell repair (Boumaza et al. 2009). Researchers also observed that MSC-derived EV mimics the effect of MSC on DC maturation, impairing antigen presentation.

2.3.5 MSCs as Cellular Vehicle for Insulin-Producing Gene Therapy

MSCs are a promising tool for cell-based gene therapy against a variety of different diseases (Hamada et al. 2005). Their high self-renewal potentiality makes them strong candidates for delivering genes and restoring function of organs and tissues. The ability to genetically modify MSCs provides durable expression of therapeutic genes.

Human insulin gene is located on chromosome 11p15.5 (Ohneda et al. 2000). Insulin synthesis and its release from islet β cells is complex and tightly regulated mechanism. Glucose affects insulin at all levels, including transcription, translation, and release. Mature insulin results from a processing pathway starting from the rough endoplasmic reticulum and ending at the Golgi apparatus. Translation of insulin mRNA yields pre-pro-insulin, cleaved by endoproteinases PC1 and PC2/PC3 to give pro-insulin first and mature insulin + C-peptide subsequently. In the secretory granule, six insulin molecules are coordinated by a Zn atom, seen under microscope by dithizone stain.

With better assays for SCs and improving the vector biology, gene transfer efficiency into MSCs has been increased. The transfected MSCs expressed the insulin gene and secreted insulin in culture media consistently. Xu et al. reported that diabetes could be relieved effectively for up to 6 weeks in mice model by intrahepatic transplantation of BM-derived murine MSCs infected with the recombinant retrovirus carrying human insulin gene. In further study, implantation of engineered cells using diabetic animal models and observed its therapeutic effect with more tests of efficacy and safety of engineered human MSCs as surrogate β cells (Xu et al. 2007). Further work was also carried out for modified herpes I virus as a vector for the human insulin gene (Calne 2005). The theoretical advantages of the herpes I virus are (i) its large capacity to accommodate a construct; (ii) its ability to infect primary and secondary cell lines in vitro; (iii) even though its entry in to the nucleus not integrate with the host DNA and functions separately as an episome which is not likely to unmask ontogenesis; (iv) most of the patients have already had contacts with the herpes I virus, which normally resides in a quiescent state in neurological tissue; (v) relatively mild immune reaction against the virus; and (vi) antiviral treatment against the herpes virus is established and available. Thus, herpes I virus could serve as a new vector for human insulin gene delivery into MSCs.

2.3.6 Minimum Requirements for Replacement β Cells

IPCs act as replacement β cells for cell therapy of T1DM which could be generated either by trans-differentiation of MSCs or by delivery of insulin gene into MSCs. These MSC-derived IPCs may solve the problem of donor shortage for islet cell transplantation and provide a cure for T1DM. Any options for primary islets of Langerhans will require some minimum essentiality. The basic requirements for surrogate β cells are:

1. Large numbers of replacement β cells will be required for significant therapeutic impact. Current transplantation protocols use up to 1×10^6 primary human islets per recipient, equivalent to approximately $2–4 \times 10^9$ β cells. As a result, the ability of MSCs to replicate and to differentiate toward pancreatic endocrine phenotype makes them attractive candidates for producing replacement β cells.
2. Replacement β cells must have synthesize, store, and release ability of insulin in response to changes in the ambient glycemia. Understanding β-cell function at the molecular level will likely facilitate the manufacturing of physiologically competent IPCs from MSCs.
3. Proliferative capacity of replacement β cells must be tightly controlled to avoid development of hyperinsulinemic hypoglycemia because of its expansion can occur in vivo. Excluding proliferative cells from the transplant material will help to overcome this problem. In the case of insulin gene transferred MSCs, the possibility of tumor formation has to be considered.
4. The transplanted replacement β cells must not be destroyed by the recipient's immune system (Burns et al. 2004).

With appropriate immunosuppressive medication, autologous MSCs-derived IPCs transplantation will circumvent the immune-rejection dilemma. On the other hand, Burt et al. indicated that HSC transplantation may reintroduce tolerance to islet cells in T1DM (Burt et al. 2002). Thus, co-transplantation of MSCs-derived IPCs and HSC from the same donor (autologous/allogeneic) could evade the risks of recurring autoimmunity. Furthermore, the pathways of β-cell differentiation were different in "in vitro" and "in vivo" (Houard et al. 2003). However, in vitro differentiation protocols can generate surrogate β cells having some phenotypic and functional similarity to authentic β cells, which is actually not the β cells. Since, MSCs-derived IPCs are developmentally and immunologically distinct from primary β cells, who may escape the recipient's autoimmune assault.

2.3.7 *Hematopoietic Stem Cells*

Multipotent SCs that give rise to all the other blood cells located in red BM and are derived from mesoderm are called HSC. BM is an important source of relatively easily accessible adult SCs. BM transplantation is considered to be effective for the treatment of autoimmune T1DM. However, there is a great debate on the issue of the fate of transplanted BM-SC. Based on animal models, autoimmune diseases have been effectively treated with combination of HSC and high-dose immunosuppression (Sykes and Nikolic 2005; Burt et al. 2008), and the very first patient was treated with HSC for autoimmune diseases in 1996 (Voltarelli et al. 2011). To date, approximately 1500 patients carrying autoimmune disease have been treated (Passweg and Tyndall 2007; Vanikar et al. 2012) with HSC transplantation because of low risk of complications. After decades of clinical use of HSC transplantation for severe and stubborn autoimmune diseases, it can be speculated that this approach may also be useful for treating T1DM. In an experiment in nonobese diabetic mice, clinically observable T1DM has been easily precluded by allogeneic HSC transplantation and not by autologous HSC. Results could be predicted by the genetic nature of the disease in this animal model (Atkinson and Leiter 1999). However, the clinically observable T1DM in nonobese diabetic mice cannot be reversed only by the allogeneic HSC transplantation but also require efficient source of pancreatic β cells (Sykes and Nikolic 2005; Kang et al. 2005). Allogeneic HSC can restore tolerance to pancreatic β cells but cannot restore the cells pool if once destroyed by autoimmune system.

The pancreatic duodenal homeobox-1 (PDX-1) gene-modified MSCs derived from the human BM can be induced to differentiate into functional IPCs (Li et al. 2007; Karnieli et al. 2007). In addition, Sun et al. demonstrated that BM-derived MSCs can differentiate into IPCs under appropriate conditions in vitro in diabetics. This study provides the information regarding the feasibility of using autologous BM-MSCs as a source of IPCs for β-cell replacement therapy (Sun et al. 2007).

2.3.8 *Umbilical Cord Blood Stem Cells*

SCs from umbilical cord can also be used for treating DM (Kucia et al. 2006). These SCs have higher regenerative potency and have low rate of rejection than BM cells after allogeneic transplantation. In human UCB, the presence of MSCs was reported, when cells isolated from UCB exhibited the characteristic immunephenotype and differentiation of BM-MSC (Lu et al. 2006; Kern et al. 2006). In animal models to prevent or reverse T1DM, MSC (Wang et al. 2011), cord blood mononuclear cells (Ende et al. 2004), combination of T-regulatory cells, and UCB-SC (McGuckin and Forraz 2008) have been used. Cord blood MSCs have been obtained from cord walls with greater efficiency than the cord blood, and these cells have the capability to treat T1DM (Anzalone et al. 2011). In an experiment, in Florida University, no significant improvements have been reported as compared to control group when 15 T1DM patients were injected with autologous mononuclear cord blood cells (Haller et al. 2009). In 1988, after first productive transplantation of UCB (Gluckman et al. 1989), it has been recognized as the ultimate source for HSC to treat blood diseases and genetic disorders. In addition to human UCB, Wharton's jelly of the UCB is rich in MSC. UCB-SC has gained substantial attention for the therapeutic options in regenerative medicine.

2.3.9 Hepatic Stem Cells

In developmental biology, endoderm gives rise to the pancreas and liver, and these contain similar progenitor cells, so it has been proposed that liver cells would be an ultimate source for β cells because of their easy availability by biopsy and their strong regenerative capability (Zaret and Grompe 2008). By using adenoviral transduction, pancreatic endocrine and exocrine gene expressions have been reported in liver cells when NGN3 and PDX1 have been expressed in mouse liver (Liu et al. 2013). Newly grown pancreatic tissues form clusters around the central veins of the liver secreting insulin and without disturbing the normal liver functions. Moreover, insulin-producing cells maintain STZ-induced hyperglycemia and glucose levels for 8-month time period (Ber et al. 2003; Yechoor et al. 2009). To date, no in vitro evidence has been reported to show that modified liver cells can be proliferated in vitro conditions. In the future, efficient methodology needs to be developed for getting in vitro expressions of trans-differentiated cells from liver.

2.3.10 Induced Pluripotent Stem Cells

The process of formation of pluripotent SC (PSC) from non-pluripotent cells is referred to as induced pluripotency. Somatic cells can be transformed to PSC under specific conditions, and such cells are called iPSCs. This ground breaking discovery led to an outburst in the studies of reprogramming of cells (Takahashi and Yamanaka 2006). Induced PSCs have the same characteristics like ESCs. These have high telomerase activity as well as gene promoters are also hypo-methylated. Human iPSCs obtained by reprogramming of human somatic cells (skin fibroblasts) can represent an alternate to human ESCs and eliminate the ethical issues pertaining to

ESCs. Therefore, these are ethically more acceptable. Induced PSCs can be the preferred cell type for the treatment of DM for autologous cell transplantation. Autoimmunity in T1DM may lead to the immune rejection of transplanted autologous iPSCs. However, this strategy may be successful for T2DM where autologous SCT is required. Second problem pertaining to iPSCs is same as with ESCs: the formation of teratoma. These PSCs are formed due to the expression of some specific factors which involve the use of genome-integrating viruses. In IPSCs, the use of DNA-based reprogramming strategy could lead to the insertional mutagenesis and the use of oncogenic reprogramming factors could add to the risk of tumor formation (Okita et al. 2013). Reprogramming of cells using nonintegrating approach may be required before it moves toward clinics. The problem of genome integration was addressed by the study of Stadtfeld et al., by introduction of four factor of pluripotency using nonintegrating adenoviral vectors (Stadtfeld et al. 2008).

2.3.11 Induction of Insulin-Producing Cells from Stem Cells by Protein Transduction Technology

Protein transduction technology has been recently emerged for induction of IPCs from SCs. A variety of peptides, like protein transduction domains (PTDs) or cell-penetrating peptides (CPPs), have been characterized for their ability to translocate into live cells. When proteins and peptides synthesized as recombinant fusion proteins or covalently cross-linked to PTDs, these can be directly internalized into cells. Biologically active full-length proteins and peptides have been delivered to cells both in vitro and in vivo. The homeodomain transcription factors like Antennapedia (Antp), HSV type-1 protein VP22, and HIV-1 trans-activator TAT protein are most commonly studied PTDs. The involved mechanism is endocytosis followed by passage from the vesicle into the cytoplasm for PTD-mediated protein transduction (Noguchi and Matsumoto 2006). This technology facilitates the differentiation of SCs into IPCs, so it can cure the T1DM. Two pancreatic endocrine transcription factors, PDX-1 protein and BETA2/NeuroD protein, have a PTD sequence in their structure. Noguchi et al. observed that PDX-1 (Noguchi et al. 2003) or BETA2/NeuroD (Noguchi et al. 2005) protein induced insulin expression in pancreatic ductal progenitor cells. Similarly, Domínguez-Bendala et al. (2005) reported that in vitro pancreatic endocrine differentiation was stimulated by TAT-mediated neurogenin-3 (ngn3) protein transduction. Gräslund's group (Kilk et al. 2001) showed that the third helix of the homeodomain of transcription factor Isl-1 internalized into cells. Thus, delivery of exogenous transcription factors (PDX-1, BETA2/Neuro D, ngn3, Isl-1, etc.) by protein transduction technology could be a novel strategy for generating IPCs from SCs/progenitor cells without requiring gene transfer technology. Thus, MSC is the strong candidate for this emerging modality.

2.3.12 Our Experience with Insulin-Secreting Cells from Adipose Tissue-Derived MSC for T1DM

In 2008, Trivedi et al. reported safe and effective treatment in five insulinopenic diabetes using intra-portal infusion of insulin-secreting AD-MSC with BM-HSC. No xenogeneic material was used in this study. There was 30–50% fall in insulin requirement with 4- to 26-fold rise in serum C-peptide levels. This effect was found to be sustained for 3 years. No further follow-up was available. No immunosuppression was used. Subtotal lymphoid irradiation of 200 cGY for 5 days followed by rabbit antithymocyte globulin, 1.5 mg/kgBW, was used for conditioning, before infusing the SC in liver, subcutaneous tissue, and thymus. No infective episodes or graft versus host disease were observed (Trivedi et al. 2008). Vanikar et al., in 2010, have generated in vitro MSC from human adipose tissue (Vanikar et al. 2010), which qualify the definition standardized by the Mesenchymal and Tissue Stem Cell Committee of the International Society for Cellular Therapy (Ya˜nez et al. 2006). Our cells fulfilled these criteria. They were further differentiated into insulin-secreting cells (ISC) under defined culture conditions; these cells were phenotypically identical to pancreatic β cells (Trivedi et al. 2008). These cells expressed transcription factors pdx-1, pax-6, and isl-1; all three are central controlling genes capable of reprogramming non-pancreatic cells to surrogate β-cell functions. Further infusion of AD-MSC-ISC with BM-HSC was carried out safely in 11 diabetics with 1 to 24 years of disease duration. Over a mean follow-up of 23 months, all patients responded, with a decreased mean exogenous insulin requirement from 1.14 to 0.63 units/kg BW/day, Hb1Ac dropped from 8.47% to 7.39%, serum C-peptide levels increased from 0.1 to 0.38 ng/mL, and all of them became free of diabetic ketoacidosis events.

In 2015, Thakkar et al. have treated 20 T1DM patients, 10 with autologous SCT and 10 with allogeneic SCT (Thakkar et al. 2015). Both groups received exactly similar treatment of AD-MSC-ISC with BM-HSC into the liver via intra-portal route, thymic circulation, and subcutaneous fat pad. Thymic infusion was carried out to achieve central tolerance (Sprent and Kishimoto 2001), and portal circulation was performed to take advantage of tolerogenicity of the liver and for better grafting (Starzl 2001). Subcutaneous tissue is an immunologically privileged site; hence, we decided to inject part of the cells into abdominal subcutaneous tissue so that it could serve as a "back-up reservoir" for insulin supply (Prokhorova et al. 2009). Results in both groups were compared, autologous SCT was found to be better than allogeneic source vis a vis long-term control of hyperglycemia. This study also established one important fact that patients with any type of DM need not search for donors; they can use their own fat reservoir for treating their own disease.

2.3.13 Lessons Learned

Still some researchers and clinicians are disappointed because of limited benefits with recent advancement of T1DM research and therapy. Large investments in terms of time, finances, and patient resources have been required for SCT, islet-cell transplantation, genetics, primary/secondary prevention, and reversal of T1DM (Greenbaum and Atkinson 2011). It is difficult to decide whether the goal is disease prevention or reversal even with therapeutic interventions using conventional or experimental agents (Staeva et al. 2013). T1DM is a multifactorial etiology that overcomes its autoimmune nature; permanent cure of the disease is not achieved and till now requires more intense research. Similarly, islet-cell transplantation depends on overcoming recurrent autoimmunity and averting alloimmunity (Vendrame et al. 2010). Even islet/pancreas transplantation have their limitations/ hurdles as described previously in this review, so investigators are focusing now on xenotransplantation, encapsulation, novel sites for cell delivery (e.g., eye), and development of surrogate insulin-producing cells (Gibly et al. 2011). T1DM has polygenic nature, in which more than 40 loci have been identified with disease susceptibility/resistance (Concanon et al. 2009), combined with environmental factors suggesting the unpredictable pathogenesis of the disease. Experiments are going on to improve understanding of the genetic risk for T1DM by genotyping at multiple susceptibility loci (Winkler et al. 2012). Genetic involvement for disease development is complex in nature regarding the immune response in T1DM. Mechanism of selective destruction/loss of pancreatic β cells remains unclear apart from antigen-specific immune response till now for this disease (Atkinson et al. 2011). Many researchers have described the putative role for adaptive rather than innate immune responses in the disease pathogenesis which is crucial for the development of improved management. Luckily, trial networks like NIH TrialNet and Immune Tolerance Network and registries like T1D Exchange can judge the ability of therapeutic agents and effects in terms of improvement of patient recruitment and increase the precision of disease prediction (Sosenko et al. 2012). Modification/ changes in clinical trial design like adaptive trial design, utilization of animal model (Atkinson 2011a, b), identifying more practical therapies, and better defining disease heterogeneity (Pozzilli 2012) could improve the applicability of T1DM research and which will be more effective. If the therapeutic intervention is applied during the natural history of the disease with silent or asymptomatic state, this may be more effective.

2.3.14 Road Map for Future Generation

The most pressing questions are: Will the recipients' immune response to infused cells destroy them eventually? Whether more number of SCs are required or repeated stem cell infusion cycles are required? Is there any better engraftment technique available? Whether requirement of more potent/supporting cells like regulatory T/B cells? Are SCs capable for producing long-term immunological tolerance? For predicting disease development, can improved markers be obtained? Can replication and neogenesis of β cell induce safely in humans? Is it possible to develop safe and effective closed-loop therapy system?

These questions make a road map of investigations for the next generation and if properly addressed, should result in substantial improvements in the quality of life of T1DM patients.

2.3.15 Current Challenges and Future Perspectives

SCs are the potentially unlimited source of functioning surrogate β cells of the pancreas; still its research is in fundamental stage (Hansson et al. 2004; Dor et al. 2004). Induced surrogate β cells from the SCs by differentiation and its infusion technique should be improved and maintained them functionally in a specialized microenvironment termed as SC niche. The niches maintain the SCs quantum, and multiple signals are required to maintain a balanced control of SC self-renewal (Hou and Singh 2008; Singh et al. 2007; Singh and Hou 2008, 2009; Scheres 2007; Fuchs 2009; Yamashita 2009; Meirelles and Nardi 2009; Discher et al. 2009). Advance technology is required for successful transplantation of β cells into suitable niches to achieve maximum therapeutic effects.

SCT has some technical and clinical limitations. Various cytoprotective strategies and agents are under investigation for improving the SC yield and outcome. Patients should get the benefit of insurance and/or reimbursement for the therapy.

Several pathological conditions have regulatory mechanism via microRNAs (Liu et al. 2008; Mishra et al. 2009; Wang and Wu 2009). This experience suggests that they may have a role in mechanisms underlying differentiation of SCs into β cells. Adenovirus genes for cell engineering reduce cell immunogenicity, allowed successful transplantation across allogeneic barriers, without immunosuppression/immune-isolation. Genetic modification can be applied in the future to cultured human islets, to derive a universal donor β cells because of current use of adenovirus genes in cell engineering. Genetically engineered β cells hold the promise of replacing exogenous administration of insulin as an accurate, convenient, and safe way for long-term control of euglycemia in T1DM (Orlando et al. 2014). Thus in vitro generated ISC from SCs to treat T1DM appears extremely promising, with bona fide hope for a complete cure.

2.3.16 Unanswered Questions

The key goal of research in T1DM is detection of ex vivo islet autoreactive T cells and their functions. This may provide the markers for detection of patients at risk and to design intervention strategies to preserve surrogate β cells.

Why islet β cells are target specific for destruction/elimination and do inherent processes for developing the disease? Is the clinical dilemma involving the autoimmune issue? How can control continuously generated autoantibodies against β cells? Understanding the innate and adaptive immune response helps in improving the therapies.

This is a review of change with respect to understanding of the epidemiology, current ongoing research in management, and prospects for curing T1DM with cell-based therapy. In hindsight, many long-held goals once thought readily achievable have been difficult to realize, and concepts regarded as dogmas have proven to be flawed.

2.3.17 Raised Issue

The major raised question is whether it is reasonable to expose diabetics to such type of therapies with some/minimal degree of untoward effects with risk, when the therapeutic option of exogenous insulin administration is effectively available.

2.4 Conclusion

SCT is a better alternative to islet/pancreas transplantation. It is safe, viable, and easily reproducible treatment modality for T1DM which has recently achieved successful graft function, with long-term better metabolic control without any untoward effects. However, intense work needs to be addressed properly before pushing the cell-based therapy from bench to bedside.

Acknowledgment Author acknowledges the immense help received from the scholars whose articles are cited and included in the references of this review chapter. The authors are also grateful to the author/editors/publishers of all those articles, journals, and books from where the literature for this chapter has been reviewed and discussed.

References

Anzalone R, Iacono ML, Loria T et al (2011) Wharton's jelly mesenchymal stem cells as candidates for beta cells regeneration: extending the differentiative and immunomodulatory benefits of adult mesenchymal stem cells for the treatment of type 1 diabetes. Stem Cell Rev Rep 7:342–363

Assady S, Maor G, Amit M, Itskovitz-Eldor J, Skorecki KL, Tzukerman M (2001) Insulin production by human embryonic stem cells. Diabetes 50:1691–1697

Atkinson MA (2011a) It's time to consider changing the rules: the rationale for rethinking control groups in clinical trials aimed at reversing type 1 diabetes. Diabetes 60:361–363

Atkinson MA (2011b) Evaluating preclinical efficacy. Sci Transl Med 3:96cm22
Atkinson MA, Leiter EH (1999) The NOD mouse model of type 1 diabetes: as good as it gets? Nat Med 5:601–604
Atkinson MA, Bluestone JA, Eisenbarth GS et al (2011) How does type 1 diabetes develop?: the notion of homicide or beta-cell suicide revisited. Diabetes 60:1370–1379
Bach JF (2011) Anti-CD3 antibodies for type 1 diabetes: beyond expectations. Lancet 378: 459–460
Balamurugan AN, Breite AG, Anazawa T, Loganathan G, Wilhelm JJ et al (2010) Successful human islet isolation and transplantation indicating the importance of class 1 collagenase and collagen degradation activity assay. Transplantation 89:954–961
Balamurugan AN, Naziruddin B, Lockridge A et al (2014) Islet product characteristics and factors related to successful human islet transplantation from the collaborative islet transplant registry (CITR) 1999–2010. Am J Transplant 14:2595–2606
Ball SG, Barber TM (2003) Molecular development of the pancreatic beta cell: implications for cell replacement therapy. Trends Endocrinol Metab 14:349–355
Ballinger WF, Lacy PE (1972) Transplantation of intact pancreatic islets in rats. Surgery 72:175–186
Banting FG (1965) Diabetes, insulin, nobel lectures, physiology or medicine 1922–1941. Elsevier, Amsterdam. http://nobelprize.org/nobelprizes/medicine/laureates/1923/banting-lecture.html
Barcala Tabarrozzi AE, Castro CN, Dewey RA (2013) Cell-based interventions to halt autoimmunity in type 1 diabetes mellitus. Clin Exp Immunol 171:135–146
Becker AJ, McCulloch EA, Till JE (1963) Cytological demonstration of the clonal nature of spleen colonies derived from transplanted mouse marrow cells. Nature 197:452–454
Bell GI, Pictet RL, Rutter WJ, Cordell B, Tischer E, Goodman HM (1980) Sequence of the human insulin gene. Nature 284:26–32
Benedum J (1999) The early history of endocrine cell transplantation. J Mol Med 77:30–35
Ber I, Shternhall K, Perl S et al (2003) Functional, persistent, and extended liver to pancreas transdifferentiation. J Biol Chem 278:31950–31957
Blanton D, Han Z, Bierschenk L et al (2011) Reduced serum vitamin D-binding protein levels are associated with type 1 diabetes. Diabetes 60:2566–2570
Blyszczuk P, Asbrand C, Rozzo A et al (2004) Embryonic stem cells differentiate into insulin-producing cells without selection of nestin-expressing cells. Int J Dev Biol 48:1095–1104
Bonner-Weir S (1991) Anatomy of the islet of Langerhans. In: Samols E (ed) The endocrine pancreas. Raven Press, New York, pp 15–27
Bonner-Weir S, Baxter LA, Schuppin GT, Smith FE (1993) A second pathway for regeneration of adult exocrine and endocrine pancreas: a possible recapitulation of embryonic development. Diabetes 42:1715–1720
Boumaza I, Srinivasan S, Witt WT, Feghali-Bostwick C, Dai Y, Garcia-Ocana A, Feili-Hariri M (2009) Autologous bone marrow-derived rat mesenchymal stem cells promote PDX-1 and insulin expression in the islets, alter T cell cytokine pattern and preserve regulatory T cells in the periphery and induce sustained normoglycemia. J Autoimmun 32:33–42
Bouwens L, Lu WG, De Krijger R (1997) Proliferation and differentiation in the human fetal endocrine pancreas. Diabetologia 40:398–404
Bruno S, Deregibus MC, Camussi G (2015) The secretome of mesenchymal stromal cells: role of extracellular vesicle in the immunomodulation. Immunol Lett 168:154–158. pii: S0165-2478(15)00105-4
Burns CJ, Persaud SJ, Jones PM (2004) Stem cell therapy for diabetes: do we need to make beta cells? J Endocrinol 183:437–443
Burt RK, Oyama Y, Traynor A, Kenyon NS (2002) Hematopoietic stem cell therapy for type 1 diabetes: induction of tolerance and islet cell neogenesis. Autoimmun Rev 1:133–138
Burt RK, Loh Y, Pearce W et al (2008) Clinical applications of blood-derived and marrow-derived stem cells for non-malignant diseases. JAMA 299:925–936

Buzzetti R, Cernea S, Petrone A et al (2011) C-peptide response and HLA genotypes in subjects with recent-onset type 1 diabetes after immunotherapy with DiaPep277: an exploratory study. Diabetes 60:3067–3072

Cabrera O, Berman DM, Kenyon NS et al (2006) The unique cyto-architecture of human pancreatic islets has implications for islet cell function. Proc Natl Acad Sci U S A 103:2334–2339

Calne R (2005) Cell transplantation for diabetes. Philos Trans R Soc Lond B Biol Sci 360:1769–1774

Canadian Diabetes Association (2003) Clinical practice guidelines for the prevention and management of diabetes in Canada. Can J Diabetes 27(suppl 2):S21–S23

Casteilla L, Planat-Benard V, Cousin B et al (2005) Plasticity of adipose tissue: a promising therapeutic avenue in the treatment of cardiovascular and blood diseases. Arch Mal Coeur Vaiss 98:922–926

Cefalu WT (2012) American diabetes association-European association for the study of diabetes position statement: due diligence was conducted. Diabetes Care 35:1201–1203

Chhabra P, Brayman KL (2013) Stem cell therapy to cure type 1 diabetes: from hype to hope. Stem Cells Transl Med 2:328–336

Chung Y, Klimanskaya I, Becker S, Marh J, Lu SJ, Johnson J, Meisner L, Lanza R (2006) Embryonic and extraembryonic stem cell lines derived from single mouse blastomeres. Nature 439:216–219

Ciancio G, Burke GW, Viciana AL et al (1996) Destructive allograft fungal arteritis following simultaneous pancreas-kidney transplantation. Transplantation 61:1172–1175

Comparison of clinical features between (juvenile) type 1 diabetes, type 2 diabetes and LADA (2006) Islets of Hope

Concanon P, Rich SS, Nepom GT (2009) Genetics of type 1A diabetes. N Engl J Med 360: 1646–1654

Cooper JD, Smyth DJ, Walker NM et al (2011) Inherited variation in vitamin D genes is associated with predisposition to autoimmune disease type 1 diabetes. Diabetes 60:1624–1631

Copland PS (2004) The roman catholic church and embryonic stem cells. J Med Ethics 30:607–608

D'Amour KA, Agulnick AD, Eliazer S, Kelly OG, Kroon E, Baetge EE (2005) Efficient differentiation of human embryonic stem cells to definitive endoderm. Nat Biotechnol 23:1534–1541

D'Amour KA, Bang AG, Eliazer S et al (2006) Production of pancreatic hormone-expressing endocrine cells from human embryonic stem cells. Nat Biotechnol 24:1392–1401

Dabelea D (2009) The accelerating epidemic of childhood diabetes. Lancet 373:1999–2000

Daneman D (2006) Type 1 diabetes. Lancet 367:847–858

Davani B, Ariely S, Ikonomou L, Oron Y, Gershengorn MC (2009) Human islet-derived precursor cells can cycle between epithelial clusters and mesenchymal phenotypes. J Cell Mol Med 13:2570–2581

Dazzi F, Marelli-Berg FM (2008) Mesenchymal stem cells for graft-versus-host disease: close encounters with T cells. Eur J Immunol 38:1479–1482

De Meyer J (1904) Sur la signification physiologique de la secretion interne du pancreas. Zbl Physiol 18:S826

Deregibus MC, Cantaluppi V, Calogero R, Lo Iacono M, Tetta C, Biancone L, Bruno S, Bussolati B, Camussi G (2007) Endothelial progenitor cell derived microvesicles activate an angiogenic program in endothelial cells by a horizontal transfer of mRNA. Blood 110:2440–2448

Discher DE, Mooney DJ, Zandstra PW (2009) Growth factors, matrices, and forces combine and control stem cells. Science 324:167–1677

Domínguez-Bendala J, Klein D, Ribeiro M, Ricordi C, Inverardi L, Pastori R, Edlund H (2005) TAT-mediated neurogenin 3 protein transduction stimulates pancreatic endocrine differentiation *in vitro*. Diabetes 54:720–726

Dominici M, Le Blanc K, Mueller I et al (2006) Minimal criteria for defining multipotent mesenchymal stromal cells. The International Society for Cellular Therapy position statement. Cytotherapy 8:315–317

Dor Y, Brown J, Martinez OI, Melton DA (2004) Adult pancreatic beta cells are formed by self-duplication rather than stem-cell differentiation. Nature 429:41–46

Drjer K (1992) The bioactivity of insulin analogues from in vitro receptor binding to in vivo glucose uptake. Diabetes Metab Rev 8:259–285

Eaton RP, Allen RC, Schade DS, Erickson KM, Standefer J (1980) Prehepatic insulin production in man: kinetic analysis using peripheral connecting peptide behavior. J Clin Endocrinol Metab 51:520–528

Eich T, Eriksson O, Lundgren T (2007) Nordic network for clinical islet transplantation. Visualization of early engraftment in clinical islet transplantation by positron-emission tomography. N Engl J Med 356:2754–2755

Ende N, Chen R, Reddi AS (2004) Effect of human umbilical cord blood cells on glycemia and insulitis in type 1 diabetic mice. Biochem Biophys Res Commun 325:665–669

Evans MJ, Kaufman MH (1981) Establishment in culture of pluripotential cells from mouse embryos. Nature 292:154–156

Fajans SS, Bell GI, Polonsky KS (2001) Molecular mechanisms and clinical pathophysiology of maturityonset diabetes of the young. N Engl J Med 345:971–980

Fandrich F, Ungefroren H (2010) Customized cell-based treatment options to combat autoimmunity and restore beta-cell function in type 1 diabetes mellitus: current protocols and future perspectives. Adv Exp Med Biol 654:641–665

Favaro E, Carpanetto A, Lamorte S, Fusco A, Caorsi C, Deregibus MC, Bruno S, Amoroso A, Giovarelli M, Porta M, Perin PC, Tetta C, Camussi G, Zanone MM (2014) Human mesenchymal stem cell-derived microvesicles modulate T cell response to islet antigen glutamic acid decarboxylase in patients with type 1 diabetes. Diabetologia 57:1664–1673

Fiorina P, Jurewicz M, Augello A et al (2009) Immunomodulatory function of bone marrow-derived mesenchymal stem cells in experimental autoimmune type 1 diabetes. J Immunol 183:993–1004

Fiorina P, Voltarelli J, Zavazava N (2011) Immunological applications of stem cells in type 1 diabetes. Endocr Rev 32:725–754

Friedenstein AJ, Piatetzky-Shapiro II, Petrakova KV (1966) Osteogenesis in transplants of bone marrow cells. J Embryol Exp Morpholog 16:381–390

Fuchs E (2009) Finding one's niche in the skin. Cell Stem Cell 2009(4):499–502

Fujikawa T, Oh SH, Pi L, Hatch HM, Shupe T, Petersen BE (2005) Teratoma formation leads to failure of treatment for type I diabetes using embryonic stem cell-derived insulin-producing cells. Am J Pathol 166:1781–1791

Gale EA (2005) Type 1 diabetes in the young: the harvest of sorrow goes on. Diabetologia 48:1435–1438

Gibly RF, Graham JG, Luo X, Lowe WL Jr, Hering BJ, Shea LD (2011) Advancing islet transplantation: from engraftment to the immune response. Diabetologia 54:2494–2505

Gluckman E, Broxmeyer HA, Auerbach AD et al (1989) Hematopoietic reconstitution in a patient with Fanconi's anemia by means of umbilical-cord blood from an HLA-identical sibling. N Engl J Med 321:1174–1178

Greenbaum C, Atkinson MA (2011) Persistence is the twin sister of excellence: an important lesson for attempts to prevent and reverse type 1 diabetes. Diabetes 60:693–694

Gruessner AC, Sutherland DE (2001) Report from the international pancreas transplant registry 2000. Transplant Proc 33:1643–1646

Gruessner AC, Sutherland DER (2005) Pancreas transplant outcomes for United States (US) and non-US cases as reported to the United Network for Organ Sharing (UNOS) and the International Pancreas Transplant Registry (IPTR) as of June 2004. Clin Transpl 19:433–455

Gruessner RW, Burke GW, Stratta R et al (1996) Multicenter analysis of the first experience with FK506 for induction and rescue therapy after pancreas transplantation. Transplantation 61:261–273

Gruss P (2003) Human ES cells in Europe. Science 301:1017

Halban PA (2004) Cellular sources of new pancreatic beta cells and therapeutic implications for regenerative medicine. Nat Cell Biol 6:1021–1025

Haller MJ, Wasserfall CH, McGrail KM et al (2009) Autologous umbilical cord blood transfusion in very young children with type 1 diabetes. Diabetes Care 32:2041–2046
Halvorsen TL, Beattie GM, Lopez AD, Hayek A, Levine F (2000) Accelerated telomere shortening and senescence in human pancreatic islet cells stimulated to divide in vitro. J Endocrinol 166: 103–109
Hamada H, Kobune M, Nakamura K, Kawano Y, Kato K, Honmou O, Houkin K, Matsunaga T, Niitsu Y (2005) Mesenchymal stem cells (MSC) as therapeutic cytoreagents for gene therapy. Cancer Sci 96:149–156
Hansson M, Tonning A, Frandsen U et al (2004) Artifactual insulin release from differentiated embryonic stem cells. Diabetes 53:2603–2609
Harjutsalo V, Sjoberg L, Tuomilehto J (2008) Time trends in the incidence of type 1 diabetes in Finnish children: a cohort study. Lancet 371:1777–1782
Hoogduijn MJ, Popp F, Verbeek R et al (2010) The immunomodulatory properties of mesenchymal stem cells and their use for immunotherapy. Int Immunopharmacol 10:1496–1500
Hori Y, Gu X, Xie X, Kim SK (2005) Differentiation of insulin-producing cells from human neural progenitor cells. PLoS Med 2:e103
Hou SX, Singh SR (2008) Germline stem cells. Springer, New York, p 450
Houard N, Rousseau GG, Lemaigre FP (2003) HNF-6-independent differentiation of mouse embryonic stem cells into insulin producing cells. Diabetologia 46:378–385
Izadpanah R, Trygg C, Patel B et al (2006) Biologic properties of mesenchymal stem cells derived from bone marrow and adipose tissue. J Cell Biochem 99:1285–1297
Jahansouz C, Kumer SC, Ellenbogen M, Brayman KL (2011) Evolution of beta-cell replacement therapy in diabetes mellitus: pancreas transplantation. Diabetes Technol Ther 13:395–418
Jahr H, Bretzel RG (2003) Insulin-positive cells in vitro generated from rat bone marrow stromal cells. Transplant Proc 35:2140–2141
Johansson H, Lukinius A, Moberg L et al (2005) Tissue factor produced by the endocrine cells of the islets of Langerhans is associated with a negative outcome of clinical islet transplantation. Diabetes 54:1755–1762
Jurewicz M, Yang S, Augello A et al (2010) Congenic mesenchymal stem cell therapy reversese-hyperglycemia in experimental type 1 diabetes. Diabetes 59:3139–3147
Kahn CR (1994) Insulin action, diabetogene and the cause of type II diabetes: banting lecture. Diabetes 43:1066–1084
Kahn HS, Morgan TM, Case LD et al (2009) Association of type 1 diabetes with month of birthamong US youth: the SEARCH for diabetes in youth study. Diabetes Care 32:2010–2015
Kang EM, Zickler PP, Burns S et al (2005) Hematopoietic stem cell transplantation prevents diabetes in NOD mice but does not contribute to significant islet cell regeneration once disease is established. Exp Hematol 33(6):699–705
Karnieli O, Izhar-Prato Y, Bulvik S, Efrat S (2007) Generation of insulin-producing cells from human bone marrow mesenchymal stem cells by genetic manipulation. Stem Cells 25: 2837–2844
Kassem SA, Ariel I, Thornton PS, Scheimberg I, Glaser B (2000) Beta-cell proliferation and apoptosis in the developing normal human pancreas and in hyperinsulinism of infancy. Diabetes 49:1325–1333
Kelly WD, Lillehei RC, Merkel FK, Idezuki Y, Goetz FC (1967) Allotransplantation of the pancreas and duodenum along with the kidney in diabetic nephropathy. Surgery 61:827–837
Kern S, Eichler H, Stoeve J et al (2006) Comparative analysis of mesenchymal stem cells from bone marrow, umbilical cord blood, or adipose tissue. Stem Cells 24:1294–1301
Kilk K, Magzoub M, Pooga M, Eriksson LE, Langel U, Gräslund A (2001) Cellular internalization of a cargo complex with a novel peptide derived from the third helix of the islet-1 homeodomain. Comparison with the penetratin peptide. Bioconjug Chem 12:911–916
Knip M, Virtanen SM, Akerblom HK (2010) Infant feeding and the risk of type 1 diabetes. Am J Clin Nutr 91:1506–1513
Kroon E, Martinson LA, Kadoya K et al (2008) Pancreatic endoderm derived from human embryonic stem cells generates glucose-responsive insulin-secreting cells in vivo. Nat Biotechnol 26:443–452

Kubo A, Shinozaki K, Shannon JM et al (2004) Development of definitive endoderm from embryonic stem cells in culture. Development 131:1651–1662
Kucia M, Halasa M, Wysoczynski M et al (2006) Morphological and molecular characterization of novel population of CXCR4+ SSEA-4+ Oct-4+ very small embryonic-like cells purified from human cord blood–preliminary report. Leukemia 21:297–303
Kukko M, Kimpimaki T, Korhonen S et al (2005) Dynamics of diabetes-associated autoantibodies in young children with human leukocyte antigen-conferred risk of type 1 diabetes recruited from the general population. J Clin Endocrinol Metab 90:2712–2717
Lakey JR, Burridge PW, Shapiro AM (2003) Technical aspects of islet preparation and transplantation. Transpl Int: Off J Eur Soc Organ Transplant 16:613–632
Landry DW, Zucker HA (2004) Embryonic death and the creation of human embryonic stem cells. J Clin Invest 114:1184–1186
Langerhans P (1869) Beiträgezurmikroskopischen Anatomie der Bauchspeicheldrüse. Inaug.-Diss, Berlin
Lee RH, Seo MJ, Reger RL et al (2006) Multipotent stromal cells from human marrow home to and promote repair of pancreatic islets and renal glomeruli in diabetic NOD/scid mice. Proc Natl Acad Sci U S A 103:17438–17443
Leibiger IB, Berggren PO (2008) Insulin signaling in the pancreatic β-cell. Annu Rev Nutr 28:233–251
Leroith D, Taylor SI, Olefky JM (eds) (2003) Diabetes mellitus. A fundamental and clinical text, 3rd edn. Lippincott, Williams Wilkins, Philadelphia
Li Y, Zhang R, Qiao H, Zhang H, Wang Y, Yuan H, Liu Q, Liu D, Chen L, Pei X (2007) Generation of insulin-producing cells from PDX-1 gene-modified human mesenchymal stem cells. J Cell Physiol 211:36–44
Liu Z, Sall A, Yang D (2008) MicroRNA: an emerging therapeutic target and intervention tool. Int J Mol Sci 9:978–999
Liu X, Wang Y, Li Y, Pei X (2013) Research status and prospect of stem cells in the treatment of diabetes mellitus. Sci China Life Sci 56:306–312
Lu LL, Liu YJ, Yang SG et al (2006) Isolation and characterization of human umbilical cord mesenchymal stem cells with hematopoiesis-supportive function and other potentials. Haematologica 91:1017–1026
Ludvigsson J, Krisky D, Casas R et al (2012) GAD65 antigen therapy in recently diagnosed type 1 diabetes mellitus. N Engl J Med 366:433–442
Lumelsky N, Blondel O, Laeng P, Velasco I, Ravin R, McKay R (2001) Differentiation of embryonic stem cells to insulin-secreting structures similar to pancreatic islets. Science 292: 1389–1394
Maahs DM, West NA, Lawrence JM, Mayer-Davis EJ (2010) Epidemiology of type 1 diabetes. Endocrinol Metab Clin N Am 39:481–497
MacCracken J, Hoel D (1997) From ants to analogues: puzzles and promises in diabetes management. Postgrad Med 101(4):138–140. commentaries
Maclaren N, Atkinson M (1992) Is insulin-dependent diabetes mellitus environmentally induced? N Engl J Med 327:348–349
Madec AM, Mallone R, Afonso G et al (2009) Mesenchymal stem cells protect NOD mice from diabetes by inducing regulatory T cells. Diabetologia 52:1391–1399
Martin GR (1981) Isolation of a pluripotent cell line from early mouse embryos cultured in medium conditioned by teratocarcinoma stem cells. Proc Natl Acad Sci U S A 78:7634–7638
Masoud MS, Anwar SS, Afzal MZ, Mehmood A, Khan SN, Riazuddin S (2012) Pre-conditioned mesenchymal stem cells ameliorate renal ischemic injury in rats by augmented survival and engraftment. J Transl Med 10:243
Matsumoto S, Okitsu T, Iwanaga Y et al (2005) Insulin independence after living-donor distal pancreatectomy and islet allotransplantation. Lancet 365:1642–1644
Matsumoto S, Okitsu T, Iwanaga Y et al (2006) Follow-up study of the first successful living donor islet transplantation. Transplantation 82:1629–1633
McGuckin CP, Forraz N (2008) Potential for access to embryonic-like cells from human umbilical cord blood. Cell Prolif 41:31–40

Meier JJ, Butler PC (2005) Insulin secretion. In: Endocrinology. Elsevier Saunders, Philadelphia
Meirelles Lda S, Nardi NB (2009) Methodology, biology and clinical applications of mesenchymal stem cells. Front Biosci 14:4281–4298
Meissner A, Jaenisch R (2006) Generation of nuclear transfer-derived pluripotent ES cells from cloned Cdx2-deficient blastocysts. Nature 439:212–215
Mering JV, Minkowski O (1889) Diabetes mellitus nach Pankreasexstirpation. Zschrklin Med 14:404–423
Mineo D, Ciancio G, Burke GW, Alejandro R, Ricordi C (2010) Islet and pancreas transplantation. In: Efrat S (ed) Stem cell therapy for diabetes. Stem cell biology and regenerative medicine. Humana Press, New York, pp 41–83. doi:10.1007/978–1–60761-366-4_2
Mishra PK, Tyagi N, Kumar M, Tyagi SC (2009) MicroRNAs as a therapeutic target for cardiovascular disease. J Cell Mol Med 13:778–789
Moberg L, Johansson H, Lukinius A et al (2002) Production of tissue factor by pancreatic islet cells as a trigger of detrimental thrombotic reactions in clinical islet transplantation. Lancet 360:2039–2045
Mokarizadeh A, Delirezh N, Morshedi A, Mosayebi G, Farshid AA, Mardani K (2012) Microvesicles derived from mesenchymal stem cells: potent organelles for induction of tolerogenic signaling. Immunol Lett 147:47–54
Moltchanova EV, Schreier N, Lammi N, Karvonen M (2009) Seasonal variation of diagnosis of type 1 diabetes mellitus in children worldwide. Diabet Med 26:673–678
Murphy R, Ellard S, Hattersley AT (2008) Clinical implications of a molecular genetics classification of monogenicβ–cell diabetes. Nat Clin Pract Endocrinol Metab 4:200
Nilsson B, Ekdahl KN, Korsgren O (2011) Control of instant blood-mediated inflammatory reaction to improve islets of Langerhans engraftment. Curr Opin Organ Transplant 16:620–626
Noguchi H, Matsumoto S (2006) Protein transduction technology offers a novel therapeutic approach for diabetes. J Hepato-Biliary-Pancreat Surg 13:306–313
Noguchi H, Kaneto H, Weir GC, Bonner-Weir S (2003) PDX-1 protein containing its own antennapedia-like protein transduction domain can transduce pancreatic duct and islet cells. Diabetes 52:1732–1737
Noguchi H, Bonner-Weir S, Wei FY, Matsushita M, Matsumoto S (2005) BETA2/neuro D protein can be transduced into cells due to an arginine- and lysinerich sequence. Diabetes 54: 2859–2866
Oakley J (2002) Democracy, embryonic stem cell research, and the roman catholic church. Med Ethics 28:22856
Oedayrajsingh-Varma M, van Ham S, Knippenberg M et al (2006) Adipose tissue-derived mesenchymal stem cell yield and growth characteristics are affected by the tissue-harvesting procedure. Cytotherapy 8:166–177
Oh SH, Muzzonigro TM, Bae SH, LaPlante JM, Hatch HM, Petersen BE (2004) Adult bone marrow-derived cells trans-differentiating into insulin-producing cells for the treatment of type I diabetes. Lab Investig 84:607–617
Ohneda K, Ee H, German M (2000) Regulation of insulin gene transcription. Semin Cell Dev Biol 11:227–233
Okita K, Yamakawa T, Matsumura Y et al (2013) An efficient nonviral method to generate integration-free human-induced pluripotent stem cells from cord blood and peripheral blood cells. Stem Cells 31:458–466
Orban T, Bundy B, Becker DJ et al (2011) Co-stimulation modulation with abatacept in patients with recent-onset type 1 diabetes: a randomised, double-blind, placebo-controlled trial. Lancet 378:412–419
Orlando G, Gianello P, Salvatori M, Stratta RJ, Soker S, Ricordi C et al (2014) Cell replacement strategies aimed at reconstitution of the beta-cell compartment in type 1 diabetes. Diabetes 63:1433–1444

Ostman J, Lonnberg G, Arnqvist HJ et al (2008) Gender differences and temporal variation in the incidence of type 1 diabetes: results of 8012 cases in the nationwide diabetes incidence study in Sweden 1983–2002. J Intern Med 263:386–394
Pap E, Pállinger E, Falus A (2011) The role of membrane vesicles in tumorigenesis. Crit Rev Oncol Hematol 79:213–223
Passweg J, Tyndall A (2007) Autologous stem cell transplantation in autoimmune diseases. Semin Hematol 44:278–285
Patterson CC, Dahlquist GG, Gyürüs E, Green A, Soltész G, EURODIAB Study Group (2009) Incidence trends for childhood type 1 diabetes in Europe during 1989–2003 and predicted new cases 2005–20: a multicentre prospective registration study. Lancet 373:2027–2033
Pescovitz MD, Greenbaum CJ, Krause-Steinrauf H et al (2009) Rituximab, B-lymphocyte depletion, and preservation of beta-cell function. N Engl J Med 361:2143–2152
Pickup JC, Willium G (2003) The history of diabetes mellitus, Textbook of diabetes, vol 1, 3rd edn. Blackwell Science Limited, Oxford
Poitout V, Robertson RP (2008) Glucolipotoxicity: fuel excess and β-cell dysfunction. Endocr Rev 29:351–366
Polonsky KS, Pugh W, Jaspan JB, Cohen DM, Karrison T, Tager HS, Rubenstein AH (1984) C-peptide and insulin secretion. Relationship between peripheral concentrations of C-peptide and insulin and their secretion rates in the dog. J Clin Invest 74:1821–1829
Posselt AM, Szot GL, Frassetto LA, Masharani U, Tavakol M et al (2010) Islet transplantation in type 1 diabetic patients using calcineurin inhibitor-free immunosuppressive protocols based on T-cell adhesion or costimulation blockade. Transplantation 90:1595–1601
Pozzilli P (2012) Type 1 diabetes mellitus in 2011: heterogeneity of T1DM raises questions for therapy. Nat Rev Endocrinol 8:78–80
Prokhorova TA, Harkness LM, Frandsen U, Ditzel N, Schroder HD, Burns JS et al (2009) Teratoma formation by human embryonic stem cells is site dependent and enhanced by the presence of Matrigel. Stem Cells Dev 18:47–54
Pybus FC (1924) Notes on suprarenal and pancreatic grafting. Lancet 204:550–551
Rackham CL, Chagastelles PC, Nardi NB et al (2011) Co-transplantation of mesenchymal stem cells maintains islet organisation and morphology in mice. Diabetologia 54:1127–1135
Rahier J, Goebbels RM, Henquin JC (1983) Cellular composition of the human diabetic pancreas. Diabetologia 24:366–371
Ramiya VK, Maraist M, Arfors KE, Schatz DA, Peck AB, Cornelius JG (2000) Reversal of insulin-dependent diabetes using islets generated in vitro from pancreatic stem cells. Nat Med 6:278–282
Ratajczak J, Miekus K, Kucia M, Zhang J, Reca R, Dvorak P, Ratajczak MZ (2006) Embryonic stem cell-derived microvesicles reprogram hematopoietic progenitors: evidence for horizontal transfer of mRNA and protein delivery. Leukemia 20:847–856
Ricordi C (ed) (1992) 1892–1992. One century of transplantation for diabetes: pancreatic islet cell. Transplantation. R.G. Landes Company, Austin, p 291
Ricordi C, Lacy PE, Scharp DW (1989) Automated islet isolation from human pancreas. Diabetes 38:140–142
Robbins PD, Morelli AE (2014) Regulation of immune responses by extracellular vesicles. Nat Rev Immunol 14:195–208
Ryan EA, Paty BW, Senior PA et al (2005) Five-year follow-up after clinical islet transplantation. Diabetes 54:2060–2069
Scharp DW, Lacy PE, Santiago JV et al (1990) Insulin independence after islet transplantation into type I diabetic patient. Diabetes 39:515–518
Scheres B (2007) Stem-cell niches: nursery rhymes across kingdoms. Nat Rev Mol Cell Biol 8:345–354
Schloot NC, Meierhoff G, Lengyel C et al (2007) Effect of heat shock protein peptide DiaPep277 on beta-cell function in paediatric and adult patients with recent-onset diabetes mellitus type 1: two prospective, randomized, double-blind phase II trials. Diabetes Metab Res Rev 23:276–285

Seaberg RM, Smukler SR, Kieffer TJ et al (2004) Clonal identification of multipotent precursors from adult mouse pancreas that generate neural and pancreatic lineages. Nat Biotechnol 22:1115–1124

Segev H, Fishman B, Ziskind A, Shulman M, Itskovitz-Eldor J (2004) Differentiation of human embryonic stem cells into insulin-producing clusters. Stem Cells 22:265–274

Shapiro AM, Lakey JR, Ryan EA et al (2000) Islet transplantation in seven patients with type 1 diabetes mellitus using a glucocorticoid-free immunosuppressive regimen. N Engl J Med 343:230–238

Shapiro AM, Ricordi C, Hering BJ et al (2006) International trial of the Edmonton protocol for islet transplantation. N Engl J Med 355:1318–1330

Sherry N, Hagopian W, Ludvigsson J et al (2011) Teplizumab for treatment of type 1 diabetes (Protege study): 1-year results from randomised, placebo-controlled trial. Lancet 378: 487–497

Siminovitch L, McCulloch EA, Till JE (1963) The distribution of colony-forming cells among spleen colonies. J Cell Physiol 62:327–336

Sims E, Evans-Molina C (2012) Stem cells as a tool to improve outcomes of islet transplantation. J Transplant 2012:736491

Singh SR, Hou SX (2008) Lessons learned about adult kidney stem cells from the malpighian tubules of Drosophila. J Am Soc Nephrol 19:660–666

Singh SR, Hou SX (2009) Multipotent stem cells in the Malpighian tubules of adult *Drosophila melanogaster*. J Exp Biol 212:413–423

Singh SR, Liu W, Hou SX (2007) The adult Drosophila malpighian tubules are maintained by multipotent stem cells. Cell Stem Cell 1:191–203

Soria B, Roche E, Berná G, Leon-Quinto T, Reig JA, Martin F (2000) Insulin-secreting cells derived from embryonic stem cells normalize glycemia in streptozotocin induced diabetic mice. Diabetes 49:157–162

Sosenko JM, Skyler JS, Mahon J et al (2012) The application of the diabetes prevention trial-type 1 risk score for identifying a preclinical state of type 1 diabetes. Diabetes Care 35:1552–1555

Spaggiari GM, Abdelrazik H, Becchetti F, Moretta L (2009) MSCs inhibit monocyte-derived DC maturation and function by selectively interfering with the generation of immature DCs: central role of MSC-derived prostaglandin E2. Blood 113:6576–6583

Sprent J, Kishimoto H (2001) The thymus and central tolerance. Philos Trans R Soc Lond B Biol Sci 356:609–616

Stadtfeld M, Nagaya M, Utikal J, Weir G, Hochedlinger K (2008) Induced pluripotent stem cells generated without viral integration. Science 322:945–949

Staeva TP, Chatenoud L, Insel R, Atkinson MA (2013) Recent lessons learned from prevention and recent-onset type 1 diabetes immunotherapy trials. Diabetes 62:9–17

Starzl TE (2001) The "privileged" liver and hepatic tolerogenicity. Liver Transpl 7:918–920

Stefan Y, Orci L, Malaisse-Lagae F, Perrelet A, Patel Y, Unger RH (1982) Quantitation of endocrine cell content in the pancreas of non-diabetic and diabetic humans. Diabetes 31:694–700

Stene LC, Rewers M (2012) Immunology in the clinic review series; focus on type 1 diabetes and viruses: the enterovirus link to type 1 diabetes: critical review of human studies. Clin Exp Immunol 168:12–23

Subramanian S, Trence DL (2007) Immunosuppressive agents: effects on glucose and lipid metabolism. Endocrinol Metab Clin N Am 36:891–905

Sun Y, Chen L, Hou XG, Hou WK, Dong JJ, Sun L et al (2007) Differentiation of bone marrow derived mesenchymal stem cells from diabetic patients into insulin-producing cells in vitro. Chin Med J 120:771–776

Suzuki A, Nakauchi H, Taniguchi H (2004) Prospective isolation of multipotent pancreatic progenitors using flow-cytometric cell sorting. Diabetes 53:2143–2152

Svoren BM, Volkening LK, Wood JR, Laffel LM (2009) Significant vitamin D deficiency in youth with type 1 diabetes mellitus. J Pediatr 154:132–134

Sykes M, Nikolic B (2005) Treatment of severe autoimmune disease by stem-cell transplantation. Nature 435:620–627

Takahashi K, Yamanaka S (2006) Induction of pluripotent stem cells from mouse embryonic and adult fibroblast cultures by defined factors. Cell 126:663–676

Tariq M, Masoud MS, Mehmood A, Khan SN, Riazuddin S (2013) Stromal cell derived factor-1alpha protects stem cell derived insulin-producing cells from glucotoxicity under high glucose conditions in-vitro and ameliorates drug induced diabetes in rats. J Transl Med 11:115

Thakkar UG, Trivedi HL, Vanikar AV, Dave SD (2015) Insulin-secreting adipose-derived mesenchymal stromal cells with bone marrow-derived hematopoietic stem cells from autologous and allogenic sources for type 1 diabetes mellitus. Cytotherapy 17:940–947

The American Diabetes Association (2009) Diagnosis and classification of diabetes mellitus. Diabetes Care 32:S62–S67

The Diabetes Control and Complications Trial Research Group (1993) The effect of intensive treatment of diabetes on the development and progression of long-term complications in insulin-dependent diabetes mellitus. N Engl J Med 329:977–986

Théry C, Ostrowski M, Segura E (2009) Membrane vesicles as conveyors of immune responses. Nat Rev Immunol 9:581–593

Thomson JA, Itskovitz-Eldor J, Shapiro SS et al (1998) Embryonic stem cell lines derived from human blastocysts. Science 282:1145–1147

Thunander M, Petersson C, Jonzon K et al (2008) Incidence of type 1 and type 2 diabetes in adults and children in Kronoberg. Sweden Diabetes Res Clin Pract 82:247–255

Trivedi HL, Vanikar AV, Thakker U et al (2008) Human adipose tissue-derived mesenchymal stem cells combined with hematopoietic stem cell transplantation synthesize insulin. Transplant Proc 40:1135–1139

Ullrich A, Shrine J, Chirgwin J (1977) Rat insulin genes construction of plasmids containing the coding sequence. Science 196:1313–1319

Urban VS, Kiss J, Kovacs J et al (2008) Mesenchymal stem cells cooperate with bone marrow cells in therapy of diabetes. Stem Cells 26:244–253

Valadi H, Ekström K, Bossiois A, Sjörstrand M, Lee JJ, Lötvall JO (2007) Exosome-mediated transfer of mRNA and mirco RNAs is a novel mechanism of genetic exchange between cells. Nat Cell Biol 9:654–659

Vanikar AV, Dave SD, Thakkar UG, Trivedi HL (2010) Co-transplantation of adipose tissue-derived insulin-secreting mesenchymal stem cells and hematopoietic stem cells: a novel therapy for insulin-dependent diabetes mellitus. Stem Cells Int 2010 :5. doi:10.4061/2010/582382Article ID 582382

Vanikar AV, Trivedi HL, Patel RD, Kanodia KV, Modi PR, Shah VR (2012) Allogenic hematopoietic stem cell transplantation in pemphigus vulgaris: a single-center experience. Indian J Dermatol 57:9–11

Vantyghem MC, Marcelli-Tourvielle S, Pattou F et al (2007) Effects of non-steroid immunosuppressive drugs on insulin secretion in transplantation. Ann Endocrinol 68:21–27

Vendrame F, Pileggi A, Laughlin E et al (2010) Recurrence of type 1 diabetes after simultaneous pancreas-kidney transplantation, despite immunosuppression, is associated with autoantibodies and pathogenic autoreactive CD4 T-cells. Diabetes 59:947–957

Voltarelli JC, Couri CE, Rodrigues MC et al (2011) Stem cell therapies for type 1 diabetes mellitus. Indian J Exp Biol 49:395–400

Walter M, Philotheou A, Bonnici F, Ziegler AG, Jimenez R (2009) No effect of the altered peptide ligand NBI-6024 on beta-cell residual function and insulin needs in new-onset type 1 diabetes. Diabetes Care 32:2036–2040

Wang V, Wu W (2009) MicroRNA-based therapeutics for cancer. Bio Drugs 23:15–23

Wang RN, Kloppel G, Bouwens L (1995) Duct- to islet-cell differentiation and islet growth in the pancreas of duct-ligated adult rats. Diabetologia 38:1405–1411

Wang HS, Shyu JF, Shen WS et al (2011) Transplantation of insulin-producing cells derived from umbilical cord stromal mesenchymal stem cells to treat NOD mice. Cell Transplant 20:455–466

Watson R (2003) Euro MPs threaten UK stem cell research. BMJ 326:838

Webb MA, Dennison AR, James RF (2012) The potential benefit of non-purified islets preparations for islet transplantation. Biotechnol Genet Eng Rev 28:101–114

Wherrett DK, Bundy B, Becker DJ et al (2011) Antigen-based therapy with glutamic acid decarboxylase (GAD) vaccine in patients with recent-onset type 1 diabetes: a randomised double-blind trial. Lancet 378:319–327

Wilkin TJ (2008) Diabetes 1 and 2, or one and the same? Progress with the accelerator hypothesis. Pediatr Diabetes 9:23–32

Winkler C, Krumsiek J, Lempainen J et al (2012) A strategy for combining minor genetic susceptibility genes to improve prediction of disease in type 1 diabetes. Genes Immun 13: 549–555

World Health Organisation (1999) Definition, diagnosis and classification of diabetes mellitus and its complications. Part 1: diagnosis and classification of diabetes mellitus. WHO/NCD/NCS/99.2. Geneva. Ref Type: Report

Xu J, Lu Y, Ding F, Zhan X, Zhu M, Wang Z (2007) Reversal of diabetes in mice by intrahepatic injection of bone-derived GFP murine mesenchymal stem cells infected with the recombinant retrovirus-carrying human insulin gene. World J Surg 31:1872–1882

Xu X, D'Hoker J, Stange G et al (2008) β-cells can be generated from endogenous progenitors in injured adult mouse pancreas. Cell 132:197–207

Yamashita YM (2009) Regulation of asymmetric stem cell division: spindle orientation and the centrosome. Front Biosci 14:3003–3011

Yañez R, Lamana ML, García-Castro J, Colmenero I, Ramírez M, Bueren JA (2006) Adipose tissue-derived mesenchymal stem cells have in vivo immunosuppressive properties applicable for the control of the graft-versus-host disease. Stem Cells 24:2582–2591

Yechoor V, Liu V, Espiritu C et al (2009) Neurogenin3 is sufficient for transdetermination of hepatic progenitor cells into neo-islets in vivo but not transdifferentiation of hepatocytes. Dev Cell 16:358–373

Yeung WC, Rawlinson WD, Craig ME (2011) Enterovirus infection and type 1 diabetes mellitus: systematic review and meta-analysis of observational molecular studies. BMJ 342:35

Yoon JW, Jun HS (2005) Autoimmune destruction of pancreatic β-cells. Am J Ther 12:580–591

Zaret KS, Grompe M (2008) Generation and regeneration of cells of the liver and pancreas. Science 322:1490–1494

Zhang J, Shehabeldin A, da Cruz LA, Butler J, Somani AK, McGavin M et al (1999) Antigen receptor-induced activation and cytoskeletal rearrangement are impaired in Wiskott-Aldrich syndrome protein-deficient lymphocytes. J Exp Med 190:1329–1342

Zimmet P, Alberti KG, Shaw J (2001) Global and societal implications of the diabetes epidemic. Nature 414:782–787

Chapter 3
New Trends in Stem Cell Transplantation in Diabetes Mellitus Type I and Type II

Alexander E. Berezin

Abbreviations

ABMSCs	Autologous bone marrow-derived mesenchymal stem cells
ADSCs	Adipose-derived stem cells
CPC	Circulating precursor cells
DM	Diabetes mellitus
EPC	Endothelial progenitor cells
ESCs	Embryonic stem cells
PSCs	Pluripotent stem cells
SCs	Stem cells
T1DM	Type one diabetes mellitus
T2DM	Type two diabetes mellitus

3.1 Introduction

Diabetes mellitus (DM) is recognized as the most common metabolic disease, which affects more than 347 million people worldwide and is reported as major cause of morbidity and mortality in general population (Scully 2012). Currently, type 1 DM (T1DM) is reported as an autoimmune disease characterized by insulin secretion deficiency due to destruction of insulin-producing β-cells (Ashcroft and Rorsman 2012). It is well established that the type 2 DM (T2DM) associates predominantly with metabolically active obese, insulin resistance, and adipocytokine production imbalance leading secondary to dysfunction/apoptosis of pancreatic β-cell (Paneni et al. 2013). Despite contemporary treatment strategy, T1DM and T2DM are frequently coexisting with major microvascular and macrovascular complications leading to target organ damages, i.e., ischemic tissue injury, retinopathy,

A.E. Berezin (✉)
Cardiology Unit, Internal Medicine Department, State Medical University,
26, Mayakovsky av., Zaporozhye, Ukraine
e-mail: dr_berezin@mail.ru

P.V. Pham (ed.), *Pancreas, Kidney and Skin Regeneration*, Stem Cells in Clinical Applications, DOI 10.1007/978-3-319-55687-1_3

nephropathy, and diabetes-related cardiomyopathy (Ali and Dayan 2009). The ability of endogenous repair system to improve ischemic damage of the tissues and to restore of innate mechanisms of cell-to-cell cooperation to attenuate tissue perfusion and endothelial function is sufficiently worse (Cade 2008). As a consequence, intensified tissue remodeling and the extended ischemic injury lead to increased cardiovascular (CV) morbidity and mortality (Berezin 2016a). Finally, DM-related mortality may approximately twofold increase when compared with death rate in DM-free patient population (Nwaneri et al. 2013). Nevertheless, not all complications of DM are considered consequence of ischemic tissue injury, and they may closely relate to metabolic memory phenomenon (Berezin 2016b). Various factors contributed to increased CV risk in DM are determined, i.e., hyperglycemia, lipotoxicity, age-related and diabetes-related comorbidity, and known CV diseases. In fact, they all may lead to malignant evolution of the diabetes associated with poor outcomes and may reflect a shortcoming of current therapies. Currently established therapies molecular targets in diabetic patients affect not only insulin secretion, glucose regulator peptides, hormones, enzymes, and transporters. However, they should also mediate improving hypoglycemia, suppression of oxidative stress and endoplasmic reticulum stress, prevention of atherosclerosis, attenuation of dyslipidemia and endothelial function, modification of coexisting CV risk factors, and adequate control in comorbidities (Howangyin and Silvestre 2014).

Although there is an understanding of the several mechanisms and different phases in pathogenesis of DM, it is unclear whether alternative translational approaches regarding tissue reparation and restoration of failing β-cell function based on attenuation of metabolic processes via stem cell transplantation are effective (Holditch et al. 2014). Indeed, clinical use of pluripotent stem cells (PSCs), as well as embryonic stem cells (ESCs) and induced PCSs (iPCSs), appears to be promising and safe in a long-term prospective (Liew 2010; Abdelalim et al. 2014). As it is expecting, various types of ESCs and iPSCs lines having a great differential potential may translate into all cell types, and they have a high potency to differentiate into insulin-secreting β-like cells without or very low risk of immune rejection (Abdelalim et al. 2014). However, the results of the recently performed studies regarding regenerative care in DM are controversial and require to be explained in detail (Chidgey et al. 2008). In this context, there are discrepancies between results received in the animal studies and data that have been obtained in the clinical investigations. The first controversy relates to some inconsistencies, which might accompany the fact that several types of stem cells were tested as prospective for regenerative strategy, and not all of stem cells were available in routine clinical practice. The second controversy attached to the patients with different types of DM at several stages of evolution of the disease failed to uniform considered candidates for stem cells transplantation, and they would probably require further investigations (Aguayo-Mazzucato and Bonner-Weir 2010; Anastasia et al. 2010). Given the conflicting evidence concerning stem cell replacement in T2DM and T1DM, the aim of the chapter was to seek, analyzes, and summarize the data to clarify actual knowledge and identify the future perspectives for regenerative therapy in diabetics.

3.2 The Regenerative Care Paradigm

The paradigm of regenerative therapy bases on novel knowledge of pathogenesis and mechanisms of endogenous reparation mechanisms (Terzic and Behfar 2014). It has been postulated that regenerative therapy based on cell care might attenuate or reverse disease progression. In this context, stem cell therapies are applied essentially as adjuvants to standard of care with the goal of furthering an otherwise limited self-renewal capacity of the disease (Qi et al. 2012).

3.3 Expected Effects of Cell Therapy

The possible effects of cell therapy have many phases, and they affect several sides of pathophysiological mechanisms of DM evolution (Berezin 2014). The possible therapeutic approaches of cell care in diabetics are presented in Fig. 3.1.

The possible approaches are:

- Restoring and renewing of pool of the functional β-cells from human stem cells
- Stimulation of the endogenous reparation processes
- A main result of stem cell transplantation is regulation of metabolic processes through production of both cytokines and growth factors

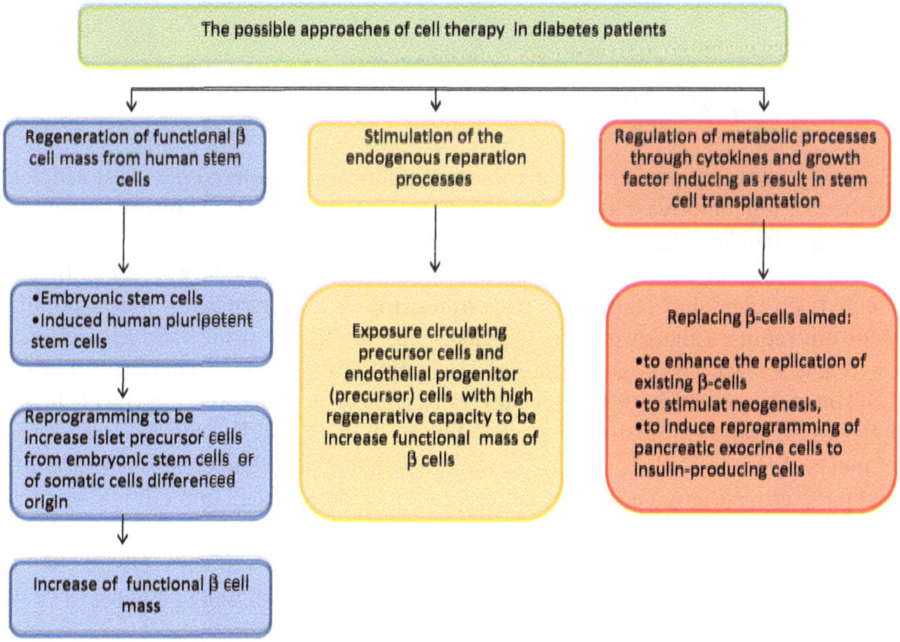

Fig. 3.1 The approaches toward improved clinical results in the diabetics enrolled for cell replacement therapy

3.3.1 Regeneration of Functional β-Cell Mass

Development of DM is characterized by the loss of functional pool of the pancreatic β-cells due to necrosis/apoptosis and negative autocrine/paracrine regulation that lead to progressive insufficiency in the insulin secretion (Holditch et al. 2014). The attempt to generate surrogate β-cells is widely used technology aim of which was to compensate the short supply of islets for transplantation to diabetics requiring daily short-acting insulin (Bhonde et al. 2014). Unfortunately, the poor availability of donor islets and high risk of rejection even within chronic immune suppression has severely restricted the broad routine clinical use of pancreas islet cell transplantation (Calafiore et al. 2014). Consequently, there is increased interest in islet cell that might obtain through neogenesis from stem cells' lines originated from embryonic or mesenchymal cells and then used as a translator for favorable responses in diabetics (Burt et al. 2002; Chen et al. 2014; Kojima 2014). The progress in our knowledge ragarding surrogate β-cells and β-cell like cells produced insulin has been mediating a use of ESCs, iPSCs, human perinatal tissues including peripheral blood mononuclears, cord blood, amnion, placenta, umbilical cord, bone marrow, pancreas, and postnatal tissues involving adipose tissue (Bhonde et al. 2014; Rabelink and Little 2013; Bhansali et al. 2009; Abraham et al. 2008; Burt et al. 2002; Lampeter et al. 1998). Although these progresses in derivation of β-cell-like cells from ESCs using reprogramming technology have taken a greater leap, the clinical implementation is limited due to controversies affecting human ESC rejection (Bhonde et al. 2014; Voltarelli et al. 2011).

3.3.2 Allogenic Pancreatic Islet Transplantation

Currently allogenic pancreatic islet transplantation is considered the most efficient method of regenerative care in DM (Nostro and Keller 2012). Transcriptional signaling mechanisms that are in vivo involved in pancreas reparation and islet pancreatic development are well-structured and closely controlled processes. In this context, up-regulated pancreatic islet transcripts in differentiating β-cell populations are required the formation of β-like cells producing insulin. Several clinical approaches toward reprogramming and differentiation of islet precursor cells from ESC base on regulation of the relevant transcription factor expression (Pdx1, Ngn3, Isl-1, etc.), some extracellular factors (Voltarelli et al. 2011), or lentiviral vectors (Jimenez-Moreno et al. 2015). These approaches might initiate creation of bioartificial pancreas, although a significant translation of similar idea into clinical application for T1DM and T2DM is not evident (Ludwig and Ludwig 2015; Khosravi-Maharlooei et al. 2015). However, the advantages in islet transplantation have exhibited a dramatic improvement in the 5-year insulin independence rates for diabetics (Khosravi-Maharlooei et al. 2015).

It was noted that there is another source for transplantation of exogenous human and nonhuman pancreas/islets or even artificial islets. However, any translational strategies regarding enhancing proliferation and maturation of endogenous β-cells, loss prevention of β-cell and β-like cells, or any methods regarding renewal of β-like-cell populations from ESCs and non-β-cells appear to be promised (Weir et al. 2011). Although currently used methods of regeneration enhancement of functional pool of β-cells from human ESCs appear as the most promising clinical approach for T1DM regenerative therapy (Lindahl et al. 2014), the cell replacement care regarding generation of unlimited sources of β-cells have been met with some sufficient limitations (Calafiore et al. 2014). As one, the purification procedure of desire cell population is a critical step to obtain enough portions of islet precursors needed to further cell-lineage selection (Soria 2001). The prevention of islets' loss in long-term prospective might need an immunosuppressant use (Jafarian et al. 2014).

3.3.3 Human Pluripotent Stem Cells

There are various ESC sources, which are considered as an appropriate resource of creating functional β-like cell generations in a safe and efficient manner (Fu 2014; Weir et al. 2011; Pandian et al. 2014; Soria 2001):

1. ESCs derived from surplus blastocysts reprogrammed for fertilization procedures in vitro
2. iPSCs created resulting in the reprogramming method of various somatic cells

The practical value of ESCs has been widely investigated during the last decade. At least two large clinical trials have recently completed, but clinical value of obtained results has become controversial (Philonenko et al. 2011). Although ECSs or adult stem cells, which were derived from various cell lines, may differentiate into β-like cells and restore the insulin production, they have represented the immune effects on the β-cells leading to their direct autoimmune destruction and destroying. (Calafiore 2014).

3.3.4 Induced Pluripotent Stem Cells

In this context, human iPSCs are discussed the most promising source for transplantation, because contemporary cellular reprogramming technology did not associate with immune modulatory ability of iPSCs did not represented immune modulatory ability (Bar-Nur et al. 2011). First iPSCs have been successfully derived from human somatic cells, i.e., dermal fibroblasts and keratinocytes (Soejitno and Prayudi 2011). It is known that redifferentiation of iPSCs into functional maturated pancreatic islets may modify disease progression through an increase of failing islet survival (Kudva et al. 2012). Importantly, the contemporary technology of iPSC derivation from adult somatic cells completely excludes the exposure of embryonic cells. Currently, iPSCs have been derived from diabetics using integrating retroviral vectors that incorporate directly in the host genome (Reiland et al. 2011; Ma et al. 2013). In fact, various reprogramming systems are different in their ability to biosafety, and integration-free reprogramming systems are more modern and hazardless (Kudva et al. 2012; Sommer et al. 2012).

The discovery of novel technologies suggests seeking of possible approaches regarding generation of patient-specific iPSCs. Various reprogramming factors have been now identified as functionally acting molecules – the main role of which was to significantly improve the results of iPSC reprogramming procedure. However, prior to a clinical implementation of the iPSCs, a strong concerns about their specificity, kinetics, and safety is required. (Bai et al. 2013).

3.3.5 Trans-differentiation Procedure

Yet one of the attractive strategies for regenerative care is the trans-differentiation procedure based on the direct conversion of the single somatic cell to another type of cell (Ma et al. 2013). Recent studies have shown a new paradigm of trans-differentiation, i.e., using specific transcriptional factors to induce novel PSC generation through trans-differentiation or induce PSC trans-differentiation through transcription factors. The trans-differentiation procedure allows generating plastic intermediates synthesis, which may attenuate iPSC reprogramming and a wide range of tissue-specific precursor cells (Ma et al. 2013; Bai et al. 2013). Clinically based evidences are required to be disseminating knowledge about novel methods of iPSC trans-differentiation on routine clinical practice.

The results of the investigations regarding an ability of surrogate cells derived from ESCs to produce insulin in vivo have appeared to be controversial. It particularly relates to absence of the commonly used pretty accurate standard protocol. The currently implemented protocol represents the requirement regarding methods of derivation of the pancreatic progenitors from PSCs (Bar-Nur et al. 2011). Moreover, there are no commonly used essential criteria in helping determine number of in vitro generated β-like cells enough to further transplantation (Naujok et al. 2011; Naujok and Lenzen 2012). Therefore, human-derived bone marrow-originated mesenchymal stem cells (hBM-MSCs) may be considered as a source for reprogramming procedure of insulin-producing β-like cells. A specific protocol regarding the generation of insulin-producing islet-like clusters derived from hBM-MSCs has now been produced. (Jafarian et al. 2014). On this way, the platelet-rich plasma might attenuate the environment for further BMSC development and differentiation (Lian et al. 2014). It has been postulated that hBM-MSCs may be considered an optimal source for an appropriate transplantation procedure compared with iPSC, while large clinical investigations are required to obtain strong evidence regarding this item.

3.3.6 The Endogenous Reparation Process Stimulation

Recent studies have shown that the development of DM-related complications, i.e., critical limb ischemia, vasculopathy, peripheral neuropathy, accelerating atherosclerosis, and neuropathic diabetic foot, closely associated with dramatic decline of number of PSCs, circulating precursor cells (CPCs), and endothelial progenitor cells (PCs) (Russ et al. 2015; Berezin 2016c). Stem cells (SCs) play a central role in the precise regulation and an appropriate provision of the organism development at the embryonic stage. Therefore, SCs represent their direct potency in tissue regeneration and an ability to regulate innate mechanisms of endogenous reparation in adult life period (Sener and Albeniz 2015). Indeed, SCs have exhibited a high potency to differentiate into a wide spectrum of the cells and also to provide several soluble circulating transcriptional factors, which are required for tissue regeneration and maintenance of repair systems (Sener and Albeniz 2015). In fact, there are evidence that the SCs exposure might be effective through modulating inflammatory changes, tissue remodeling, extracellular matrix renewal, attenuation of cell migration, and maintenance of angiogenesis/neogenesis. It is reported that EPCs with immune phenotypes CD34+KDR+(VEGFR1) and CD31+133+ might have high therapeutic value in the healing process of neuropathic and ischemic lesions in DM (Sambataro et al. 2014; Shi and VandeBerg 2015). As it is expected, direct derivation of embryonic SCs to CD34+ cells might give a source for regenerative care in the future (Shi and VandeBerg 2015). The favorable effects of EPCs on target cells may associate with stimulation of the endogenous repair systems especially affecting the endothelium. The restoring of the endothelium structure and function may be deemed as a basis to improve natural evolution of DM and clinical outcomes in DM-related diseases, such as peripheral neuropathy, atherosclerosis, and critical limb ischemia (Berezin 2016d). In this context, the modulation of EPC-related signaling pathways may be useful for supporting trans-differentiation of the endogenous human SCs into functional β-like cells and mature endothelial cells (Mayhew and Wells 2010).

3.3.7 Regulation of Metabolic Processes via SCs

Replacing β-cells is frequently considered a simple way to enhance the population of pre-existing β-cells, stimulate angiogenesis, and reprogram pancreatic exocrine cells to β-like cells with ability to produce insulin (Berezin 2014). Cellular approaches based on SCs transplantation may also be useful for restoring the immune system response in T1DM or to attenuate insulin resistance and adipocytokines' abnormalities in T2DM. It has been predisposed that identification of novel transcription factors and development of strategies for their modulation could lead to effective regeneration of functioning pool of pancreatic β-cells (Soria 2001; Pandian et al. 2014; Weir et al. 2011). Other promising soluble circulating factors, which could translate the effects on SCs, are cytokines and growth factors, but their clinical significance in diabetics is not yet clear and requires more investigations.

3.4 Results of Preclinical Studies of Stem Cell-Based Therapy

The first experience in the treatment of DM with using cell technologies was based on employment of the native SCs, as well as unfractionated or enriched in subpopulation PCs, while the next generations of cell delivery, i.e., directly reprogramming SCs, human bone marrow-derived mononuclear cells; lineage-specified PCs, are discussed as a more prospective cell resource, which appears to be much more promised for cell therapy.

3.4.1 Reprogramming Stem Cells

Therapeutic cloning of cells has now entered a new era in cell recruiting and reprogramming procedures in clinical settings. The novel approach has become affordable and assessable in the treatment of DM (Kang et al. 2010). By now commonly used essential requirements and regulatory approvals regarding clear design of SC use and recruitment of SCs for further reprogramming have been created (Zhou and Ding 2010). Apart from SCs suitable for reprogramming, it discussed ESCs and multipotent adult SCs/PCs derived from various tissues, i.e., pancreas, peripheral blood, intestine, liver, bone marrow, brain, etc. (Nsair and MacLellan 2011).

There is a large body of evidence regarding clinical exposure of some recombinant proteins and/or several pharmacological-active drugs that might initiate and then support the reprogramming process in the target cell population (Burns et al. 2006; Tancos et al. 2012). Various approaches including nonintegrating, nonviral, and nongenetic methods have been developed for generating clinically compatible iPSCs (Tancos et al. 2012). The are some conditions that have now been determined in vitro in which cell pluripotency is maintained, and even an ability to differentiate to specific somatic cells is desired (Lu and Zhao 2013). Hindley and Philpott (2013), summarizing our knowledge of the possible links between the core cell cycle machinery and the maintenance of iPSCs pluripotency, have emphasized that some advantages of therapeutic non-β-cell cloning includes low rate of the autoimmune reaction after transplantation (Teng et al. 2013). Although a strict similarity of iPSCs with ESCs is determined, the efficacy of reprogramming procedure is now low. Recent study performed by Soejitno and Prayudi (2011) has revealed that a typical reprogramming event may count only 0.01–0.1% of the entire cultured cell population. In fact, the establishment and design of SCs bank is discussed widely (Taylor et al. 2011). However, human iPSCs have demonstrated enormous clinical potency, which may affect their unique capability to self-renew and their innate ability to self-differentiate into all cell types when compared with human ESCs

(Yeo amd Ng 2011). However, whether or not reprogrammed iPSCs have remained fully pluripotent at long time, it is not yet clear (Kang et al. 2010).

Little is known about the importance of the abovementioned advances of iPSCs for developing a new treatment strategy in DM (Fu and Xu 2012; Jiang et al. 2014). It has been postulated that the nature of the pluripotency is under a tight mutual counter-regulated control of genetic and epigenetic mechanisms. It is unclear whether the pluripotency is essential for an increase in the efficacy of cell transplantation or there is no usefulness in pluripotency in safety at long time (Kao et al. 2008). All these unresolved issues should be addressed to further large investigations.

3.4.2 Bone Marrow-Derived Mesenchymal Stem Cell Replacement

Since the first derivation of ESCs and iPSCs, the safety of clinical application of the cell therapy has been come under watchful gaze. Recent benefits in reprogramming regarding transgene-free iPSC have taken into consideration the potential of implementation of the PSC differentiated from different populations (Jung et al. 2012). Care with mesenchymal-originated PSCs is an extremely fast-growing method of regenerative medicine that has now proven their high safety and efficacy in the therapy of various states and diseases. Human bone marrow mesenchymal SCs (hBM-MSCs) are a self-renewing pool of the multipotent cells that is able to migrate to the sites of the pathological process and then mediate regenerative effect in situ. Animal model of T2DM showed that a 6-week period after successful hBM-MSC transplantation was associated with a sufficient decrease of the fasting blood glucose and lipid plasma levels. Additionally, the circulating C-peptide level was significantly increased (Pan et al. 2014). El-Tantawy and Haleem (2014) have reported that use of the autologous BM-MSCs has significantly prevented alterations of tissue and markedly attenuated alloxan-induced oxidative stress in albino rats with DM. Authors believed that BM-MSCs may be helpful in the prevention of diabetic complications associated with oxidative stress.

Tang et al. (2014) have studied the effect of autologous BM-MSCs in miniature pigs with established streptozotocin-induced DM. The obtained results have showed that BM-MSCs transplantation may prevent natural evolution of DM in animals that is associated with restoring blood glucose, serum insulin, and C-peptide levels, as well as attenuation of the oral glucose tolerance test and an increase in the islet numbers. These data suggested the implantation of autologous BM-MSCs for T1DM may partially improve a glucose homeostasis through restoration of the pool/function of β-cells and attenuation of the pancreatic microcirculation. Overall, the majority of the investigators believe that the BM-MSC transplantation is a safe and an effective procedure with great long-term prospective regarding clinical evolution of DM (El-Tantawy and Haleem 2014; Tang et al. 2014).

3.5 Clinical Efficacy of Stem Cell-Based Regenerative Therapy

There are controversial results regarding clinical efficacy of SC therapy in T1DM and T2DM patients (Matveyenko and Vella 2015; Moore et al. 2015). In prospective, open-labeled, two-armed study in T1DM ($n = 20$), Thakkar et al. (2015) have infused allogenic and autologous adipose-derived insulin-secreting mesenchymal stromal cells (IS-AD-MSCs) in combination with bone marrow-derived hematopoietic stem cells (BM-HSCs). Authors have concluded that co-infusion with the use of autologous IS-AD-MSCs and BM-HSCs have appeared to be a better method for long-term control of hyperglycemia when compared with isolated allogenic SC therapy. Surprisingly, allogenic SCs infusion has exhibited very varied effects on fasting glucose level, while safety of the treatment was good. In similar small studies, it was shown that the implantation of mesenchymal stem cells may lower glucose levels via paracrine-mediated influences rather than through direct transdifferentiation of transplanted precursors into β-like insulin-producing cells (Katuchova et al. 2015; Dave 2014; Ezquer et al. 2008). As it is expected, BM-HSCs may represent pro-angiogenic and immunomodulatory effects that might be useful to improve metabolic control in DM, especially when there is coadministrated transplantation BM-HSCs with pancreatic islets (Trivedi et al. 2008). Probably similar approaches might have more efficacy and safety.

3.6 Limitation of the Regenerative Therapy

Because T1DM is considered a chronic metabolic disorder characterized by targeted autoimmune-mediated β-cell destruction, there are some limitations regarding successful transplantation of both pancreatic islets and SCs (Xiao et al. 2014). By now, the islet transplantation in T1DM has been determined as the curative therapy only. Yet, there are several technical limitations regarding donor shortages and cellular damage that may appear within the isolation process and critically limit the further exposure of cultured cells (Jun et al. 2014). However, there is a method created for successful islet transplantation and based on coculturing single primary islet cells with adipose-derived stem cells (ADSCs) in concave micro wells. It has suggested that ADSCs may protect islet cells from damage and increase their survival in the culture prior to transplantation. In animal model xenotransplantation of microfiber-encapsulated spheroids has shown that coculture-transplanted mice maintained their blood glucose levels better than monoculture-transplanted mice. Moreover, it needed sufficiently less islet mass to reverse DM. Jun et al. (2014) have reported that the method for culturing islet spheroids might became a novel step in creating bioartificial pancreas. However, exaggerated immunogenic capacity and potent tumor-induced capability of transplanted cells have remained serious reasons for clinical implementation, while iPSCs, ADSCs, and BM-MSCs might be considered as a future of regenerative care (Schuetz and Markmann 2015).

3.7 Expectancies of Cell Replacement Care in Diabetics: From Bench to Bedside

It is well known that all forms of DM associate with the loss of pool of insulin-produced cells. In this context, the replacement of β-like cells, pancreatic islet, and iPSCs might be argued as a way to attenuate the natural evolution of the DM through improvement of metabolic control (Schroeder 2012). On the other hand, the metabolic abnormalities in DM do not limit a loss of insulin production. The results of the preclinical studies have supported an idea that the cell replacement might improve the entire regenerative potency including islet restoration and prevention of the metabolic memory phenomenon in target cells. Interestingly, the regenerative paradigm has been involved in various clinical settings appearing to be related to cultured SC platforms (Kojima 2014), while contemporary techniques for human ESC generation are known as genetically diverse, patient-specific, or disease-related SCs (Deng 2010). However, the efficacy of the several methods regarding standardization of cell isolation, the conversional nuclear transfer and delivery protocols have now assayed as very low, and the safety of the procedure has remained under discussion and evoked a serious concern even for iPSCs (Hao et al. 2009; Naujok and Lenzen 2012). Moreover, Soejitno and Prayudi (2011), thinking about SC application in several clinical setting, have determined limiting factors, which explain an increased risk of complications after cell therapy. Consequently, the use of retroviral or lentiviral vectors, coadministration of cMyc oncogene, may associate with low efficiency of reprogramming, allogeneic immune rejection, other autoimmune response, and tumor formation resulting in pre-existing epigenetic signature of the target cells (Fu 2014). The resolve of these issues might be mitigated by a breakthrough in the contemporary technology in iPSCs (Fu 2014; Li et al. 2010).

3.8 Future Perspectives of Regenerative Therapy

The shortcoming benefits of the cell therapy in DM relate to implement in routine clinical practice the own patient cells, which are directly differentiated into β-like cells under influence of the specifically created reprogramming technique (Chen et al. 2014). On this way, the correct choosing of the cell source (i.e., adult cells of intestine, pancreas, liver, bone morrow) is discussed as a limiting factor which contributed in successful transplantation of functional insulin-producing cells because no transplantable pancreatic islets were now found (Kojima 2014). The next expectation is creating appropriate techniques that could help to seek the personally patient-specific transcriptional factors required for islet regeneration and trans-differentiation of the target cells into β-like cells suitable for adequate insulin production (Jang et al. 2012; Boland et al. 2012). Currently synthetic DNA-based small molecule, which do not directly affect genome manipulation and allow to regulate and trigger epigenetic mechanisms, i.e., epigenetic enzymes or signaling pathways, of trans-differentiation of own cells into β-like cells with desired phenotype, have been found and broadly investigated (Sohn et al. 2012; Zou et al. 2012; Dadheech et al. 2015). Finally, protection of transplanted cells or renewal β-cells/pancreatic islets from destruction by immune reaction via discover of potent pharmacological agents appears to be promised (Lysy et al. 2012; Kim 2014).

3.9 Conclusion

Cell replacement therapy is considered a promising approach in the combined therapy of DM at the different stages of its evolution. The implementation of the novel technologies regarding isolation, sorting, culture, reprogramming, and trans-differentiation of SCs open serious prospective for achieving adequate control under metabolic abnormalities in all types of the DM, while several coexisting ethical and practical concerns require resolving in large clinical trials in the future.

References

Abdelalim EM, Bonnefond A, Bennaceur-Griscelli A, Froguel P (2014) Pluripotent stem cells as a potential tool for disease modelling and cell therapy in diabetes. Stem Cell Rev 10(3): 327–337

Abraham NG, Li M, Vanella L et al (2008) Bone marrow stem cell transplant into intra-bone cavity prevents type 2 diabetes: role of heme oxygenase-adiponectin. J Autoimmun 30:128–135

Aguayo-Mazzucato C, Bonner-Weir S (2010) Stem cell therapy for type 1 diabetes mellitus. Nat Rev Endocrinol 6:139–148

Ali MA, Dayan CM (2009) The importance of residual endogenous beta-cell preservation in type 1 diabetes. Br J Diabetes Vasc Dis 9:6

Anastasia L, Pelissero G, Venerando B, Tettamanti G (2010) Cell reprogramming: expectations and challenges for chemistry in stem cell biology and regenerative medicine. Cell Death Differ 17:1230–1237

Ashcroft FM, Rorsman P (2012) Diabetes mellitus and the beta cell: the last ten years. Cell 148:1160–1171

Bai Q, Desprat R, Klein B et al (2013) Embryonic stem cells or induced pluripotent stem cells? A DNA integrity perspective. Curr Gene Ther 13(2):93–98

Bar-Nur O, Russ HA, Efrat S et al (2011) Epigenetic memory and preferential lineage-specific differentiation in induced pluripotent stem cells derived from human pancreatic islet beta cells. Cell Stem Cell 9(1):17–23

Berezin AE (2014) Diabetes mellitus and cellular replacement therapy: expected clinical potential and perspectives. World J Diabetes 5(6):777–786

Berezin AE (2016a) Endothelial progenitor cells dysfunction and impaired tissue reparation: the missed link in diabetes mellitus development. Diabetes Metab Syndr Clin Res Rev. [ahead of print]. doi:10.1016/j.dsx.2016.08.007

Berezin AE (2016b) Metabolic memory phenomenon in diabetes mellitus: achieving and perspectives. Diabetes Metab Syndr Clin Res Rev 10(2 Suppl 1):S176–S183. doi:10.1016/j.dsx.2016.03.016

Berezin A (2016c) Biomarkers for cardiovascular risk in diabetic patients. Heart. [epub ahead of print]. doi:10.1136/heartjnl-2016-310197

Berezin (2016d) Endothelial repair in diabetes: the causative role of progenitor cells dysfunction? J Clin Epigenet 2(2):22–24

Bhansali A, Upreti V, Khandelwal N et al (2009) Efficacy of autologous bone marrow-derived stem cell transplantation in patients with type 2 diabetes mellitus. Stem Cells Dev 18:1407–1415

Bhonde RR, Sheshadri P, Sharma S et al (2014) Making surrogate β-cells from mesenchymal stromal cells: perspectives and future endeavors. Int J Biochem Cell Biol 46:90–102

Boland MJ, Hazen JL, Nazor KL et al (2012) Generation of mice derived from induced pluripotent stem cells. J Vis Exp 69:e4003

Burns CJ, Persaud SJ, Jones PM (2006) Diabetes mellitus: a potential target for stem cell therapy. Curr Stem Cell Res Ther 1(2):255–266

Burt RK, Oyama Y, Traynor A et al (2002) Hematopoietic stem cell therapy for type 1 diabetes: induction of tolerance and islet cell neogenesis. Autoimmune Rev 1:133–138

Cade WT (2008) Diabetes-related microvascular and macrovascular diseases in the physical therapy setting. Phys Ther 88(11):1322–1335

Calafiore R, Montanucci P, Basta G (2014) Stem cells for pancreatic β-cell replacement in diabetes mellitus: actual perspectives. Curr Opin Organ Transplant 19(2):162– 168.

Chen YJ, Finkbeiner SR, Weinblatt D et al (2014) De novo formation of insulin-producing “neo-β cell islets” from intestinal crypts. Cell Rep 6(6):1046–1058

Chidgey AP, Layton D, Trounson A, Boyd RL (2008) Tolerance strategies for stem-cell-based therapies. Nature 453:330–337

Dadheech N, Srivastava A, Paranjape N et al (2015) Swertisin an anti-diabetic compound facilitate islet neogenesis from pancreatic stem/progenitor cells via p-38 MAP kinase-SMAD pathway: an in-vitro and in-vivo study. PLoS One 10(6):e0128244

Dave S (2014) Mesenchymal stem cells derived in vitro transdifferentiated insulin-producing cells: a new approach to treat type 1 diabetes. Adv Biomed Res 3:266

Deng W (2010) Exploiting pluripotency for therapeutic gain. Panminerva Med 52(2):167–173

El-Tantawy WH, Haleem EN (2014) Therapeutic effects of stem cell on hyperglycemia, hyperlipidemia, and oxidative stress in alloxan-treated rats. Mol Cell Biochem 391(1–2):193–200

Ezquer FE, Ezquer ME, Parrau DB et al (2008) Systemic administration of multipotent mesenchymal stromal cells reverts hyperglycemia and prevents nephropathy in type 1 diabetic mice. Biol Blood Marrow Transplant 14:631–640

Fu X (2014) The immunogenicity of cells derived from induced pluripotent stem cells. Cell Mol Immunol 11(1):14–16

Fu X, Xu Y (2012) Challenges to the clinical application of pluripotent stem cells: towards genomic and functional stability. Genome Med 4(6):55

Hao J, Zhu W, Sheng C et al (2009) Human parthenogenetic embryonic stem cells: one potential resource for cell therapy. Sci China C Life Sci 52(7):599–602

Hindley C, Philpott A (2013) The cell cycle and pluripotency. Biochem J 451(2):135–143

Holditch SJ, Terzic A, Ikeda Y (2014) Concise review: pluripotent stem cell-based regenerative applications for failing β-cell function. Stem Cells Transl Med 3(5):653–661

Howangyin KY, Silvestre JS (2014) Diabetes mellitus and ischemic diseases: molecular mechanisms of vascular repair dysfunction. Arterioscler Thromb Vasc Biol 34(6):1126–1135

Jafarian A, Taghikhani M, Abroun S et al (2014) Generation of high-yield insulin producing cells from human bone marrow mesenchymal stem cells. Mol Biol Rep 41(7):4783–4794

Jang J, Yoo JE, Lee JA et al (2012) Disease-specific induced pluripotent stem cells: a platform for human disease modeling and drug discovery. Exp Mol Med 44(3):202–213

Jiang Z, Han Y, Cao X (2014) Induced pluripotent stem cell (iPSCs) and their application in immunotherapy. Cell Mol Immunol 11(1):17–24

Jimenez-Moreno CM, de Gracia Herrera-Gomez I, Lopez-Noriega L et al (2015) A simple high efficiency intra-islet transduction protocol using lentiviral vectors. Curr Gene Ther 15 (4): 436–446
Jun Y, Kang AR, Lee JS et al (2014) Microchip-based engineering of super-pancreatic islets supported by adipose-derived stem cells. Biomaterials 35(17):4815–4826
Jung Y, Bauer G, Nolta JA (2012) Concise review: induced pluripotent stem cell-derived mesenchymal stem cells: progress toward safe clinical products. Stem Cells 30(1):42–47
Kang L, Kou Z, Zhang Y et al (2010) Induced pluripotent stem cells (iPSCs) – a new era of reprogramming. J Genet Genomics 37(7):415–421
Kao CF, Chuang CY, Chen CH et al (2008) Human pluripotent stem cells: current status and future perspectives. Chin J Physiol 51(4):214–225
Katuchova J, Harvanova D, Spakova T et al (2015) Mesenchymal stem cells in the treatment of type 1 diabetes mellitus. Endocr Pathol 26(2):95–103
Khosravi-Maharlooei M, Hajizadeh-Saffar E, Tahamtani Y, Basiri M, Montazeri L, Khalooghi K, et al (2015) THERAPY OF ENDOCRINE DISEASE: Islet transplantation for type 1 diabetes: so close and yet so far away. Eur J Endocrinol 173(5):R165– R183
Kim C (2014) Disease modeling and cell based therapy with iPSC: future therapeutic option with fast and safe application. Blood Res 49(1):7–14
Kojima N (2014) In vitro reconstitution of pancreatic islets. Organogenesis 10(2):225–230
Kudva YC, Ohmine S, Greder LV et al (2012) Transgene-free disease-specific induced pluripotent stem cells from patients with type 1 and type 2 diabetes. Stem Cells Transl Med 1(6): 451–461
Lampeter EF, McCann SR, Kolb H (1998) Transfer of insulin-dependent diabetes by bone marrow transplantation. Lancet 351:568–569
Li M, Chen M, Han W et al (2010) How far are induced pluripotent stem cells from the clinic? Ageing Res Rev 9(3):257–264
Lian Z, Yin X, Li H et al (2014) Synergistic effect of bone marrow-derived mesenchymal stem cells and platelet-rich plasma in streptozotocin-induced diabetic rats. Ann Dermatol 26(1): 1–10
Liew CG (2010) Generation of insulin-producing cells from pluripotent stem cells: from the selection of cell sources to the optimization of protocols. Rev Diabet Stud 7(2):82–92
Lindahl M, Danilova T, Palm et al (2014) MANF is indispensable for the proliferation and survival of pancreatic β cells. Cell Rep 7(2):366–375
Lu X, Zhao T (2013) Clinical therapy using iPSCs: hopes and challenges. Genomics Proteomics Bioinforma 11(5):294–298
Ludwig B, Ludwig S (2015) Transplantable bioartificial pancreas devices: current status and future prospects. Langenbecks Arch Surg 400(5):531–540
Lysy PA, Weir GC, Bonner-Weir S (2012) Concise review: pancreas regeneration: recent advances and perspectives. Stem Cells Transl Med 1(2):150–159
Ma T, Xie M, Laurent T, Ding S (2013) Progress in the reprogramming of somatic cells. Circ Res 112(3):562–574
Matveyenko A, Vella A (2015) Regenerative medicine in diabetes. Mayo Clin Proc 90(4): 546–554
Mayhew CN, Wells JM (2010) Converting human pluripotent stem cells into beta-cells: recent advances and future challenges. Curr Opin Organ Transplant 15(1):54–60
Moore SJ, Gala-Lopez BL, Pepper AR et al (2015) Bioengineered stem cells as an alternative for islet cell transplantation. World J Transplant 5(1):1–10. doi:10.5500/wjt.v5.i1.1
Naujok O, Lenzen S (2012) Pluripotent stem cells for cell replacement therapy of diabetes. Dtsch Med Wochenschr 137(20):1062–1066
Naujok O, Burns C, Jones PM et al (2011) Insulin-producing surrogate β-cells from embryonic stem cells: are we there yet? Mol Ther 19(10):1759–1768
Nostro MC, Keller G (2012) Generation of beta cells from human pluripotent stem cells: potential for regenerative medicine. Semin Cell Dev Biol 23(6):701–710
Nsair A, MacLellan WR (2011) Induced pluripotent stem cells for regenerative cardiovascular therapies and biomedical discovery. Adv Drug Deliv Rev 63(4–5):324–330

Nwaneri C, Cooper H, Bowen-Jones D (2013) Mortality in type 2 diabetes mellitus: magnitude of the evidence from a systematic review and meta-analysis. Brit J Diabetes Vasc Dis 13(4): 192–207

Pan XH, Song QQ, Dai JJ et al (2014) Transplantation of bone marrow mesenchymal stem cells for the treatment of type 2 diabetes in a macaque model. Cells Tissues Organs 198(6): 414–427

Pandian GN, Taniguchi J, Sugiyama H (2014) Cellular reprogramming for pancreatic β-cell regeneration: clinical potential of small molecule control. Clin Translat Med 3:6

Paneni F, Beckman JA, Creager MA et al (2013) Diabetes and vascular disease: pathophysiology, clinical consequences, and medical therapy: part I. Eur Heart J 34(31):2436–2443

Philonenko ES, Shutova MV, Chestkov IV et al (2011) Current progress and potential practical application for human pluripotent stem cells. Int Rev Cell Mol Biol 292:153–196

Qi SD, Smith PD, Choong PF (2012) Nuclear reprogramming and induced pluripotent stem cells: a review for surgeons. ANZ J Surg 84(6):E1–11

Rabelink TJ, Little MH (2013) Stromal cells in tissue homeostasis: balancing regeneration and fibrosis. Nat Rev Nephrol 9(12):747–753

Reiland S, Salekdeh GH, Krijgsveld J (2011) Defining pluripotent stem cells through quantitative proteomic analysis. Expert Rev Proteomics 8(1):29–42

Russ HA, Parent AV, Ringler JJ et al (2015) Controlled induction of human pancreatic progenitors produces functional beta-like cells in vitro. EMBO J 34:1759–1772

Sambataro M, Seganfreddo E, Canal F et al (2014) Prognostic significance of circulating and endothelial progenitor cell markers in type 2 diabetic foot. Int J Vasc Med 2014:589412

Schroeder IS (2012) Potential of pluripotent stem cells for diabetes therapy. Curr Diab Rep 12(5):490–498

Schuetz C, Markmann JF (2015) Immunogenicity of β-cells for autologous transplantation in type 1 diabetes. Pharmacol Res 98:60–68

Scully T (2012) Diabetes in numbers. Nature 485:S2–S3

Sener LT, Albeniz I (2015) Challenge of mesenchymal stem cells against diabetic foot ulcer. Curr Stem Cell Res Ther 10:530–534

Shi Q, VandeBerg JL (2015) Experimental approaches to derive CD34+ progenitors from human and nonhuman primate embryonic stem cells. Am J Stem Cells 4(1):32–37

Soejitno A, Prayudi PK (2011) The prospect of induced pluripotent stem cells for diabetes mellitus treatment. Ther Adv Endocrinol Metab 2(5):197–210

Sohn YD, Han JW, Yoon YS (2012) Generation of induced pluripotent stem cells from somatic cells. Prog Mol Biol Transl Sci 111:1–26

Sommer AG, Rozelle SS, Sullivan S et al (2012) Generation of human induced pluripotent stem cells from peripheral blood using the STEMCCA lentiviral vector. J Vis Exp 68:pii: 4327

Soria B (2001) In-vitro differentiation of pancreatic beta-cells. Differentiation 68(4–5):205–219

Tancos Z, Nemes C, Polgar Z et al (2012) Generation of rabbit pluripotent stem cell lines. Theriogenology 78(8):1774–1786

Tang K, Xiao X, Liu D et al (2014) Autografting of bone marrow mesenchymal stem cells alleviates streptozotocin-induced diabetes in miniature pigs: real-time tracing with MRI in vivo. Int J Mol Med 33(6):1469–1476

Taylor CJ, Bolton EM, Bradley JA (2011) Immunological considerations for embryonic and induced pluripotent stem cell banking. Philos Trans R Soc Lond Ser B Biol Sci 366(1575): 2312–2322

Teng S, Liu C, Krettek C et al (2013) The application of induced pluripotent stem cells for bone regeneration: current progress and prospects. Tissue Eng Part B Rev 20(4):328–339

Terzic A, Behfar A (2014) Regenerative heart failure therapy headed for optimization. Eur Heart J 35:1231–1234

Thakkar UG, Trivedi HL, Vanikar AV et al (2015) Insulin-secreting adipose-derived mesenchymal stromal cells with bone marrow-derived hematopoietic stem cells from autologous and allogenic sources for type 1 diabetes mellitus. Cytotherapy 17(7):940–947

Trivedi HL, Vanikar AV, Thakker U et al (2008) Human adipose tissue-derived mesenchymal stem cells combined with hematopoietic stem cell transplantation synthesize insulin. Transplant Proc 40:1135–1139

Voltarelli JC, Couri CE, Oliveira MC et al (2011) Stem cell therapy for diabetes mellitus. Kidney Int 1(3):94–98

Weir GC, Cavelti-Weder C, Bonner-Weir S (2011) Stem cell approaches for diabetes: towards beta cell replacement. Genome Med 3(9):61

Xiao F, Ma L, Zhao M et al (2014) Ex vivo expanded human regulatory T cells delay islet allograft rejection via inhibiting islet-derived monocyte chemoattractant protein-1 production in CD34+ stem cells-reconstituted NOD-scid IL2rγnull mice. PLoS One 9(3):e90387

Yeo JC, Ng HH (2011) Transcriptomic analysis of pluripotent stem cells: insights into health and disease. Genome Med 3(10):68

Zhou H, Ding S (2010) Evolution of induced pluripotent stem cell technology. Curr Opin Hematol 17(4):276–280

Zou C, Chou BK, Dowey SN et al (2012) Efficient derivation and genetic modifications of human pluripotent stem cells on engineered human feeder cell lines. Stem Cells Dev 21(12): 2298–2311

Chapter 4
New Advances in Stem Cell Therapy for Diabetes Mellitus

Lei Lei and Yuanjie Mao

4.1 Introduction

Diabetes mellitus (DM) is one of the most pandemic chronic diseases that prevail worldwide, and the prevalence has continued to be growing in recent decades. The patients with DM manifest a hyperglycemic state induced by impairments in insulin secretion (type 1 and at the late phase of type 2), insulin sensitivity (type 2), or both. Type 1 diabetes mellitus (T1DM), which accounts for less than 10% of patients with DM, occurs through mechanisms of an immune-mediated damage and destruction of pancreatic beta cells in the pancreatic islets of Langerhans, leading to absolute insulin deficiency (American Diabetes A 2011). Type 2 diabetes mellitus (T2DM), which accounts for more than 90% of patients with DM, is characterized by insulin resistance in peripheral tissues and relative insulin deficiency (Groop and Eriksson 1992; D'Souza et al. 2013). Initially, patients with T2DM do not require insulin treatment; however as population and function of pancreatic beta cells declines over time, eventually exogenous insulin supplementation will be required at a late phase. DM is often complicated with major organ damage such as retinopathy, nephropathy, and neuropathy, as well as macrovascular diseases including coronary, cerebral, or peripheral vascular atherosclerosis (Gispen and Biessels 2000).

The cure of DM relies on regenerating functional beta cells, restoring insulin secretion, and relieving abnormal autoimmunity (Wu and Mahato 2014). Allologous whole pancreas or islet transplant has been used to treat DM. The data from the

L. Lei
Department of Endocrinology, Aerospace Center Hospital, Peking University Affiliate, Beijing 100049, China
e-mail: Leileihospital@gmail.com

Y. Mao (✉)
Department of Medicine, Cleveland Clinic Akron General Medical Center, Akron, OH 44307, USA
e-mail: yuanjiemao@gmail.com

P.V. Pham (ed.), *Pancreas, Kidney and Skin Regeneration*, Stem Cells in Clinical Applications, DOI 10.1007/978-3-319-55687-1_4

International Pancreatic Transplant Registry showed promising results that clinical hyperglycemia was rapidly controlled in the recipients after transplantation, with consequentially discontinued supplementation of exogenous insulin (Cefalu 2012). However, its clinical application is significantly limited by the lack of organ donors, high risk of major surgery compilations, and the need for lifelong immunosuppressive therapy to prevent graft rejection. In terms of these limitations, currently the allologous whole pancreas or islet transplant is only recommended to insulin-dependent diabetic patients with end-stage renal disease who require kidney co-transplant (Cefalu 2012). Achievement of the cure of DM demands more efficacious and feasible methods.

Stem cells, characterized by the potential of multi-lineage differentiation and self-renewal, have demonstrated their unique roles in functional insulin-producing cell (IPC) regeneration, immune modulation, and other fields of regenerative studies. Compared to organ or tissue transplant, stem cell therapy has the following advantages in the treatment of DM (Lo and Parham 2009; Calafiore and Basta 2015; Matveyenko and Vella 2015). Autologous stem cells can accommodate a long-term stable source and are not limited by the source of donors. Stem cells are also able to secret numerous cytokines, modulate the local inflammation of pancreatic islets, and further improve the microenvironment and autoimmunity. Moreover, stem cell-derived IPCs can avoid graft rejection and eliminate the necessity of immunosuppressive therapy. Prospectively, novel attempts to replenish pancreatic beta cells in DM, with special regard to IPCs for transplant purposes, could be substituted beta cells by allo- or autografted stem cells. Hereby, we reviewed and discussed the recent advances in stem cell therapy for DM in attempt to clarify where we are and how we may go to reach the final goal of the cure of DM.

4.2 Stem Cell Therapy for Diabetes Mellitus

4.2.1 Sources of Stem Cells

A variety of stem cells have been isolated from different tissues. Based on sources of origin, biological features, and biochemical markers, they can be grossly classified into embryonic stem cells (ESCs), adult stem cells (ASCs), and induced pluripotent stem cells (iPSCs).

ESCs and ASCs each have advantages and disadvantages in terms of potential use for stem cell therapy (Lo and Parham 2009; Calafiore and Basta 2015; Matveyenko and Vella 2015). One major difference is cell types that they can differentiate into. Theoretically ESCs can differentiate into any cell type since they are pluripotent, whereas ASCs are limited to becoming into cell types of their tissue of origin. Besides, ESCs can be isolated easily and expanded quickly in culture, whereas ASCs are rare in adult tissues, hence isolating them and expanding their numbers are very challenging. Moreover, the likelihood of being rejected after transplantation is different in tissues derived from ESCs and ASCs. Tissues derived from ASCs are less likely to be rejected after transplantation (Lo and Parham 2009; Calafiore and Basta 2015; Matveyenko and Vella 2015), since these cells are isolated from a patient's own tissue, expanded in culture into presuming a specific cell type, and reintroduced into the same patient.

By being induced to express genes crucial for keeping the properties of ESCs, iPSCs are genetically reprogrammed adult cells maintained in an ESC-like state. Although these cells meet the defining criteria for pluripotent stem cells, it is still unknown if iPSCs and ESCs differ significantly in clinical application. The breakthrough discovery of iPSCs has created a new powerful tool to "dedifferentiate" adult cells and to harvest a large number of target stem cells. Moreover, tissues derived from iPSCs are almost identically matched to the cell donor, thus reducing the likelihood of rejection after transplantation (Lo and Parham 2009; Calafiore and Basta 2015; Matveyenko and Vella 2015). However, the vectors currently being used to introduce the reprogramming factors into adult cells are viruses; therefore, this process must be carefully controlled and evaluated before this technique can be applied on humans.

4.2.2 Embryonic Stem Cells (ESCs)

Mouse ESCs were first differentiated into IPCs, and upon transplantation in streptozotocin (STZ)-induced diabetic mice, those differentiated IPCs led to reversal of hyperglycemia (Soria et al. 2000). Although ESCs have the highest potential of differentiation into IPCs, only a small amount of pancreatic beta-like cells (1–3%) can be identified in vitro (Mfopou et al. 2010a). Subsequently, the findings of D'Amour et al. are a milestone for the differentiation protocol of ESC-derived IPCs. In the study, the differentiation process of ESCs to IPCs was first described in five stages referred to the developmental biology of the pancreas: definitive endoderm, foregut, hindgut, pancreatic endoderm, and then endocrine cells (D'Amour et al. 2006). Since then, strategies of improving the efficiency of differentiation have been developed. Chen et al. found that the differentiation of human ESCs into PDX1-positive cells can be promoted by a small molecule, indolactam V, both in vitro and in vivo, and a large number of ESC-derived cells can be obtained via this method (Chen et al. 2009). Similarly, a new protocol via a nestin expression step has been developed to obtain IPCs from mouse ESCs (Lumelsky et al. 2001).

The ESC-derived IPCs were able to synthesize insulin and expressed voltage-activated calcium channels; however, without the presence of insulin-containing secretary granules, they did not show the exclusive response of insulin secretion to high-glucose stimulation (Sipione et al. 2004). By using stepwise differentiation protocols, several other studies also successfully generated IPCs from human ESCs in vitro, though these generated pancreatic beta-like cells have very low function in terms of glucose responsiveness (Zhang et al. 2009; Nostro et al. 2011; Teo et al. 2013; Kroon et al. 2008; Shim et al. 2007; Mfopou et al. 2010b; Kelly et al. 2011;

Xu et al. 2011). The in vivo microenvironment has been proposed as a key element for maturation of pancreatic beta cells intended for transplant. Hence the essential function of beta cells, i.e., the ability to secrete insulin in response to high-glucose stimulation, remains an issue of ESC-derived IPCs (Rolletschek et al. 2006).

Further, the potential tumorigenesis of ESCs would be another hurdle in clinical application. ESCs have the characteristics such as rapid proliferation, self-renewal, lack of contact inhibition, and telomerase activity, resembling cancer cells (Kooreman and Wu 2010). Actually, ESCs were associated with occurrence of teratomas and teratocarcinomas in humans (Blum and Benvenisty 2009). The accumulation of potentially oncogenic chromosomal abnormalities may result from the multiple rounds of cell replication before transplantation (Knoepfler 2009). Increase differentiation status and commitment to the cell type of interest before transplant into patients might reduce the risk of tumor development in future.

4.2.3 Adult Stem Cells (ASCs)

4.2.3.1 Bone Marrow: Derived Mesenchymal Stem Cells (BM-MSCs)

The bone marrow contains two different types of stem cells, namely, mesenchymal stem cells (MSCs) and hematopoietic stem cells (HSCs). Because the bone marrow can be easily obtained by simple procedures, bone marrow stem cells have become one of the focuses of stem cell therapy research for DM (Pileggi 2012). Bone marrow-derived mesenchymal stem cells (BM-MSCs) are multipotent progenitor cells, capable of self-renewal and differentiation into adipogenic, chondrogenic, and osteogenic cell lineages (Oswald et al. 2004). The cells can be isolated from the bone marrow in a low-density cell culture method by removal of nonadherent cells (Pittenger and Martin 2004). BM-MSCs express a typical set of surface markers including CD29, CD44, CD49e, CD51, CD54, CD59, CD71, CD73, CD90, CD105, CD166, and CD200 (Hung et al. 2002; Delorme et al. 2008). Unlike HSCs, BM-MSCs do not express CD14, CD31, CD34, CD45, CD79, CD86, CD117, and glycophorin A (Reger et al. 2008; Turnovcova et al. 2009). In addition, BM-MSCs express markers of class I major histocompatibility complex (MHC) but not class II, which may be very advantageous graft-wise (Weiss et al. 2006).

It has been demonstrated that BM-MSCs are able to differentiate into IPCs in vivo and in vitro, and normalize high blood glucose in diabetic mice (Kim et al. 2012; Ho et al. 2012). Experiments aimed at inducing BM-MSCs to differentiate into IPCs were attempted to properly reprogram these cells by activating "ad hoc" differentiation pathways. Oh et al. suggested that rat BM-MSCs could transdifferentiate into IPCs when cultured in a high-glucose medium. These cells grew into a mixed population of islet-like cells, possessed granules with relatively low insulin content, and expressed typical pancreatic endocrine genes such as insulin, glucagon, and somatostatin. Transplanted these cells into diabetic mice normalized blood glucose levels for over 3 months (Oh et al. 2004). Xie et al. demonstrated a three-step differentiation protocol from human BM-MSCs to IPCs, which activin A as a key differentiating agent was added at the final step. The achievement of IPCs was confirmed by morphological analyses and the expression of typical pancreatic beta cell genes such as Beta2/NeuroD, Glut2, Isl-1, nestin, ngn3, Nkx6.1, Pax6, Pdx-1, insulin, glucagon, and C-peptide. Notably, these IPCs secreted insulin in a glucose-dependent manner and could control hyperglycemia in STZ-induced diabetic rats for more than 1 month (Xie et al. 2009). Further, their ability to escape immune recognition and immunomodulatory potentials is discussed in a different section.

However, because only a small portion of the generated cells was originated from differentiated bone marrow cells, the differentiating ability of BM-MSCs into IPCs is questioned (Hess et al. 2003). Besides, after transplantation, increased levels of insulin and decreased levels of plasma glucose had been observed before the presence of MSC-derived IPCs. Therefore, the success of BM-MSCs treatment is less likely to be obtained by direct differentiation into IPCs but probably by paracrine or endocrine mechanisms (Hess et al. 2003). Finally, BM-MSCs have limitations regarding procured cell mass, requiring in vitro expansion which may increase the risk of microbial contamination, losing stemness properties, and inducing artifactual chromosomal changes (Baksh et al. 2007).

4.2.3.2 Adipose Tissue-Derived Mesenchymal Stem Cells (AT-MSCs)

MSCs are multipotent cells existing in several tissues including adipose tissue (Kern et al. 2006). As AT-MSCs can be easily isolated from a patient's own tissue ex vivo, expanded, differentiated into IPCs, and transplanted back to the same patient, adipose tissue has been gaining increased attention for cell therapy as a primary source of MSCs (Zuk et al. 2002; Schaffler and Buchler 2007). Adipose tissue produces a number of bioactive molecules named adipokines, modulating fat mass, nutrient homeostasis, lipid and glucose metabolism, and blood pressure. Adiponectin, leptin, and visfatin are well-recognized adipokines and play crucial roles in insulin sensitivity and glucose regulation (Kojima et al. 2004). It has also been reported that adipocytes from the carp secreted insulin and the proliferative population of AT-MSCs expressed the transcription factor ISL-1 and PAX-6, which are involve in pancreatic endocrine development (Timper et al. 2006; Chandra et al. 2009). Moreover, several studies have revealed that AT-MSCs have even greater potencies in proliferation, differentiation, and immunomodulatory compared to BM-MSCs (Kern et al. 2006; Kim et al. 2007; Pendleton et al. 2013; Melief et al. 2013; Lee et al. 2004). All these features render AT-MSCs a prominent candidate in stem cell therapies for DM.

Chandra et al. showed that AT-MSCs from murine epididymis could differentiate into IPCs under a 10-day inductive protocol. Differentiated cells expressed Glut2, NeuroD, Ngn3, Pax4, PDX1 and secreted insulin and C-peptide in response to glucose levels. Secretory granules in the cell cytoplasm were confirmed by electron microscopy. Normoglycemic state was restored 2 weeks after intraperitoneal transplant into diabetic mice (Chandra et al. 2009). A recent study demonstrated that

AT-MSCs differentiate into IPCs after 38-day co-culture with islet cells. Insulin and C-peptide production were confirmed by ELISA and immunoassaying. After co-transplant of IPCs and islet cells under kidney capsule, hyperglycemic state was recovered in diabetic rats (Karaoz et al. 2013). Moreover, it has been shown that combination of differentiated AT-MSCs and islets resulted in better recovery from diabetes compared to islet transplant alone or co-transplant of islets and differentiated BM-MSCs (Karaoz et al. 2013).

4.2.3.3 Wharton Jelly-Derived Mesenchymal Stem Cells (WJ-MSCs)

Recent studies suggest that the postpartum umbilical cord-extracted Wharton jelly (WJ) contains adult MSCs that can be successfully expanded ex vivo, cryopreserved, and differentiated into ectodermal, mesodermal, and endodermal cellular lineages (Romanov et al. 2003). The gene expression profile of WJ-derived MSCs (WJ-MSCs) is similar to that of BM-MSCs, although it also expresses additional markers (e.g., CD117) (La Rocca et al. 2009; Montanucci et al. 2011). The immune features of WJ-MSCs resemble those of BM-MSCs, since both do not express type II MHC. Moreover, both also express key molecules associated with immunomodulatory properties (Weiss et al. 2008; Deuse et al. 2011). These characteristics clearly indicated the potential of WJ-MSCs that they can transdifferentiate into IPCs or ancillary cells that may assist IPCs, and they can be transplanted and stay functionally active in a diabetic recipient (Ricordi et al. 2012). As an important requisite for allogeneic graft is low immunogenicity, WJ-MSCs is very competent in allologous cell transplantation. Recently, a study demonstrated long-term effects of WJ-MSC therapy in newly onset T1DM (Hu et al. 2013). All of these evidences supported that WJ-MSCs may be an efficient allologous cell candidate for the cure of DM.

Chao et al. successfully differentiated WJ-MSCs into IPCs through a four-step protocol. They transplanted the IPCs into the liver of diabetic mice (Chao et al. 2008). The results showed insulin and C-peptide secretion in response to plasma physiological glucose levels and pancreas-specific gene expression such as Glut-2, HLXB-9, Nkx2.2, and PDX1 in the transplanted IPCs (Chao et al. 2008; Palmer 2009). In a study by He et al. after infected with PDX1 gene-carrying recombinant adenovirus and then treated with inductive factors, WJ-MSCs differentiated into IPCs in vitro. It showed that the differentiated IPCs expressed beta cell-related genes like PDX1, Ngn3, Glut2, and Nkx6.1 and were able to respond to high-glucose stimulation (He et al. 2011). The beta cell-related genes were expressed in both differentiated cells and beta like-cells transplanted into the liver of STZ-induced diabetic rats through the portal vein. As a result, blood glucose levels were significantly reduced 4 weeks after transplantation (Tsai et al. 2012). Wang et al. also differentiated WJ-MSCs into IPCs with an inductive medium. They confirmed that differentiated IPCs responded to the glucose challenge test in vitro. After retro-orbital injection of IPCs into nonobese diabetic (NOD) mice, they found that IPCs containing human nuclei and human C-peptide were located in the liver. They concluded that differentiated IPCs from human WJ-MSCs can alleviate hyperglycemia in NOD mice (Wang et al. 2011). These promising data indicated that WJ-MSCs possess the ability to differentiate into IPCs, both in vivo and in vitro. With respect to the outstanding differentiation and immunomodulatory capacities, WJ-MSCs should be considered as a potential cell therapy option.

4.2.3.4 Pancreatic Stem Cells

It has been shown that pancreatic tissues of adult rat can regenerate efficiently even after resection of 90% of the pancreas (Bonner-Weir et al. 1993). Physiological changes such as pregnancy and obesity also promote pancreatic cell proliferation in great numbers. These suggested that pancreatic cells are capable of self-renewal. However, it remains uncertain whether new regenerated cells are derived from the differentiation of pancreatic stem cells or the proliferation of existing mature cells. By cell sorting and lineage tracing, Simon et al. isolated PDX1+/insulin+/GLUT2 cells from rodent pancreatic islets and pancreatic duct (Smukler et al. 2011). In vitro these cells could expand clonally and differentiate into different types of pancreatic cells. Transplant of these cells reduced high blood glucose levels in STZ-induced diabetic mice. Therefore, this group of cells was considered as pancreatic stem cells (Smukler et al. 2011). However, several experiments were performed to stimulate pancreatic regeneration in vivo, and no pancreatic stem cells could be traced (Furuyama et al. 2011). Bonner et al. concluded that the process of cell growth and regeneration is likely from slow replication of mature cells rather than pancreatic stem cells (Bonner-Weir et al. 2010; Brennand et al. 2007).

4.2.3.5 Hepatic Stem Cells

It has been assumed that liver cells can be an alternative source of IPCs, because in developmental biology, both the liver and pancreas originate from the endoderm, and they share common progenitor cells (Zaret and Grompe 2008). After exogenous PDX1 and NGN3 were transducted into the mouse liver, pancreatic endocrine and exocrine gene expression was substantially induced in liver cells. These transdifferentiated cells were able to survive in the liver and form new pancreatic tissue clusters around the central veins. Besides, they did not affect the normal liver function but could secret insulin, which had normalized blood glucose levels for 8 months in STZ-induced diabetic mice (Ber et al. 2003; Yechoor et al. 2009). Yang et al. further demonstrated that the differentiation of PDX1-reprogrammed liver tissue into functional IPCs only occurred under a state of high-glucose stimulation in vivo and in vitro (Yang 2006). However, so far no experiment could obtain sufficient number of functional IPCs for transplantation via the in vitro proliferation of reprogrammed liver cells.

4.2.4 *Induced Pluripotent Stem Cells (iPSCs)*

Induced pluripotent stem cells (iPSCs) are ESC-like cells from reprogrammed adult somatic cells by the introduction of embryonic genes. Thus, a large number of cells specific to the donor can be obtained, thereby reducing the likelihood of rejection when these cells are transplanted back. Human iPSCs were successfully achieved by reprogrammed somatic cells like fibroblasts (Takahashi et al. 2007), hepatocytes and stomach cells (Aoi et al. 2008), blood cells, and keratinocytes (Hanna et al. 2008), with a cocktail of key transcription factors including cMYC, KLF4, LIN28, NANOG, OCT4, and SOX2 (Group CR 2009). IPSCs are able to differentiate into IPCs by stepwise differentiation protocols including SOX17-positive cells, PDX1-positive cells (pancreatic progenitors), and NGN3-positive cells (endocrine progenitors) (D'Amour et al. 2006; Maehr et al. 2009; Hua et al. 2013; Thatava et al. 2013), which are similar to those applied to ESCs (D'Amour et al. 2006; Maehr et al. 2009). However, like human ESCs, it is still under investigation by which methods iPSCs can be committed to proper cells of interest effectively and reproducibly.

The differentiation of human iPSCs into pancreatic beta cells was first reported in 2008 (Tateishi et al. 2008). By using a four-stage differentiation protocols, IPCs with glucose responsiveness were differentiated from skin fibroblast-derived iPSCs (Thatava et al. 2013). Furthermore, human iPSC clones showed the variations in pancreatic differentiation abilities into IPCs, which was more prominent at the final stage of differentiation (Thatava et al. 2011; Liew et al. 2008). However it is controversial in terms of the differentiation of iPSCs into fully functional IPCs with glucose responsiveness (Teo et al. 2013). The iPSC-derived IPCs had the expression of multi-hormones such as insulin, C-peptide, and glucagon, but they did not express the specific markers for mature pancreatic beta cells such as MAFA and NKX6-1 (Tateishi et al. 2008).

Transplantation of pancreatic progenitor cells or immature pancreatic beta cells into experimental animals may lead to the maturation of iPSC-derived IPCs. In both T1DM and T2DM mouse models, the transplanted iPSC-derived beta cells were able to proficiently secrete insulin with glucose responsiveness and improve hyperglycemia (Alipio et al. 2010). Furthermore, transplantation of monkey iPSC-derived beta cells into diabetic mouse models could correct their hyperglycemia (Zhu et al. 2011). IPCs from the differentiation of beta cell- derived iPSCs can release insulin upon glucose stimulation and after transplantation into T2DM mice can normalize high blood glucose and reduce glycated hemoglobin levels (Bar-Nur et al. 2011). The transplantation of iPSC-derived IPCs obtained from pancreatic epithelial cells into a kidney led to a functional response to glucose stimulation in NOD mice (Jeon et al. 2012). These findings indicated that in vivo maturation is the key of the functionality of iPSC-derived IPCs.

Currently, viral transfection of transcription factors is a major step in the reprogramming of somatic cells into iPSCs. The viral backbone and transgenes are permanently integrated into the genome of transfected cells, and this integration may expose the iPSCs to the risk of mutations, tumorigenesis, dysfunction, or reduced differentiation ability (Okita et al. 2007). With the modification of differentiating protocols, the safety of iPSCs has been greatly improved. By using an adenoviral reprogramming protocol to generate iPSCs, exogenous genes (Oct4, Sox2, Klf4, and c-Myc) are highly and transiently expressed within the cells, but adenovirus does not integrate into the genome (Stadtfeld et al. 2008; Okita et al. 2008). Further, some transfected genes used to generate iPSCs are tumorigenic genes like MYC, with which iPSCs were granted pluripotency and tumorigenicity (Yamanaka 2007; Knoepfler 2008). Recently, to exclude the oncogene MYC so as to prevent genetic modifications, new techniques have been developed by the use of microribonucleic acids (Nakagawa et al. 2008; Wernig et al. 2008). By using only three factors (OCT4, SOX2, and KLF4), and excluding the oncogene MYC, iPSCs have also been produced from patients with T1DM (Maehr et al. 2009). Another method to generated iPSCs is the supplementation of appropriate growth factors into the medium of cultured spermatogonial stem cells that can reprogram iPSCs into cells with three germ layers (Golestaneh et al. 2009). As exogenous genes are not used in the reprogramming process, related risks to exogenous gene integration are eliminated. Certainly, to enhance the efficiency of these methods, further studies are warranted.

4.3 Immune-Modulation of Stem Cell Therapy

MSCs have been shown to have presumptive plasticity potential to differentiate into multiple lineages, and their ability to escape immune recognition and potent immunomodulatory properties have received great interest in regenerative medicine (Anzalone et al. 2011; Abdi et al. 2008; Larijani et al. 2012; Fiorina et al. 2011). Currently, research focusing on the treatment of diabetes with MSCs has led to the following findings. Several experimental studies showed that allogeneic or syngeneic BM-MSCs could prevent or reverse autoimmune diabetes in diabetic animals (Hess et al. 2003; Lee et al. 2006; Ezquer et al. 2008; Boumaza et al. 2009; Zhao et al. 2008; Jurewicz et al. 2010; Fiorina et al. 2009). Urban et al. demonstrated that the immunomodulation of MSCs is one of the mechanisms that support cell regeneration (Urban et al. 2008). In mice with STZ/radiation-induced diabetes, a mixture of HSCs and BM-MSCs were injected into the bone marrow, where these cells inhibited the proliferation of pancreatic cell-specific T cells, reduced the damage caused by T cells, and increased new cell regeneration to a certain extent. The transplantation successfully enabled the blood glucose control in the mice (Urban et al. 2008). Bassi et al. reported that administrated AT-MSCs reversed hyperglycemia in NOD mice. The underlying mechanisms were induction of regulatory T cells and reduction of CD4+ Type 1 T helper cell response as well as decrease in interferon gamma levels (Bassi et al. 2012). Madec et al. revealed that MSCs could induce interleukin 10-producing regulatory T cells and suppress beta cell- specific T cell responses in vitro and in NOD mice (Madec et al. 2009).

Moreover, BM-MSCs have been shown to inhibit both alloimmunization and autoimmunization (Hashemian et al. 2015). Combining pancreatic islets with MSCs transplantation enables the graft to escape immune surveillance and improves the survival rate of the graft (Hashemian et al. 2015). Ding et al. also exploited this feature of MSCs to transplant pancreatic islets and BM-MSCs, the transplanted islets had prolonged survival time, and blood glucose levels were normalized for a significant period of time (Ding et al. 2009). It was speculated that synthesis and secretion of matrix metalloproteinase 2 and 9 into the extracellular matrix play a role in the immune evasion effects of MSCs (Ding et al. 2009).

However, in different microenvironments, MSCs utilize different mechanisms to exert immunosuppressive function. Thus, it should be taken into consideration that animal models are not exactly equal to their human counterparts, and, in different species, MSCs have different functions (Ren et al. 2009). Once safety issues of iPSC were solved, combined transplantation of recipient-specific iPSC and MSCs could be promising (Calafiore and Basta 2015). Another ultimate frontier to cure T1DM is, by proper stimuli of MSCs, to eliminate the autoimmune destruction of pancreatic beta cells; hence the original beta cell reservoir could be regenerated to reconstitute a sufficient mass of IPCs (Calafiore and Basta 2015).

4.4 Future Direction

There still have a number of challenges for the real clinical application of stem cell therapy for DM. One of these challenges is the cell source. Each transplantation needs approximately 100,000 pancreatic islets, and a larger number of alternative cells have to be produced in vitro (Liu et al. 2013). Although a variety of methods to differentiate various stem cells into IPCs have been developed, in terms of clinical application, these methods have relatively low differentiation efficiency, and their generated cell numbers are far from meeting transplant requirements. Therefore, a more effective method for enhancing the differentiation of stem cells into functional IPCs should be developed.

Second, they have several potential critical risks associated with stem cell therapy. Kroon et al. have shown that after human ESC-derived IPCs were transplanted into mice, a 15% of generated cells had components of teratomas or other tissue (Kroon et al. 2008). In vitro proliferation of MSCs was shown to increase the risk of tumor formation and metastasis (Tang et al. 2012; Vajdic and van Leeuwen 2009). These safety concerns must be cautiously evaluated and prevented before stem cell therapy can be used in humans.

Finally, the immune rejection is an issue associated with stem cell therapy (Halban et al. 2010). After receiving allogeneic cell therapy, patients must receive lifelong immunosuppression to avoid graft rejection. Immunosuppressive agents may inhibit insulin secretion or exacerbate insulin resistance, thus counteract the insulin-producing effects of implanting cells. Meanwhile, immunosuppressive agents may increase the risk of malignancy. Further, for the treatment of T1DM, even the autologous cells can be rapid damaged by autoimmunity. In the scenario of the dysfunction of beta cells that is related to genetic abnormalities or changes, transplantation of these autologous cells may not be able to be functioning as well (Liu et al. 2013).

4.5 Conclusion

To find a cure of DM, various scientific areas of research have been extensively explored, with stem cell therapy being one of them. In the present chapter, we have reviewed the current basic and clinical research regarding the development of stem cell therapy for DM. An effective cell transplantation and therapeutic immunomodulatory strategy are required in stem cell therapy for DM (Li and Ikehara 2013). Various stem cells are capable to generate functional IPCs and to improve diabetes in animals and humans. MSCs can inhibit the T cell- mediated autoimmune against newly generated IPCs and prevent destruction of beta cells in DM (Li and Ikehara 2013). Therefore, a blended strategy that combines reliable existing therapies such as islet and pancreas transplantation, the latest bioengineering techniques, and novel immunosuppressive and immunomodulatory agents, with an effective and safe stem cell protocol would secure an optimistic approach for successful translation of stem cell therapy into a cure of DM (Chhabra and Brayman 2013). Overall, for clinical application of stem cell therapy in DM, more studies with various sources of stem cells, larger population of patients undergoing transplantation, and longer monitoring duration are needed to testify the efficacy and safety of this auspicious therapeutic approach.

Disclosure Summary The authors have nothing to disclose.

References

Abdi R, Fiorina P, Adra CN, Atkinson M, Sayegh MH (2008) Immunomodulation by mesenchymal stem cells: a potential therapeutic strategy for type 1 diabetes. Diabetes 57(7):1759–1767

Alipio Z, Liao W, Roemer EJ, Waner M, Fink LM, Ward DC et al (2010) Reversal of hyperglycemia in diabetic mouse models using induced-pluripotent stem (iPS)-derived pancreatic beta-like cells. Proc Natl Acad Sci U S A 107(30):13426–13431

American Diabetes A (2011) Diagnosis and classification of diabetes mellitus. Diabetes Care 34(Suppl 1):S62–S69

Anzalone R, Lo Iacono M, Loria T, Di Stefano A, Giannuzzi P, Farina F et al (2011) Wharton's jelly mesenchymal stem cells as candidates for beta cells regeneration: extending the differentiative and immunomodulatory benefits of adult mesenchymal stem cells for the treatment of type 1 diabetes. Stem Cell Rev 7(2):342–363

Aoi T, Yae K, Nakagawa M, Ichisaka T, Okita K, Takahashi K et al (2008) Generation of pluripotent stem cells from adult mouse liver and stomach cells. Science 321(5889):699–702

Baksh D, Yao R, Tuan RS (2007) Comparison of proliferative and multilineage differentiation potential of human mesenchymal stem cells derived from umbilical cord and bone marrow. Stem Cells 25(6):1384–1392

Bar-Nur O, Russ HA, Efrat S, Benvenisty N (2011) Epigenetic memory and preferential lineage-specific differentiation in induced pluripotent stem cells derived from human pancreatic islet beta cells. Cell Stem Cell 9(1):17–23

Bassi EJ, Moraes-Vieira PM, Moreira-Sa CS, Almeida DC, Vieira LM, Cunha CS et al (2012) Immune regulatory properties of allogeneic adipose-derived mesenchymal stem cells in the treatment of experimental autoimmune diabetes. Diabetes 61(10):2534–2545

Ber I, Shternhall K, Perl S, Ohanuna Z, Goldberg I, Barshack I et al (2003) Functional, persistent, and extended liver to pancreas transdifferentiation. J Biol Chem 278(34):31950–31957

Blum B, Benvenisty N (2009) The tumorigenicity of diploid and aneuploid human pluripotent stem cells. Cell Cycle 8(23):3822–3830

Bonner-Weir S, Baxter LA, Schuppin GT, Smith FE (1993) A second pathway for regeneration of adult exocrine and endocrine pancreas. A possible recapitulation of embryonic development. Diabetes 42(12):1715–1720

Bonner-Weir S, Li WC, Ouziel-Yahalom L, Guo L, Weir GC, Sharma A (2010) Beta-cell growth and regeneration: replication is only part of the story. Diabetes 59(10):2340–2348

Boumaza I, Srinivasan S, Witt WT, Feghali-Bostwick C, Dai Y, Garcia-Ocana A et al (2009) Autologous bone marrow-derived rat mesenchymal stem cells promote PDX-1 and insulin expression in the islets, alter T cell cytokine pattern and preserve regulatory T cells in the periphery and induce sustained normoglycemia. J Autoimmun 32(1):33–42

Brennand K, Huangfu D, Melton D (2007) All beta cells contribute equally to islet growth and maintenance. PLoS Biol 5(7):e163

Calafiore R, Basta G (2015) Stem cells for the cell and molecular therapy of type 1 diabetes mellitus (T1D): the gap between dream and reality. Am J Stem Cells 4(1):22–31

Cefalu WT (2012) American diabetes association-European association for the study of diabetes position statement: due diligence was conducted. Diabetes Care 35(6):1201–1203

Chandra V, Swetha G, Phadnis S, Nair PD, Bhonde RR (2009) Generation of pancreatic hormone-expressing islet-like cell aggregates from murine adipose tissue-derived stem cells. Stem Cells 27(8):1941–1953

Chao KC, Chao KF, Fu YS, Liu SH (2008) Islet-like clusters derived from mesenchymal stem cells in Wharton's Jelly of the human umbilical cord for transplantation to control type 1 diabetes. PLoS One 3(1):e1451

Chen S, Borowiak M, Fox JL, Maehr R, Osafune K, Davidow L et al (2009) A small molecule that directs differentiation of human ESCs into the pancreatic lineage. Nat Chem Biol 5(4):258–265

Chhabra P, Brayman KL (2013) Stem cell therapy to cure type 1 diabetes: from hype to hope. Stem Cells Transl Med 2(5):328–336

D'Amour KA, Bang AG, Eliazer S, Kelly OG, Agulnick AD, Smart NG et al (2006) Production of pancreatic hormone-expressing endocrine cells from human embryonic stem cells. Nat Biotechnol 24(11):1392–1401

D'Souza DM, Al-Sajee D, Hawke TJ (2013) Diabetic myopathy: impact of diabetes mellitus on skeletal muscle progenitor cells. Front Physiol 4:379

Delorme B, Ringe J, Gallay N, Le Vern Y, Kerboeuf D, Jorgensen C et al (2008) Specific plasma membrane protein phenotype of culture-amplified and native human bone marrow mesenchymal stem cells. Blood 111(5):2631–2635

Deuse T, Stubbendorff M, Tang-Quan K, Phillips N, Kay MA, Eiermann T et al (2011) Immunogenicity and immunomodulatory properties of umbilical cord lining mesenchymal stem cells. Cell Transplant 20(5):655–667

Ding Y, Xu D, Feng G, Bushell A, Muschel RJ, Wood KJ (2009) Mesenchymal stem cells prevent the rejection of fully allogenic islet grafts by the immunosuppressive activity of matrix metalloproteinase-2 and -9. Diabetes 58(8):1797–1806

Ezquer FE, Ezquer ME, Parrau DB, Carpio D, Yanez AJ, Conget PA (2008) Systemic administration of multipotent mesenchymal stromal cells reverts hyperglycemia and prevents nephropathy in type 1 diabetic mice. Biol Blood Marrow Transplant: J Am Soc Blood Marrow Transplant 14(6):631–640

Fiorina P, Jurewicz M, Augello A, Vergani A, Dada S, La Rosa S et al (2009) Immunomodulatory function of bone marrow-derived mesenchymal stem cells in experimental autoimmune type 1 diabetes. J Immunol 183(2):993–1004

Fiorina P, Voltarelli J, Zavazava N (2011) Immunological applications of stem cells in type 1 diabetes. Endocr Rev 32(6):725–754

Furuyama K, Kawaguchi Y, Akiyama H, Horiguchi M, Kodama S, Kuhara T et al (2011) Continuous cell supply from a Sox9-expressing progenitor zone in adult liver, exocrine pancreas and intestine. Nat Genet 43(1):34–41

Gispen WH, Biessels GJ (2000) Cognition and synaptic plasticity in diabetes mellitus. Trends Neurosci 23(11):542–549

Golestaneh N, Kokkinaki M, Pant D, Jiang J, DeStefano D, Fernandez-Bueno C et al (2009) Pluripotent stem cells derived from adult human testes. Stem Cells Dev 18(8):1115–1126

Groop LC, Eriksson JG (1992) The etiology and pathogenesis of non-insulin-dependent diabetes. Ann Med 24(6):483–489

Group CR (2009) 2007 update on allogeneic islet transplantation from the Collaborative Islet Transplant Registry (CITR). Cell Transplant 18(7):753–767

Halban PA, German MS, Kahn SE, Weir GC (2010) Current status of islet cell replacement and regeneration therapy. J Clin Endocrinol Metab 95(3):1034–1043

Hanna J, Markoulaki S, Schorderet P, Carey BW, Beard C, Wernig M et al (2008) Direct reprogramming of terminally differentiated mature B lymphocytes to pluripotency. Cell 133(2):250–264

Hashemian SJ, Kouhnavard M, Nasli-Esfahani E (2015) Mesenchymal stem cells: rising concerns over their application in treatment of type one diabetes mellitus. J Diabetes Res 2015:675103

He D, Wang J, Gao Y, Zhang Y (2011) Differentiation of PDX1 gene-modified human umbilical cord mesenchymal stem cells into insulin-producing cells in vitro. Int J Mol Med 28(6):1019–1024

Hess D, Li L, Martin M, Sakano S, Hill D, Strutt B et al (2003) Bone marrow-derived stem cells initiate pancreatic regeneration. Nat Biotechnol 21(7):763–770

Ho JH, Tseng TC, Ma WH, Ong WK, Chen YF, Chen MH et al (2012) Multiple intravenous transplantations of mesenchymal stem cells effectively restore long-term blood glucose homeostasis by hepatic engraftment and beta-cell differentiation in streptozotocin-induced diabetic mice. Cell Transplant 21(5):997–1009

Hu J, Yu X, Wang Z, Wang F, Wang L, Gao H et al (2013) Long term effects of the implantation of Wharton's jelly-derived mesenchymal stem cells from the umbilical cord for newly-onset type 1 diabetes mellitus. Endocr J 60(3):347–357

Hua H, Shang L, Martinez H, Freeby M, Gallagher MP, Ludwig T et al (2013) iPSC-derived beta cells model diabetes due to glucokinase deficiency. J Clin Invest 123(7):3146–3153

Hung SC, Chen NJ, Hsieh SL, Li H, Ma HL, Lo WH (2002) Isolation and characterization of size-sieved stem cells from human bone marrow. Stem Cells 20(3):249–258

Jeon K, Lim H, Kim JH, Thuan NV, Park SH, Lim YM et al (2012) Differentiation and transplantation of functional pancreatic beta cells generated from induced pluripotent stem cells derived from a type 1 diabetes mouse model. Stem Cells Dev 21(14):2642–2655

Jurewicz M, Yang S, Augello A, Godwin JG, Moore RF, Azzi J et al (2010) Congenic mesenchymal stem cell therapy reverses hyperglycemia in experimental type 1 diabetes. Diabetes 59(12):3139–3147

Karaoz E, Okcu A, Unal ZS, Subasi C, Saglam O, Duruksu G (2013) Adipose tissue-derived mesenchymal stromal cells efficiently differentiate into insulin-producing cells in pancreatic islet microenvironment both in vitro and in vivo. Cytotherapy 15(5):557–570

Kelly OG, Chan MY, Martinson LA, Kadoya K, Ostertag TM, Ross KG et al (2011) Cell-surface markers for the isolation of pancreatic cell types derived from human embryonic stem cells. Nat Biotechnol 29(8):750–756

Kern S, Eichler H, Stoeve J, Kluter H, Bieback K (2006) Comparative analysis of mesenchymal stem cells from bone marrow, umbilical cord blood, or adipose tissue. Stem Cells 24(5):1294–1301

Kim Y, Kim H, Cho H, Bae Y, Suh K, Jung J (2007) Direct comparison of human mesenchymal stem cells derived from adipose tissues and bone marrow in mediating neovascularization in response to vascular ischemia. Cell Physiol Biochem: Int J Exp Cell Physiol Biochem Pharmacol 20(6):867–876

Kim SJ, Choi YS, Ko ES, Lim SM, Lee CW, Kim DI (2012) Glucose-stimulated insulin secretion of various mesenchymal stem cells after insulin-producing cell differentiation. J Biosci Bioeng 113(6):771–777

Knoepfler PS (2008) Why myc? An unexpected ingredient in the stem cell cocktail. Cell Stem Cell 2(1):18–21

Knoepfler PS (2009) Deconstructing stem cell tumorigenicity: a roadmap to safe regenerative medicine. Stem Cells 27(5):1050–1056

Kojima H, Fujimiya M, Matsumura K, Nakahara T, Hara M, Chan L (2004) Extrapancreatic insulin-producing cells in multiple organs in diabetes. Proc Natl Acad Sci U S A 101(8):2458–2463

Kooreman NG, Wu JC (2010) Tumorigenicity of pluripotent stem cells: biological insights from molecular imaging. J R Soc Interface/R Soc 7(Suppl 6):S753–S763

Kroon E, Martinson LA, Kadoya K, Bang AG, Kelly OG, Eliazer S et al (2008) Pancreatic endoderm derived from human embryonic stem cells generates glucose-responsive insulin-secreting cells in vivo. Nat Biotechnol 26(4):443–452

La Rocca G, Anzalone R, Corrao S, Magno F, Loria T, Lo Iacono M et al (2009) Isolation and characterization of Oct-4+/HLA-G+ mesenchymal stem cells from human umbilical cord matrix: differentiation potential and detection of new markers. Histochem Cell Biol 131(2):267–282

Larijani B, Esfahani EN, Amini P, Nikbin B, Alimoghaddam K, Amiri S et al (2012) Stem cell therapy in treatment of different diseases. Acta Med Iran 50(2):79–96

Lee RH, Kim B, Choi I, Kim H, Choi HS, Suh K et al (2004) Characterization and expression analysis of mesenchymal stem cells from human bone marrow and adipose tissue. Cell Physiol Biochem: Int J Exp Cell Physiol Biochem Pharmacol 14(4–6):311–324

Lee RH, Seo MJ, Reger RL, Spees JL, Pulin AA, Olson SD et al (2006) Multipotent stromal cells from human marrow home to and promote repair of pancreatic islets and renal glomeruli in diabetic NOD/scid mice. Proc Natl Acad Sci U S A 103(46):17438–17443

Li M, Ikehara S (2013) Bone marrow stem cell as a potential treatment for diabetes. J Diabetes Res 2013:329596

Liew CG, Shah NN, Briston SJ, Shepherd RM, Khoo CP, Dunne MJ et al (2008) PAX4 enhances beta-cell differentiation of human embryonic stem cells. PLoS One 3(3):e1783

Liu X, Wang Y, Li Y, Pei X (2013) Research status and prospect of stem cells in the treatment of diabetes mellitus. Sci China Life Sci 56(4):306–312

Lo B, Parham L (2009) Ethical issues in stem cell research. Endocr Rev 30(3):204–213

Lumelsky N, Blondel O, Laeng P, Velasco I, Ravin R, McKay R (2001) Differentiation of embryonic stem cells to insulin-secreting structures similar to pancreatic islets. Science 292(5520):1389–1394

Madec AM, Mallone R, Afonso G, Abou Mrad E, Mesnier A, Eljaafari A et al (2009) Mesenchymal stem cells protect NOD mice from diabetes by inducing regulatory T cells. Diabetologia 52(7):1391–1399

Maehr R, Chen S, Snitow M, Ludwig T, Yagasaki L, Goland R et al (2009) Generation of pluripotent stem cells from patients with type 1 diabetes. Proc Natl Acad Sci U S A 106(37):15768–15773

Matveyenko A, Vella A (2015) Regenerative medicine in diabetes. Mayo Clin Proc 90(4):546–554

Melief SM, Zwaginga JJ, Fibbe WE, Roelofs H (2013) Adipose tissue-derived multipotent stromal cells have a higher immunomodulatory capacity than their bone marrow-derived counterparts. Stem Cells Transl Med 2(6):455–463

Mfopou JK, Chen B, Sui L, Sermon K, Bouwens L (2010a) Recent advances and prospects in the differentiation of pancreatic cells from human embryonic stem cells. Diabetes 59(9):2094–2101

Mfopou JK, Chen B, Mateizel I, Sermon K, Bouwens L (2010b) Noggin, retinoids, and fibroblast growth factor regulate hepatic or pancreatic fate of human embryonic stem cells. Gastroenterology 138(7):2233–2245. 45 e1–14

Montanucci P, Basta G, Pescara T, Pennoni I, Di Giovanni F, Calafiore R (2011) New simple and rapid method for purification of mesenchymal stem cells from the human umbilical cord Wharton jelly. Tissue Eng Part A 17(21–22):2651–2661

Nakagawa M, Koyanagi M, Tanabe K, Takahashi K, Ichisaka T, Aoi T et al (2008) Generation of induced pluripotent stem cells without Myc from mouse and human fibroblasts. Nat Biotechnol 26(1):101–106

Nostro MC, Sarangi F, Ogawa S, Holtzinger A, Corneo B, Li X et al (2011) Stage-specific signaling through TGFbeta family members and WNT regulates patterning and pancreatic specification of human pluripotent stem cells. Development 138(5):861–871

Oh SH, Muzzonigro TM, Bae SH, LaPlante JM, Hatch HM, Petersen BE (2004) Adult bone marrow-derived cells trans-differentiating into insulin-producing cells for the treatment of type I diabetes. Lab Invest; J Tech Methods Pathol 84(5):607–617

Okita K, Ichisaka T, Yamanaka S (2007) Generation of germline-competent induced pluripotent stem cells. Nature 448(7151):313–317

Okita K, Nakagawa M, Hyenjong H, Ichisaka T, Yamanaka S (2008) Generation of mouse induced pluripotent stem cells without viral vectors. Science 322(5903):949–953

Oswald J, Boxberger S, Jorgensen B, Feldmann S, Ehninger G, Bornhauser M et al (2004) Mesenchymal stem cells can be differentiated into endothelial cells in vitro. Stem Cells 22(3):377–384

Palmer JP (2009) C-peptide in the natural history of type 1 diabetes. Diabetes Metab Res Rev 25(4):325–328

Pendleton C, Li Q, Chesler DA, Yuan K, Guerrero-Cazares H, Quinones-Hinojosa A (2013) Mesenchymal stem cells derived from adipose tissue vs bone marrow: in vitro comparison of their tropism towards gliomas. PLoS One 8(3):e58198

Pileggi A (2012) Mesenchymal stem cells for the treatment of diabetes. Diabetes 61(6):1355–1356

Pittenger MF, Martin BJ (2004) Mesenchymal stem cells and their potential as cardiac therapeutics. Circ Res 95(1):9–20

Reger RL, Tucker AH, Wolfe MR (2008) Differentiation and characterization of human MSCs. Methods Mol Biol 449:93–107

Ren G, Su J, Zhang L, Zhao X, Ling W, L'Huillie A et al (2009) Species variation in the mechanisms of mesenchymal stem cell-mediated immunosuppression. Stem Cells 27(8):1954–1962

Ricordi C, Inverardi L, Dominguez-Bendala J (2012) From cellular therapies to tissue reprogramming and regenerative strategies in the treatment of diabetes. Regen Med 7(6 Suppl):41–48

Rolletschek A, Kania G, Wobus AM (2006) Generation of pancreatic insulin-producing cells from embryonic stem cells – 'proof of principle', but questions still unanswered. Diabetologia 49(11):2541–2545

Romanov YA, Svintsitskaya VA, Smirnov VN (2003) Searching for alternative sources of postnatal human mesenchymal stem cells: candidate MSC-like cells from umbilical cord. Stem Cells 21(1):105–110

Schaffler A, Buchler C (2007) Concise review: adipose tissue-derived stromal cells – basic and clinical implications for novel cell-based therapies. Stem Cells 25(4):818–827

Shim JH, Kim SE, Woo DH, Kim SK, Oh CH, McKay R et al (2007) Directed differentiation of human embryonic stem cells towards a pancreatic cell fate. Diabetologia 50(6):1228–1238

Sipione S, Eshpeter A, Lyon JG, Korbutt GS, Bleackley RC (2004) Insulin expressing cells from differentiated embryonic stem cells are not beta cells. Diabetologia 47(3):499–508

Smukler SR, Arntfield ME, Razavi R, Bikopoulos G, Karpowicz P, Seaberg R et al (2011) The adult mouse and human pancreas contain rare multipotent stem cells that express insulin. Cell Stem Cell 8(3):281–293

Soria B, Roche E, Berna G, Leon-Quinto T, Reig JA, Martin F (2000) Insulin-secreting cells derived from embryonic stem cells normalize glycemia in streptozotocin-induced diabetic mice. Diabetes 49(2):157–162

Stadtfeld M, Nagaya M, Utikal J, Weir G, Hochedlinger K (2008) Induced pluripotent stem cells generated without viral integration. Science 322(5903):945–949

Takahashi K, Tanabe K, Ohnuki M, Narita M, Ichisaka T, Tomoda K et al (2007) Induction of pluripotent stem cells from adult human fibroblasts by defined factors. Cell 131(5):861–872

Tang DQ, Wang Q, Burkhardt BR, Litherland SA, Atkinson MA, Yang LJ (2012) In vitro generation of functional insulin-producing cells from human bone marrow-derived stem cells, but long-term culture running risk of malignant transformation. Am J Stem Cells 1(2):114–127

Tateishi K, He J, Taranova O, Liang G, D'Alessio AC, Zhang Y (2008) Generation of insulin-secreting islet-like clusters from human skin fibroblasts. J Biol Chem 283(46):31601–31607

Teo AK, Wagers AJ, Kulkarni RN (2013) New opportunities: harnessing induced pluripotency for discovery in diabetes and metabolism. Cell Metab 18(6):775–791

Thatava T, Nelson TJ, Edukulla R, Sakuma T, Ohmine S, Tonne JM et al (2011) Indolactam V/GLP-1-mediated differentiation of human iPS cells into glucose-responsive insulin-secreting progeny. Gene Ther 18(3):283–293

Thatava T, Kudva YC, Edukulla R, Squillace K, De Lamo JG, Khan YK et al (2013) Intrapatient variations in type 1 diabetes-specific iPS cell differentiation into insulin-producing cells. Mol Ther: J Am Soc Gene Ther 21(1):228–239

Timper K, Seboek D, Eberhardt M, Linscheid P, Christ-Crain M, Keller U et al (2006) Human adipose tissue-derived mesenchymal stem cells differentiate into insulin, somatostatin, and glucagon expressing cells. Biochem Biophys Res Commun 341(4):1135–1140

Tsai PJ, Wang HS, Shyr YM, Weng ZC, Tai LC, Shyu JF et al (2012) Transplantation of insulin-producing cells from umbilical cord mesenchymal stem cells for the treatment of streptozotocin-induced diabetic rats. J Biomed Sci 19:47

Turnovcova K, Ruzickova K, Vanecek V, Sykova E, Jendelova P (2009) Properties and growth of human bone marrow mesenchymal stromal cells cultivated in different media. Cytotherapy 11(7):874–885

Urban VS, Kiss J, Kovacs J, Gocza E, Vas V, Monostori E et al (2008) Mesenchymal stem cells cooperate with bone marrow cells in therapy of diabetes. Stem Cells 26(1):244–253

Vajdic CM, van Leeuwen MT (2009) Cancer incidence and risk factors after solid organ transplantation. Int J Cancer J Int Cancer 125(8):1747–1754

Wang HS, Shyu JF, Shen WS, Hsu HC, Chi TC, Chen CP et al (2011) Transplantation of insulin-producing cells derived from umbilical cord stromal mesenchymal stem cells to treat NOD mice. Cell Transplant 20(3):455–466

Weiss ML, Medicetty S, Bledsoe AR, Rachakatla RS, Choi M, Merchav S et al (2006) Human umbilical cord matrix stem cells: preliminary characterization and effect of transplantation in a rodent model of Parkinson's disease. Stem Cells 24(3):781–792

Weiss ML, Anderson C, Medicetty S, Seshareddy KB, Weiss RJ, Vander Werff I et al (2008) Immune properties of human umbilical cord Wharton's jelly-derived cells. Stem Cells 26(11):2865–2874

Wernig M, Meissner A, Cassady JP, Jaenisch R (2008) c-Myc is dispensable for direct reprogramming of mouse fibroblasts. Cell Stem Cell 2(1):10–12

Wu H, Mahato RI (2014) Mesenchymal stem cell-based therapy for type 1 diabetes. Discov Med 17(93):139–143

Xie QP, Huang H, Xu B, Dong X, Gao SL, Zhang B et al (2009) Human bone marrow mesenchymal stem cells differentiate into insulin-producing cells upon microenvironmental manipulation in vitro. Differ Res Biol Divers 77(5):483–491

Xu X, Browning VL, Odorico JS (2011) Activin, BMP and FGF pathways cooperate to promote endoderm and pancreatic lineage cell differentiation from human embryonic stem cells. Mech Dev 128(7–10):412–427

Yamanaka S (2007) Strategies and new developments in the generation of patient-specific pluripotent stem cells. Cell Stem Cell 1(1):39–49

Yang LJ (2006) Liver stem cell-derived beta-cell surrogates for treatment of type 1 diabetes. Autoimmun Rev 5(6):409–413

Yechoor V, Liu V, Espiritu C, Paul A, Oka K, Kojima H et al (2009) Neurogenin3 is sufficient for transdetermination of hepatic progenitor cells into neo-islets in vivo but not transdifferentiation of hepatocytes. Dev Cell 16(3):358–373

Zaret KS, Grompe M (2008) Generation and regeneration of cells of the liver and pancreas. Science 322(5907):1490–1494

Zhang D, Jiang W, Liu M, Sui X, Yin X, Chen S et al (2009) Highly efficient differentiation of human ES cells and iPS cells into mature pancreatic insulin-producing cells. Cell Res 19(4):429–438

Zhao W, Wang Y, Wang D, Sun B, Wang G, Wang J et al (2008) TGF-beta expression by allogeneic bone marrow stromal cells ameliorates diabetes in NOD mice through modulating the distribution of CD4+ T cell subsets. Cell Immunol 253(1–2):23–30

Zhu FF, Zhang PB, Zhang DH, Sui X, Yin M, Xiang TT et al (2011) Generation of pancreatic insulin-producing cells from rhesus monkey induced pluripotent stem cells. Diabetologia 54(9):2325–2336

Zuk PA, Zhu M, Ashjian P, De Ugarte DA, Huang JI, Mizuno H et al (2002) Human adipose tissue is a source of multipotent stem cells. Mol Biol Cell 13(12):4279–4295

Chapter 5
Therapeutic Potential of Mesenchymal Stem Cells and miRNAs in Diabetes

Arefeh Jafarian and Masoud Soleimani

5.1 Introduction

Diabetes mellitus (DM) is a chronic, multifactorial metabolic disorder affecting 2–5% of the population and is a major challenge in health. DM type I (T1DM) or juvenile-onset Diabetes is characterized by autoimmune selective destruction of insulin-producing pancreatic β cells, which result in an absolute deficiency of insulin required for glucose metabolism ultimately resulting in the loss of insulin production and secretion that leads to increase in blood glucose level. So, patients are dependent on exogenous insulin for their blood glucose control. Usually 60–80% of the β-cell mass have been destroyed at the time of diagnosis. Insulin replacement therapy by either insulin pump or multiple daily injections is intensive and often associated with severe hypoglycemic episodes. Pathogenesis of DM type 2 (T2DM) is related to genetic, environmental, and lifestyle factors resulting from insulin resistance in target tissues such as liver, skeletal muscles, and adipose tissues. So, β cells are unable to sustain the increased demand for insulin, which therefore leads to chronic hyperglycemia and the onset of T2DM. Both types of diabetes have the serious long-term complications in different organs such as liver, kidneys, eyes, heart, nerves, and blood vessels (*Diagnosis and classification of diabetes mellitus* 2014). Edmonton protocol is the most reliable approach to the treatment of T1DM by transplantation of whole pancreas or isolated islets. However, the scarcity of

A. Jafarian, PhD (✉)
Immunology, Asthma and Allergy Research Institute (IAARI), Tehran University of Medical Sciences, Tehran, Iran
e-mail: ajafaryan@modares.ac.ir; Arefe_jafarian@yahoo.com

M. Soleimani, PhD
Department of Hematology, Faculty of Medical Sciences, Tarbiat Modares University, Tehran, Iran

P.V. Pham (ed.), *Pancreas, Kidney and Skin Regeneration*, Stem Cells in Clinical Applications, DOI 10.1007/978-3-319-55687-1_5

human donors and the need for lifelong immunosuppressant to prevent immune rejection are considered major obstacles to transplantation of islets (Gruessner et al. 2012; Jamiolkowski et al. 2012). Therefore, new therapeutic strategies are needed to preserve or even promote regeneration of the β-cell mass. Due to the limitation of using embryonic stem cells (ES) and induced puleripotent stem cells (IPS) in the clinic, recently cell-based therapy has been focused on mesenchymal stem cells (MSCs). MSCs have remarkable immunomodulatory properties, and they can be isolated from adipose tissue and can be differentiated into insulin-producing cells (IPC). Therefore, they have possible applications in the treatment of type 1 diabetes (Liu and Han 2008; Wei et al. 2013). In addition, in the past 15 years, a family of endogenous small noncoding RNAs known as microRNAs (miRNAs) has been discovered as new players in regulation of protein coding genes. They are a novel class of endogenous small nc-RNAs, of ~20–30 nucleotides in length that were first discovered in 1993 in *Caenorhabitis elegans* and *Drosophila* and later identified in many species (Ambros 2004).These nc-RNAs are encoded by up to 3% of all genes, and approximately 30% of the genes are supposed to be regulated by small RNA species that regulate gene expression posttranscriptionally (Zhang and Farwell 2008). In mammalians, miRNAs are transcriptional repressor and have inhibitory effects on RNA stability by base pairing between 3′ untranslated regions (UTRs) of target mRNAs and miRNA "seed region." Each miRNA may have multiple targets and therefor have multiple effects on physiological and pathological processes (van Rooij 2011). Recently, several miRNAs have been identified that have potential roles in pancreas development, islet function, insulin secretion, and diabetic complications (Zhang and Farwell 2008; Kantharidis et al. 2011).We also discussed important role of this miRNAs in diabetes and its complications.

5.2 MSCs: Cell Sources and Phenotypic Properties

MSCs are multipotent adult stem cells that have diverse distribution in the body. They localized in multiple fetal tissues including bone marrow, adipose tissue, umbilical cord blood, adult muscle, lung, and pancreas (Lu et al. 2006; Stenderup et al. 2003). They were first identified by Friedenstein in rodent bone marrow in 1976 (Friedenstein et al. 1976). They are morphologically plastic-adherent cells, and they have the potential capacity to differentiate into multiple of cells, such as osteoblasts, chondrocytes, and adipocytes. It is generally agreed by international society for cellular therapy (ISCT) guidelines that MSCs must express CD105 (SH2), CD73 (SH3/4), CD44, CD166 (vascular cell adhesion protein), CD54/CD102, CD49, CD90, and stromal antigen 1 and must be negative for hematopoietic cell surface markers including CD34, CD14, CD45, *CD14*, CD11a/LFA-1, and CD31(Bernardi et al. 2012). They have selfrenewal capacity and great multiplication potency; they can be expanded in culture for more than 60 doublings without losing their native characteristics (De Keyser 2005; Pittenger et al. 1999). Among all types of stem cells, MSCs have resived special attention because of their easily isolation from bone marrow and adipose tissue, less potent to induce teratoma or other malignant transformation and also in vitro large-scale expansion (Patel et al. 2013). Therefore, MSCs appear to be a very promising tool for gene therapy and cell-based regenerative therapy for T1DM.

5.3 MSCs: Immunomodulatory Properties

MSCs have a powerful capacity of regulating immune responses and are considered to behave hypoimmunogenic phenotype because they have low expression level of histocompatability complex (MHC) class I antigens and lack of MHC class II including co-stimulatory molecule expressions (Liu and Han 2008). They have important roles in modulation of innate and adaptive immune cell responses by intraction with immune system particularly dendritic cells, T cells, and NK cells (Aggarwal and Pittenger 2005). MSCs have anti-inflamatory effects; they could generate and expand regulatory T cells (Tregs). Tregs have important roles in modulating the immune responses. MSCs inhibit T and B cell proliferation and cytokine secretion by unknown mechanism, but paracrine factors secreted by MSCs may be involved in these effects. MSCs have inhibitory roles in immunoglobulin synthesis by inhibition of monocyte differentiation into dendritic cells. It was evidence that T-cell proliferation suppression and enhancement of anti-inflammatory cytokine secretion stimulated by prostaglandin E2 (PGE-2), transforming growth factor beta (TGF-β), hepatocyte growth factor (HGF), indoleamine 2,3-dioxygenase (IDO), and leukemia inhibitory factor (LIF) (Newman et al. 2009). In addition, inhibition of NK cell proliferation and dendritic cell maturation was contributed by paracrine factors such as IDO, TGF-β, and PGE-2.Taken together, all these immunomodulatory activities *make them* good *therapeutic* candidates for cell-based therapy in clinical application.

5.4 MSCs: Multipotent Differentiation

MSCs have the ability to adhere to plastic culture flasks and differentiate into osteoblasts, adipocytes, and chondroblasts (Dominici et al. 2006). MSC cross-differentiation into different lineages can be regulated by some regulatory genes and transcription factors. Indeed growth factors, cytokines, induction chemicals, and microenvironment conditions can also provide MSCs with appropriate proliferation and differentiation (Ding et al. 2011). Human MSCs have the ability to differentiate into ectodermal, mesodermal, and endodermal lineage cells and repair tissue without exogenous gene introduction (Wakao et al. 2012). hMSCs derived from mesoderm but have the potential to transdifferentiate into neural cells. Transdifferentiation into ectodermal lineages (neuronal cells) induced upon exposure to neural induction cocktail media supplemented with some growth factors like FGF and EGF, and neural-like obtained cells have neuronal-specific phenotypes (Naghdi et al. 2009; Kang et al. 2003). Several factors were reported that are effective in the trans-differentiation of MSCs into neuronal protein expressing cells such as neurogenin-1, LIM homoeobox transcription factor 1 α (LMX1a) (Barzilay et al. 2009), forskolin, cAMP (Rooney et al. 2009), β-Mercaptoethanol (BME), nerve growth factor (NGF) (Naghdi et al. 2009), glial cell line-derived neurotrophic growth factors (GNDF), brain-derived neurotrophic factors (BDNF), retinoic acid, 5-azacytidine, isobutyl-methylxanthine (IBMX), indomethacin (Pavlova et al. 2012), and telomerase reverse transcriptase (TERT) (Zhao et al. 2014). Therefore, MSCs can be a promising approach in the regenerative cell therapy in treating many neurological disorders. MSCs are able to easily differentiate into our embryonic origin. The in vitro differentiation into adipocytes, osteocytes, and contracts is best studded. Differentiation of MSCs into adipocytes is stimulated by a mixture of dexamethasone (Dex), isobutylmethylxanthine (IBMX), indomethacin (IM), and insulin in culture medium for 3 weeks. Lipid droplet formation have been observed by oil-red staining and peroxisome proliferator-activated receptors-γ2 (PPAR-γ2), adipocyte protein 2 (ap2), and lipoprotein lipase (LPL) genes analysis (Ding et al. 2011). Osteocytes differentiation is induced by classical method consisting of dexamethasone, β-glycerol phosphate, and ascorbic acid-2 phosphate for 3 weeks in culture media. Osteogenic differentiation was established by production of intracellular calcium deposition and increase in alkaline phosphatase activity. The mineralized cells have been observed by alizarin red staining. Numerous signaling pathways and transcription factors are involved in osteogenic differentiation. Runt-related transcription factor 2 (Runx2) and caveolin-1 are considered the most important indicating factors for osteogenic differentiation. Indeed, bone morphogenetic proteins (BMPs) , especially BMP-2, BMP-6, and BMP-9, and osteonectin are involved in this process (Hwang et al. 2009). In standard protocol, several inductive agents were used for differentiation of MSCs into chondrocytes such as insulin transferrin selenium, linoleic acid, pyrovate, ascorbic acid-2 phosphate, dexamethasone, and transforming factor β (TGF-β). Fibroblastic growth factor-2 (FGF-2) can enhance early differentiation of MSC into chondrocytes (Cheng et al. 2012). Chondrogenic differentiation were analyzed by expression of types I and II collagens and adhesion molecules such as TGF-β family (Teven et al. 2011). In addition, MSCs can cross-differentiate into endoderm lineage cells in vitro including IPCs by transient transfection of Pdx1 (Fedyunina et al. 2011) and hepatocyte-like cells using HGF and FGF-4 (Zhang et al. 2009). Therefore, generated IPCs from MSCs could be a new source for drug discovery and for IPCs replacement in diabetes (Gabr et al. 2015).

5.5 MSCs: The Therapeutic Potential in Diabetes

Differentiation of MSCs into insulin-producing cells (IPCs) has been demonstrated (Sheng et al. 2008). For the first time, nestin-positive cells were isolated from human pancreas and transplanted to diabetic nonobese diabetic/severe combined immunodeficiency (NOD-SCID) mice (Zulewski et al. 2001). These cells improved hyperglycemic condition. In addition, MSCs can be isolated from another source to treat T1DM. Transplantation of islet cells together with MSCs into streptozotocin-treated diabetic rat model improved the efficacy and survival rate of engraftment and are found useful for treating non-insulin-dependent patients in T1DM human BMMSCs which have high potential to differentiating into glucose-sensitive pancreatic lineage in vitro as well as in vivo (Phadnis et al. 2011). These mobilized cells have regenerative and reparative properties and could selectively migrate to sites of injury. So, they are appropriate candidates to regenerate IPCs. However, most of the recent protocols successfully differentiated MSCs with different sources into the insulin-producing cells, but there is a critical need for the development of an effective protocol for MSC differentiation into IPCs.

5.6 MSCs: In Vitro Differentiation into Insulin-Producing Cells

In vitro induction of human and murine MSCs into pancreatic islets are well documented. *Among adult stem cells*, the significant therapeutic potential of MSCs to generate cells with ability to adopt a pancreatic endocrine phenotype is apparent. Multistep differentiation protocols are used to generate IPCs from MSCs with combination of nicotinamide, activin A, retinoic acid, epidermal growth factor (EGF), taurine, and exendin-4 in high-glucose medium. At the end of the culture, differentiated cells show an islet-like cluster morphology with PDX-1, insulin and glucagon gene expression, and insulin secretion in a glucose-dependent manner (Jafarian et al. 2014; Parekh et al. 2009). Several in vitro studies described multiple experimental strategies, including different growth or differentiating factors, cytokines, and supplements to achieve better culture conditions for direct differentiation of MSCs into IPCs. Therefore, further efforts are needed to generate highly efficient method for differentiation into functional IPCs.

5.7 MSCs: In Vivo Studies in Diabetes

Different sources of MSCs are used to generate insulin-secreting granules and transplanted into diabetic animals. In vivo studies demonstrated that transplantation of bone marrow-derived MSCs in noneobese diabetic (NOD) mice increased levels of serum insulin and reversed insulin-dependent diabetes (Patel et al. 2013). Transplantation of these cells in streptozotocin-induced diabetic mice ameliorated hyperglycemia and increased serum level of insulin by inducing regeneration of endogenous pancreatic islets. However, the mechanisms underlying their therapeutic effects have not been clearly defined. Histological analysis of the pancreas in the transplanted animals showed that MSCs could migrate to the site of injury to promote formation of IPCs (Tang et al. 2004; Chang et al. 2007). On the other hand, it is suggested that antidiabetic effect of MSCs is related to their immonomodulatory effects on immune response or may inhibit apoptosis of injured pancreatic β cells and increase in regeneration of endogenous pancreatic progenitor cells and subsequently β-cell mass through paracrine effects (Xu et al. 2008). In the some studies, it was shown that transplanted BMMSCs to diabetic animals by direct injections into pancreas increased insulin levels and formed new islets that are smaller than normal islets in the histological studies (Chang et al. 2009). Recent studies used MSCs derived from adipose tissue (AD-MSCs) to generate IPCs. These cells are easily accessible and have differentiation capacities similar to BMMSCs (Zuk et al. 2002). They have high capacity in proliferation, differentiation, and immunomodulatory effects compared with BMMSCs (Kim et al. 2007; Pendleton et al. 2013; Melief et al. 2013). AD-MSCs could differentiate into IPCs under a stepwise inductive protocol. Differentiated cells expressed endocrine and β-cell-specific markers and secreted C-peptide in response to glucose with different concentrations. Intraperitoneal transplantation of these cells could restore normoglycemia in diabetic mice (Chandra 2009).

5.8 MSCs: Clinical Trials

MSCs have been shown to decrease inflammatory-mediated immune reactions and reduce inflammation and assist in tissue repair. In recent clinical trials, long-term follow-up of MSCs generated very promising findings in treatment of T1DM. MSCs have been used in several studies to generate IPCs (Karnieli et al. 2007) and treat diabetic foot ulcers and limb ischemia (Lu et al. 2011). Data published from Hu et al. showed that long-term injection of Wharton's jelly-derived MSCs in T1DM patients led to tight glycemic control and increased levels of C-peptide after 2 years from injection compared to T1DM patients having the same age and intensive insulin therapy (Hu et al. 2013). MSCs have been used in several diabetes-related complications like cardiomyopathy, nephropathy, polyneuropathy, and diabetic wounds (Volarevic et al. 2011). The results of these studies will decide the future of cell-based treatment for the most devastating degenerative conditions. Completed or ongoing clinical trials using MSCs are listed in Table 5.1. One of them has multiple administrations of ROCHYMAL® (allogeneic bone marrow-derived MSCs from healthy volunteer donors with commercial formulation) to decreasing the immune response and repair damaged pancreatic tissue, leading to restore normal glycemia in diabetic patients. The results of this trial showed that intravenous infusion of these cells could normalized hyperglycemia in newly diagnosed type 1 diabetes patients (Sàrl 2008). MSCs have a greater biosafety profile and lower risk of tumorigenicity in comparison with other kind of stem cells. Recently, some studies found that the transplantation of bone marrow-derived MSCs into STZ-induced diabetic mice often induced tumor formation at the injection site (Jeong et al. 2011).

Table 5.1 Clinical trials of mesenchymal stem cell-based therapy in T1D

Study	Year	Identifier	Sponsor	Status
Phase II, PROCHYMAL® (Human Mesenchymal Stem Cells) for the Treatment of Recently Diagnosed Type 1 Diabetes Mellitus (T1DM)	2008	NCT00690066	Mesoblast International Sàrl	Completed
Phase I, Comparison of Autologous Mesenchymal Stem Cells and Mononuclear Cells on Diabetic Critical Limb Ischemia and Foot Ulcer	2009	NCT00955669	Third Military Medical University	Completed
Phase II, Induced Wound Healing by Application of Expanded Bone Marrow Stem Cells in Diabetic Patients With Critical Limb Ischemia	2010	NCT01065337	Ruhr University of Bochum	Completed
Phase I/II, Treatment of Patients With Newly Onset of Type 1 Diabetes With Mesenchymal Stem Cells	2010	NCT01068951	Uppsala University Hospital	Completed
Safety Study of Stem Cells Treatment in Diabetic Foot Ulcers	2012	NCT01686139	Sheba Medical Center	Incomplete
Mesenchymal Stem Cells to Intervene in the Development of Type 1 Diabetes: a Blinded Randomized Study	2014	NCT02057211	Uppsala University Hospital	Incomplete

To overcome limitations of cell therapy procedures, cells were encapsulated by using many different polymers, methods, and sizes. This barrier protects cells from an immune attack and allows the passage of nutrients without larger molecules, antibodies, or cells (Hernandez et al. 2010). But, further clinical trials will be critical to evaluating clinical benefits and risk of assuming adverse effects for this kind of treatment.

5.9 Roles of miRNAs in Diabetes

5.9.1 Biogenesis of miRNAs and Mechanisms of Gene Silencing

microRNAs (miRNAs) a new class of small noncoding RNAs that are involved in regulation of genes are involved in developmental timing, proliferation, differentiation, carcinogenesis, and apoptosis (Gauthier and Wollheim 2006). The miRNAs have inhibitory effects on RNA stability and mRNA translation by binding to the 3′-untranslated regions (3′UTR) of target messenger RNAs (mRNAs) (Carthew and Sontheimer 2009). They are 21–23 nucleotides that regulate target genes at the posttranscriptional level. In mammals, primary transcripts of miRNAs (pri-miRNAs) are transcribed from the genome by ribonucleases DROSHA (a nuclear RNase III enzyme) and DICER (a cytoplasmic RNase III enzyme) to generate 21–22-nucleotide double-stranded RNAs (Fig. 5.1). Then, the active single-stranded miRNA enters the RNA-induced silencing complex (RISC) and interacts with target mRNAs or can be released by the cells. In addition, the miRNAs can be loaded into high-density lipoprotein (HDL) or proteins and released into the extracellular space during plasma membrane blebbing or after fusion of multivesicular bodies (MVBs) with the plasma membrane (Fig. 5.1). A number of studies show that miRNAs play an important role in the pancreas development, insulin secretion, and DM and its complications. A number of studies indicate that miRNAs may serve as potential biomarkers for the diagnosis and prognosis of diabetes (Chen et al. 2014).

5.10 miRNA Function in Pancreatic Cell Fate and Pancreas Development

Islets of Langerhans consist of different cell types including α, β, δ, ε, and PP that produce and secrete glucagon, insulin, somatostatin, ghrelin, and pancreatic polypeptide, respectively. Decrease in number of β cells leads to an imbalance in insulin and glucose resulting in diabetes. The miR-375, one of the most widely studied and highly enriched in pancreatic islets, is essential for the formation of insulin-secreting pancreatic islets (Kloosterman et al. 2007) (Fig. 5.2). It is essential to maintain the

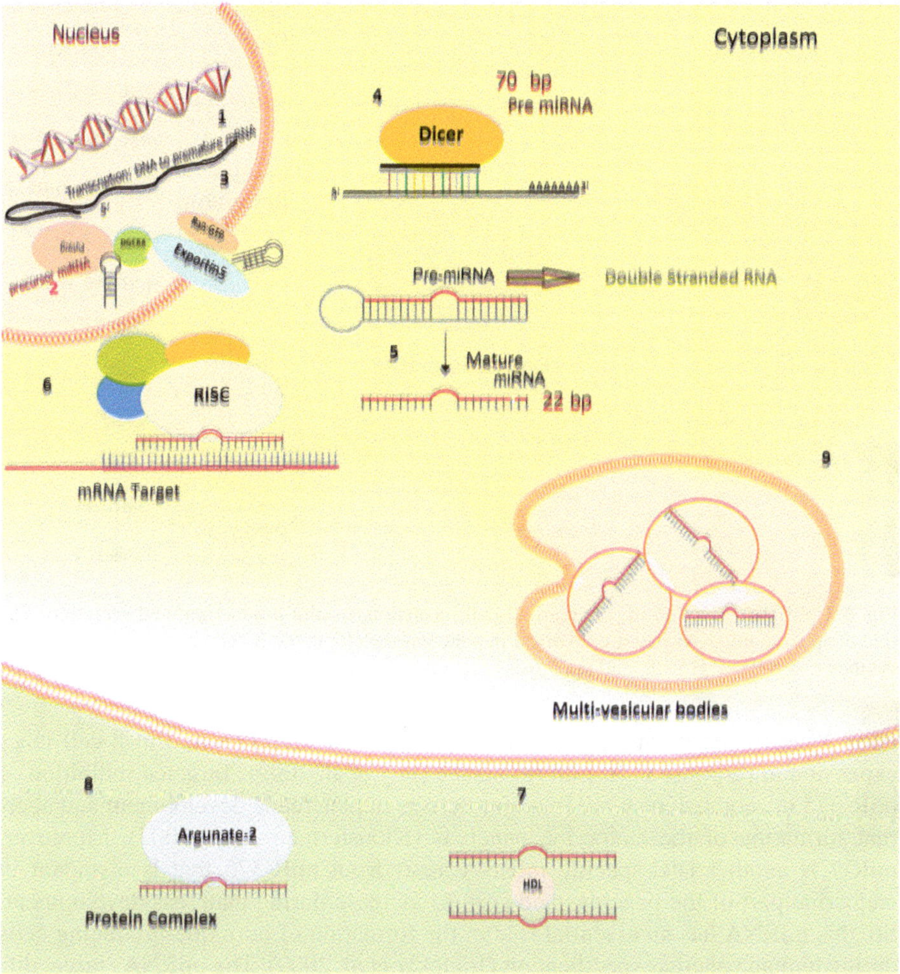

Fig. 5.1 Biogenesis of miRNAs and mechanisms of gene silencing. Pre-miRNAs are generated in the nucleus by the ribonuclease III enzyme Drosha after cleavage of pri-miRNAs (*1*, *2*). The pre-miRNAs are then transported in the cytoplasm by an active process involving exportin-5 and the GTP-binding protein Ran (*3*) and further processed by Dicer to yield 21–23 nucleotide duplexes (*4*). One strand of the miRNA duplex can either associate to the RISC complex and guide translational repression of target mRNAs (*6*) or be released by the cells. Then, mature miRNA (*5*) binds to RNA-binding proteins such as Argonaute-2 (*8*) or to lipoproteins (*7*). Alternatively, the miRNAs can be loaded in microvesicles formed by plasma membrane blebbing or in exosomes that are released into the extracellular space upon exocytic fusion of multivesicular bodies with the plasma membrane (*9*). Mature miRNAs regulate gene expression by guiding the RISC complex to their target complementary mRNA, causing mRNA degradation and repression of translation initiation. *Abbreviations*: *miRNA* microRNA, *pre-miRNA* miRNA precursor, *pri-miRNA* primary miRNA transcript, *RISC* RNA-induced silencing complex

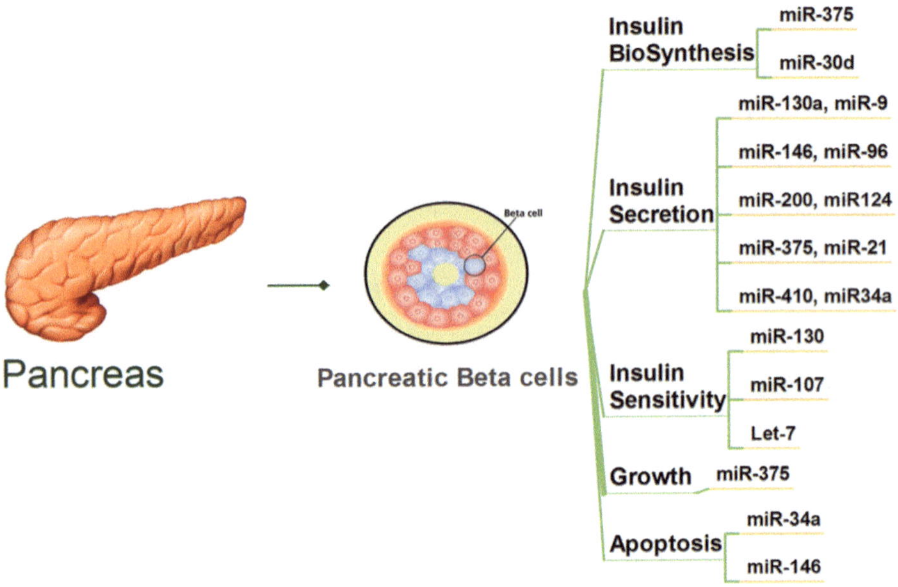

Fig. 5.2 miRNA function in pancreatic β-cell formation, insulin production, and secretion. The role of specific miRNAs that are involved in pancreas development, β-cell formation, insulin biosynthesis, secretion, growth, and apoptosis

normal pancreatic α- and β-cell mass and hormone secretion and in β-cell mass expansion in response to insulin resistance (Poy et al. 2009). Targeted inhibition of miR-375 in zebrafish resulted in major defects in pancreatic development and aberrant formation of the endocrine pancreas (Kloosterman et al. 2007). Moreover, miR-7 is another islet-specific miRNA apart from miR-375 that is abundant in endocrine part of the pancreas and seen to increase during pancreas development. So, this miRNA has an essential role in the formation of hormone-producing cells during human pancreas specification (Joglekar et al. 2009). The miRNAs target different transcription factors that control pancreas development and maintenance, but some of them remain to be identified.

5.11 miRNAs in Insulin Production and Insulin Secretion

In the pancreas islets, miR-375 regulates insulin exocytosis and homeostasis (Gauthier and Wollheim 2006). miR-375 targets 3′-phosphoinositide-dependent protein kinase-1 (PDK1) and decreases glucose-induced insulin gene expression and protein synthesis (Chen et al. 2014). The myotrophin is a transcription factor and another target of miR-375 (Mtpn) that is involved in actin depolymerization and

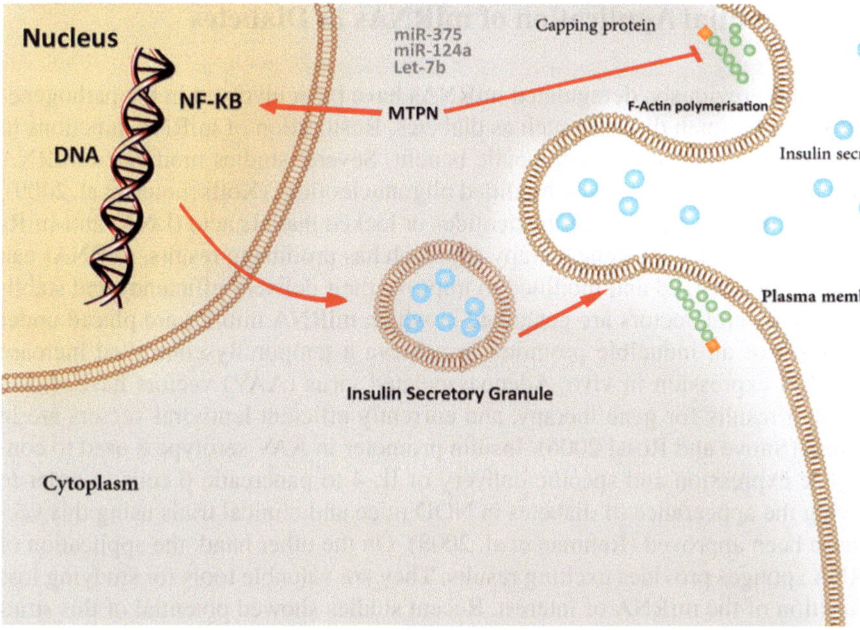

Fig. 5.3 Possible role of miRNA-375 in insulin secretion. The microRNAs miR-375, miR-124, and let-7b suppress the translation of mRNA encoding myotrophin (MTPN). MTPN interacts with cytoskeletal elements and is thought to open the F-actin mesh, allowing access of the secretory insulin granules to exocytotic sites. Furthermore, MTPN may activate the transcription factor NF-κB in the nucleus. NF-κB has an important role in insulin activation and secretion, possibly by activating genes that control vesicle trafficking and exocytosis. *RISC* RNA-induced silencing complex, *ANK* ankyrin motifs, *GLP-1* glucagon-like peptide-1

insulin granule fusion, thereby reducing insulin exocytosis (Poy et al. 2009) (Fig. 5.3). In addition, miR-124a, miR-9, miR-96a, and miR-33 have been identified to regulate insulin secretion. miR-124a negatively regulates insulin exocytosis insulin secretion in pancreatic β cells by direct targeting of *Rab27a* expression (Lovis et al. 2008). miR-9 is another miRNA that has been involved in the control of insulin exocytosis by direct targeting of ONECUT2 (a Granuphilin gene repressor), mRNA. The granuphilin has been well characterized as a negative regulator of insulin secretion. Its expression is downregulated in insulin-producing cells (Ramachandran et al. 2011). miR-9 by increasing the level of granuphilin (SLP4) facilitates mobilizing of insulin and fusing with the plasma membrane *during* exocytosis processing (Chen et al. 2014). miR-34a is upregulated in db/db mice and affects insulin secretion by targeting vesicl-associated membrane protein 2 (Vamp2), a key player involved in insulin exocytosis from β-cell membranes (Lovis et al. 2008). miR-30d induces insulin expression in β-cells through the downregulation of mitogen-activated protein 4 kinase 4 (MAP4K4) expression, the negative regulator of the insulin transcription factor, MAFA (Zhao et al. 2012).

5.12 Potential Application of miRNAs in Diabetes

As outlined previously, deregulated miRNAs have been involved in the pathogenesis of several human diseases such as diabetes. Restoration of miRNA functions to normal levels may provide therapeutic benefit. Several studies modulated miRNA expression by using chemically modified oligonucleotides (Kolfschoten et al. 2009). The use of miRNA mimic oligonucleotides or locked nucleic acid (LNA) anti-miRs and morpholinos for the gene therapy approach has promising results. miRNAs can be readily synthesized and modified to improve their delivery efficiency and stability. Today several vectors are generated in which miRNA mimics are placed under the control of an inducible promoter to achieve a temporally controlled increase of miRNA expression in vivo. Adenoassociated virus (AAV) vectors have shown promising results for gene therapy, and currently efficient lentiviral vectors are in progress (Snove and Rossi 2006). Insulin promoter in AAV serotype 8 used to control gene expression and specific delivery of IL-4 to pancreatic β cells in order to delaying the appearance of diabetes in NOD mice and clinical trials using this vector have been approved (Rehman et al. 2008). On the other hand, the application of miRNA sponges provides exciting results. They are valuable tools for studying loss of function of the miRNA of interest. Recent studies showed potential of this strategy to decrease the activity of miR-31 or of the miR-15a/16-1 cluster stably in vivo their role in cancer development (Valastyan et al. 2009). Engineered mesenchymal stem cells with pancreas-specific miRNAs may enhance their capacity for differentiation to insulin-producing cells and insulin secretion with glucose-dependent manner and improved their transplantation efficiency (Jafarian et al. 2015). In vivo delivery of therapeutic miRNA in β cells and insulin-target tissues is important to create new strategies for treating the disease (Guay and Regazzi 2013). However, evaluation of safety and cost-effectiveness is needed to achieve appropriate levels of miRNA for the treatment of diabetes. In addition to the alteration in insulin-producing cells and insulin-target tissues, diabetes is also associated with distinct changes in the blood miRNA profile. In the future, measurements of the level of specific miRNAs may become useful tools to identify individuals at risk for developing type 1 or type 2 diabetes, hopefully preventing the development of the disease.

5.13 Potential miRNA Biomarkers in T1D

miRNAs have been found in high concentration in body flouids. They are new biomarkers for detecting several diseases including various forms of cancers, autoimmune disease, and sepsis (Guay and Regazzi 2013). Circulating miR-375 levels have been shown to be a biomarker of β-cell death and predict the development of T1DM in animal models. In NOD mice, an autoimmune model of diabetes, it was significantly increased at 2 weeks before onset of diabetes (Erener et al. 2013). miR-326 is another important biomarker that elevated levels of expression which was found in peripheral blood lymphocytes of T1D patients with ongoing islet autoimmunity (Sebastiani et al. 2011). Diminished expression of miR-21a and miR-93 was seen in peripheral blood mononuclear cells (PBMC) of patients with T1DM, and clinically upregulation of 12 serum miRNAs (miR-152, miR-30a-5p, miR-181a, miR-24, miR-148a, miR-210, miR-27a, miR-29a, miR-26a, miR-27b, miR-25, and miR-200a) was detected in these patients, and it is cleared that miR-25 was negatively associated with β-cell function.

5.14 Potential miRNA Biomarkers in T2D

Seven miRNAs (miR-9, miR-29a, miR-30d, miR-34a, miR-124a, miR-146a, and miR-375) were significantly elevated in a serum analysis in T2D patients. miR-9, miR-29a, miR-34a, miR-146a, and miR-375 were significantly upregulated in individuals with normal glucose tolerance in comparison to prediabetes individuals. Importantly, a decrease in miR-15a, miR-29b, miR-126, and miR-223 and an increase in miR-28-3p levels in plasma were valuable biomarkers for predicting T2D. Three serum miRNAs (miR-138, miR-376a, and miR-15b) are potential biomarkers for distinguishing obese patients from obese T2D and T2D patients. It was revealed that three serum miRNAs (miR-132, miR-29a, and miR-222) can predict gestational DM with high sensitivity and specificity (Chen et al. 2014), but the potential clinical use of miRNAs as diabetic biomarkers still needs further investigation.

5.15 Conclusion and Prospects

MSCs are one of the most valuable candidates for cell-based therapy in various human diseases. They have gained interest based on their immunomodulatory properties and parecrine secretion activity. They are successfully explored in animal models of in tissue engineering, transplantation, and treatment of autoimmune disease. MSCs seem to be a promising approach in treatment of type 1 diabetes. They are relatively isolated from different organs, expanded, and transduced efficiently with viral vectors for treatment of genetic diseases. MSCs are a good cells for gene therapy, including the transfer of manipulated miRNAs into MSCs, and detect their roles in the differentiation of MSCs into insulin-secreting cells. miRNA presence in extracellular fluids in remarkably stable forms has led to the idea of using them as biomarkers for a variety of diseases, given the numerous associations between miRNAs and disease. They are directly involved in β-cell survival and insulin exocytosis, glucose homeostasis, and insulin resistance. These properties have also attracted the attention of diabetologists who have initiated the search for miRNAs that enable early detection of T1DM and T2DM and their associated complications. So, these small RNAs may serve as a potential biomarkers in early detection and treatment of diabetes, and they will soon serve as valuable new blood parameters that will help physicians to refine their therapeutic interventions.

References

Aggarwal S, Pittenger MF (2005) Human mesenchymal stem cells modulate allogeneic immune cell responses. Blood 105(4):1815–1822

Ambros V (2004) The functions of animal microRNAs. Nature 431(7006):350–355

Barzilay R, Ben-Zur T, Bulvik S, Melamed E, Offen D (2009) Lentiviral delivery of LMX1a enhances dopaminergic phenotype in differentiated human bone marrow mesenchymal stem cells. Stem Cells Dev 18(4):591–601

Bernardi S, Severini GM, Zauli G, Secchiero P (2012) Cell-based therapies for diabetic complications. Exp Diabetes Res 872504(10):9

Carthew RW, Sontheimer EJ (2009) Origins and mechanisms of miRNAs and siRNAs. Cell 136(4):642–655

Chandra V, G S, Phadnis S, Nair PD, Bhonde RR (2009) Generation of pancreatic hormone-expressing islet-like cell aggregates from murine adipose tissue-derived stem cells. Stem Cells 27(8):1941–1953

Chang C, Niu D, Zhou H, Li F, Gong F (2007) Mesenchymal stem cells contribute to insulin-producing cells upon microenvironmental manipulation in vitro. Transplant Proc 39(10): 3363–8

Chang C, Wang X, Niu D, Zhang Z, Zhao H, Gong F (2009) Mesenchymal stem cells adopt beta-cell fate upon diabetic pancreatic microenvironment. Pancreas 38(3):275–281

Chen H, Lan HY, Roukos DH, Cho WC (2014) Application of microRNAs in diabetes mellitus. J Endocrinol 222(1):R1–R10

Cheng T, Yang C, Weber N, Kim HT, Kuo AC (2012) Fibroblast growth factor 2 enhances the kinetics of mesenchymal stem cell chondrogenesis. Biochem Biophys Res Commun 426(4): 544–550

De Keyser J (2005) Autologous mesenchymal stem cell transplantation in stroke patients. Ann Neurol 58(4):653–654. author reply 654–5

Diagnosis and classification of diabetes mellitus (2014) Diabetes Care 37(1):dc14–S081

Ding DC, Shyu WC, Lin SZ (2011) Mesenchymal stem cells. Cell Transplant 20(1):5–14

Dominici M, Le Blanc K, Mueller I, Slaper-Cortenbach I, Marini F, Krause D et al (2006) Minimal criteria for defining multipotent mesenchymal stromal cells. The International Society for Cellular Therapy position statement. Cytotherapy 8(4):315–317

Erener S, Mojibian M, Fox JK, Denroche HC, Kieffer TJ (2013) Circulating miR-375 as a biomarker of beta-cell death and diabetes in mice. Endocrinology 154(2):603–608

Fedyunina IA, Rzhaninova AA, Kirienko EE, Goldshtein DV (2011) Isolation of insulin-producing cells from different populations of multipotent stromal cells of the umbilical cord and human adipose tissue. Bull Exp Biol Med 151(1):114–120

Friedenstein AJ, Gorskaja JF, Kulagina NN (1976) Fibroblast precursors in normal and irradiated mouse hematopoietic organs. Exp Hematol 4(5):267–274

Gabr MM, Zakaria MM, Refaie AF, Khater SM, Ashamallah SA, Ismail AM et al (2015) Differentiation of human bone marrow-derived mesenchymal stem cells into insulin-producing cells: evidence for further maturation in vivo. Biomed Res Int 575837(10):12

Gauthier BR, Wollheim CB (2006) MicroRNAs: 'ribo-regulators' of glucose homeostasis. Nat Med 12(1):36–38

Gruessner AC, Sutherland DE, Gruessner RW (2012) Long-term outcome after pancreas transplantation. Curr Opin Organ Transplant 17(1):100–105

Guay C, Regazzi R (2013) Circulating microRNAs as novel biomarkers for diabetes mellitus. Nat Rev Endocrinol 9(9):513–521

Hernandez RM, Orive G, Murua A, Pedraz JL (2010) Microcapsules and microcarriers for in situ cell delivery. Adv Drug Deliv Rev 62(7–8):711–730

Hu J, Yu X, Wang Z, Wang F, Wang L, Gao H et al (2013) Long term effects of the implantation of Wharton's jelly-derived mesenchymal stem cells from the umbilical cord for newly-onset type 1 diabetes mellitus. Endocr J 60(3):347–357

Hwang NS, Zhang C, Hwang YS, Varghese S (2009) Mesenchymal stem cell differentiation and roles in regenerative medicine. Wiley Interdiscip Rev Syst Biol Med 1(1):97–106
Jafarian A, Taghikhani M, Abroun S, Pourpak Z, Allahverdi A, Soleimani M (2014) Generation of high-yield insulin producing cells from human bone marrow mesenchymal stem cells. Mol Biol Rep 41(7):4783–94
Jafarian A, Taghikani M, Abroun S, Allahverdi A, Lamei M, Lakpour N et al (2015) The generation of insulin producing cells from human mesenchymal stem cells by MiR-375 and anti-MiR-9. PLoS One 10(6):e0128650
Jamiolkowski RM, Guo LY, Li YR, Shaffer SM, Naji A (2012) Islet transplantation in type I diabetes mellitus. Yale J Biol Med 85(1):37–43
Jeong JO, Han JW, Kim JM, Cho HJ, Park C, Lee N et al (2011) Malignant tumor formation after transplantation of short-term cultured bone marrow mesenchymal stem cells in experimental myocardial infarction and diabetic neuropathy. Circ Res 108(11):1340–1347
Joglekar MV, Joglekar VM, Hardikar AA (2009) Expression of islet-specific microRNAs during human pancreatic development. Gene Expr Patterns 9(2):109–113
Kang SK, Lee DH, Bae YC, Kim HK, Baik SY, Jung JS (2003) Improvement of neurological deficits by intracerebral transplantation of human adipose tissue-derived stromal cells after cerebral ischemia in rats. Exp Neurol 183(2):355–366
Kantharidis P, Wang B, Carew RM, Lan HY (2003) Improvement of neurological deficits by intracerebral transplantation of human adipose tissue-derived stromal cells after cerebral ischemia in rats. Exp Neurol 183(2):355–366
Karnieli O, Izhar-Prato Y, Bulvik S, Efrat S (2007) Generation of insulin-producing cells from human bone marrow mesenchymal stem cells by genetic manipulation. Stem Cells 25(11): 2837–44
Kim Y, Kim H, Cho H, Bae Y, Suh K, Jung J (2007) Direct comparison of human mesenchymal stem cells derived from adipose tissues and bone marrow in mediating neovascularization in response to vascular ischemia. Cell Physiol Biochem 20(6):867–876
Kloosterman WP, Lagendijk AK, Ketting RF, Moulton JD, Plasterk RH (2007) Targeted inhibition of miRNA maturation with morpholinos reveals a role for miR-375 in pancreatic islet development. PLoS Biol 5(8):e203
Kolfschoten IG, Roggli E, Nesca V, Regazzi R (2009) Role and therapeutic potential of microRNAs in diabetes. Diabetes Obes Metab 4:118–129
Liu M, Han ZC (2008) Mesenchymal stem cells: biology and clinical potential in type 1 diabetes therapy. J Cell Mol Med 12(4):1155–1168
Lovis P, Gattesco S, Regazzi R (2008) Regulation of the expression of components of the exocytotic machinery of insulin-secreting cells by microRNAs. Biol Chem 389(3):305–312
Lu LL, Liu YJ, Yang SG, Zhao QJ, Wang X, Gong W et al (2006) Isolation and characterization of human umbilical cord mesenchymal stem cells with hematopoiesis-supportive function and other potentials. Haematologica 91(8):1017–1026
Lu D, Chen B, Liang Z, Deng W, Jiang Y, Li S et al (2011) Comparison of bone marrow mesenchymal stem cells with bone marrow-derived mononuclear cells for treatment of diabetic critical limb ischemia and foot ulcer: a double-blind, randomized, controlled trial. Diabetes Res Clin Pract 92(1):26–36
Melief SM, Zwaginga JJ, Fibbe WE, Roelofs H (2013) Adipose tissue-derived multipotent stromal cells have a higher immunomodulatory capacity than their bone marrow-derived counterparts. Stem Cells Transl Med 2(6):455–463
Naghdi M, Tiraihi T, Namin SA, Arabkheradmand J (2009) Transdifferentiation of bone marrow stromal cells into cholinergic neuronal phenotype: a potential source for cell therapy in spinal cord injury. Cytotherapy 11(2):137–152
Newman RE, Yoo D, LeRoux MA, Danilkovitch-Miagkova A (2009) Treatment of inflammatory diseases with mesenchymal stem cells. Inflamm Allergy Drug Targets 8(2):110–123
Parekh VS, Joglekar MV, Hardikar AA (2009) Differentiation of human umbilical cord blood-derived mononuclear cells to endocrine pancreatic lineage. Differentiation 78(4):232–240

Patel DM, Shah J, Srivastava AS (2013) Therapeutic potential of mesenchymal stem cells in regenerative medicine. Stem Cells Int 496218(10):19
Pavlova G, Lopatina T, Kalinina N, Rybalkina E, Parfyonova Y, Tkachuk V et al (2012) In vitro neuronal induction of adipose-derived stem cells and their fate after transplantation into injured mouse brain. Curr Med Chem 19(30):5170–7
Pendleton C, Li Q, Chesler DA, Yuan K, Guerrero-Cazares H, Quinones-Hinojosa A (2013) Mesenchymal stem cells derived from adipose tissue vs bone marrow: in vitro comparison of their tropism towards gliomas. PLoS ONE 8(3):12
Phadnis SM, Joglekar MV, Dalvi MP, Muthyala S, Nair PD, Ghaskadbi SM et al (2011) Human bone marrow-derived mesenchymal cells differentiate and mature into endocrine pancreatic lineage in vivo. Cytotherapy 13(3):279–293
Pittenger MF, Mackay AM, Beck SC, Jaiswal RK, Douglas R, Mosca JD et al (1999) Multilineage potential of adult human mesenchymal stem cells. Science 284(5411):143–147
Poy MN, Hausser J, Trajkovski M, Braun M, Collins S, Rorsman P et al (2009) miR-375 maintains normal pancreatic alpha- and beta-cell mass. Proc Natl Acad Sci U S A 106(14):5813–8
Ramachandran D, Roy U, Garg S, Ghosh S, Pathak S, Kolthur-Seetharam U (2011) Sirt1 and mir-9 expression is regulated during glucose-stimulated insulin secretion in pancreatic beta-islets. FEBS J 278(7):1167–1174
Rehman KK, Trucco M, Wang Z, Xiao X, Robbins PD (2008) AAV8-mediated gene transfer of interleukin-4 to endogenous beta-cells prevents the onset of diabetes in NOD mice. Mol Ther 16(8):1409–1416
van Rooij E (2011) The art of microRNA research. Circ Res 108(2):219–234
Rooney GE, Howard L, O'Brien T, Windebank AJ, Barry FP (2009) Elevation of cAMP in mesenchymal stem cells transiently upregulates neural markers rather than inducing neural differentiation. Stem Cells Dev 18(3):387–398
Sàrl MI (2008) PROCHYMAL® (human adult stem cells) for the treatment of recently diagnosed Type 1 Diabetes Mellitus (T1DM). Clinical Trials.gov
Sebastiani G, Grieco FA, Spagnuolo I, Galleri L, Cataldo D, Dotta F (2011) Increased expression of microRNA miR-326 in type 1 diabetic patients with ongoing islet autoimmunity. Diabetes Metab Res Rev 27(8):862–866
Sheng H, Wang Y, Jin Y, Zhang Q, Zhang Y, Wang L et al (2008) A critical role of IFNgamma in priming MSC-mediated suppression of T cell proliferation through up-regulation of B7-H1. Cell Res 18(8):846–857
Snove O Jr, Rossi JJ (2006) Expressing short hairpin RNAs in vivo. Nat Methods 3(9):689–695
Stenderup K, Justesen J, Clausen C, Kassem M (2003) Aging is associated with decreased maximal life span and accelerated senescence of bone marrow stromal cells. Bone 33(6):919–926
Tang DQ, Cao LZ, Burkhardt BR, Xia CQ, Litherland SA, Atkinson MA et al (2004) In vivo and in vitro characterization of insulin-producing cells obtained from murine bone marrow. Diabetes 53(7):1721–1732
Teven CM, Liu X, Hu N, Tang N, Kim SH, Huang E et al (2011) Epigenetic regulation of mesenchymal stem cells: a focus on osteogenic and adipogenic differentiation. Stem Cells Int 201371(10):11
Valastyan S, Reinhardt F, Benaich N, Calogrias D, Szasz AM, Wang ZC et al (2009) A pleiotropically acting microRNA, miR-31, inhibits breast cancer metastasis. Cell 137(6):1032–1046
Volarevic V, Arsenijevic N, Lukic ML, Stojkovic M (2011) Concise review: mesenchymal stem cell treatment of the complications of diabetes mellitus. Stem Cells 29(1):5–10
Wakao S, Kuroda Y, Ogura F, Shigemoto T, Dezawa M (2012) Regenerative effects of mesenchymal stem cells: contribution of muse cells, a novel pluripotent stem cell type that resides in mesenchymal cells. Cell 1(4):1045–1060
Wei X, Yang X, Han ZP, Qu FF, Shao L, Shi YF (2013) Mesenchymal stem cells: a new trend for cell therapy. Acta Pharmacol Sin 34(6):747–754
Xu YX, Chen L, Wang R, Hou WK, Lin P, Sun L et al (2008) Mesenchymal stem cell therapy for diabetes through paracrine mechanisms. Med Hypotheses 71(3):390–393

Zhang B, Farwell MA (2008) microRNAs: a new emerging class of players for disease diagnostics and gene therapy. J Cell Mol Med 12(1):3–21

Zhang YN, Lie PC, Wei X (2009) Differentiation of mesenchymal stromal cells derived from umbilical cord Wharton's jelly into hepatocyte-like cells. Cytotherapy 11(5):548–558

Zhao X, Mohan R, Ozcan S, Tang X (2012) MicroRNA-30d induces insulin transcription factor MafA and insulin production by targeting mitogen-activated protein 4 kinase 4 (MAP4K4) in pancreatic beta-cells. J Biol Chem 287(37):31155–64

Zhao F, Qu Y, Liu H, Du B, Mu D (2014) Umbilical cord blood mesenchymal stem cells co-modified by TERT and BDNF: a novel neuroprotective therapy for neonatal hypoxic-ischemic brain damage. Int J Dev Neurosci 38:147–154

Zuk PA, Zhu M, Ashjian P, De Ugarte DA, Huang JI, Mizuno H et al (2002) Human adipose tissue is a source of multipotent stem cells. Mol Biol Cell 13(12):4279–95

Zulewski H, Abraham EJ, Gerlach MJ, Daniel PB, Moritz W, Muller B et al (2001) Multipotential nestin-positive stem cells isolated from adult pancreatic islets differentiate ex vivo into pancreatic endocrine, exocrine, and hepatic phenotypes. Diabetes 50(3):521–533

Chapter 6
Novel Biomarkers at Risk Stratification of Diabetes Mellitus Patients

Alexander E. Berezin

Abbreviations

CAD	Coronary artery disease
CV	Cardiovascular
GDF-15	Growth differentiation factor-15
HbA1c	Glycated hemoglobin
hs-CRP	High-sensitivity C-reactive protein
hs-cTnT	High-sensitivity cardiac troponin T
IFG	Impaired fasting glucose
IGT	Impaired glucose tolerance
IL	Interleukin
NPs	Natriuretic peptides
ST2	A member of the interleukin-1 receptor family protein of tumorigenicity
T2DM	Type 2 diabetes mellitus

6.1 Introduction

Type 2 diabetes mellitus (T2DM) remains a substantial health problem and one of the most prevalent metabolic diseases worldwide, which exhibits relentless growth in general population (Kharroubi and Darwish 2015). Recent observation and epidemiological and clinical investigations have revealed a strong evidence regarding sufficient association between T2DM with cardiovascular (CV) mortality and morbidity (Achelrod et al. 2016). The impact of T2DM on CV mortality and morbidity

This research received no specific grant from any funding agency in the public, commercial, or not-for-profit sectors.

A.E. Berezin (✉)
Cardiology Unit, Internal Medicine Department, State Medical University,
26, Mayakovsky av., Zaporozhye, Ukraine
e-mail: dr_berezin@mail.ru

P.V. Pham (ed.), *Pancreas, Kidney and Skin Regeneration*, Stem Cells in Clinical Applications, DOI 10.1007/978-3-319-55687-1_6

is embedded through a nature evolution of the disease and effect of sociodemographic, health-lifestyle, and coexisting traditional CV risk factors including hypertension, obesity, and dyslipidemia, which are more prevalent in diabetic patients compared with general population (Raffield et al. 2015; Yen et al. 2016). Moreover, asymptomatic clinical conditions, i.e., ectopic vascular calcifications, atherosclerosis, worse kidney function, prolonged QTc interval on ECG, poor glycemic control, low-intense inflammation, and albuminuria, are most prevalent even at early stage of diabetes' development, and they frequently associate well with higher incidence of macrovascular and microvascular complications of the disease (Duffy and Hameed 2015). Consequently, a manifestation of T2DM has been entailing a 1.5- to 3.5-fold increased risk for CV events and disease (Raffield et al. 2015).

On the last decades, they have focused on investigations of innate accurate molecular mechanisms underlying diabetic-induced vascular complications that directly affected altered gene expression, worse cellular signaling, and cellular metabolism, which lead to various tissue and organ injuries (Bozkurt et al. 2016; Wende 2015). All these pathogenetic mechanisms are not just involved into nature evolution of T2DM, but they contribute in CV risk, a medical care response, individual prognosis (Sörensen et al. 2016).

In this context, more pretty accurate measurement of an initial CV risk and prediction of medical care response depending on drug strategy in diabetics and even in prediabetics might have an important clinical value (Mellbin et al. 2010). To improve our knowledge, peak concentration and serial measurements of various circulating biomarkers reflecting several stages of T2DM nature evolution might be useful to risk assessment and probability of clinical outcomes. There are well-established diabetes-related biomarkers, which are widely used in routine clinical practice, i.e., impaired fasting glucose (IFG), impaired glucose tolerance (IGT), and those with a glycated hemoglobin (HbA1c), glycated albumin, creatinine, and the endogenous secretory receptor for advanced glycation end products (Dixon 2016; Di Pino et al. 2016). They may modulate cardiometabolic risk status including asymptomatic atherosclerosis, diabetes-related vasculopathy, ectopic vascular calcifications, and nephropathy (Yang et al. 2015; Carlsson et al. 2016). Although recent clinical trials have shown the important role of these biomarkers as predictive tools for both the risk of developing T2DM in prediabetes and T2DM-related CV risk, the discovery of predictive biomarkers with improving ability to identification of the patients having an increased risk for CV disease is extremely promised. On this way there are several controversies associated with not fully established strong evidence regarding the role of cardiac biomarkers in predicting of mortality in diabetics. The first controversy has been relating to the ability to routine use of cardiac biomarkers to identify a risk in diabetics without known CV diseases (Alexander et al. 2015). The second controversy affects the possibility of cardiac biomarkers to assay and appraise CV risk and risk of the CV complications (i.e., arrhythmia, heart failure, or chronic kidney disease) in diabetic population with established CV disease (Martín-Timón et al. 2014). Whether novel cardiac biomarkers could exhibit much more pretty accurate CV risk than traditional biomarkers is not fully understood. The aim of the chapter is to summarize our knowledge regarding predictive role of alternate cardiac biomarkers in T2DM and their ability to improve traditional predictive scores.

6.2 Biomarkers of Cardiac Biomechanical Stress and Injury

As well known a prominent attribute of diabetic-induced CV complications is accelerated atherosclerosis, endothelial dysfunction, low-intense inflammation, and worsening of tissue repair as resulted in insulin resistance, hyperglycemia, and oxidative stress (Altabas 2015; Imai et al. 2013; Johnson 2014; Mannarino and Pirro 2008). All these factors contribute a development of coronary artery disease, diabetic cardiomyopathy, heart failure, arrhythmia, thromboembolism, and, suddenly, cardiac death (Torremocha et al. 2001; Wannamethee et al. 2011; Scholte et al. 2009; Zaccardi et al. 2014). Appropriately, biomarkers might reflect appropriate faces of multifactorial pathogenesis of disease and predict CV mortality (Berezin 2016a).

6.2.1 Natriuretic Peptides

Although natriuretic peptides (NP) mediate various cardiovascular and metabolic effects and exhibit an ability to be also the markers of biomechanical cardiac stress, their role in the diabetes nature evolution is not fully understood and appears to be strongly controversial. On the one hand, secretion of NPs is resulting in stretch of the cardiac wall and volume overload of cardiac cavities (Iwanaga et al. 2006). On the other hand, recent epidemiological and preclinical/clinical studies have shown that the NP system acts in deficiency in obesity and T2DM patients and is due to worse clearance of NP receptors and neutral endopeptidases (Moro 2016). Consequently, NP system in obese and diabetics is not able to mediate a wide spectrum of cardiovascular and metabolic protective effects (i.e., vasodilation, natriuresis, diuresis, lipolysis, weight loss, lusitropy, lipid peroxidation, and also improvement of mitochondrial respiration and insulin sensitivity) (McKie et al. 2006).

However, according contemporary clinical guidelines for heart failure (HF), acute coronary syndrome/myocardial infarction, and cardiac cytotoxicity, the NPs are recognized as powerful biomarkers with much more pretty accurate diagnostic and predictive values (McMurray et al. 2012; Yancy et al. 2013). Moreover, elevated circulating level of NPs in a large community-based cohort free of HF has strongly predicted the risk of newly diagnosed CV diseases regardless of obesity and T2DM (Ramos et al. 2015; McKie et al. 2006). However, the concentration of NPs in peripheral blood of HF patients with obesity and T2DM is extremely lower than in HF subjects without these metabolic comorbidities; despite overall results, the circulating levels of NPs in both cohorts of HF individuals are higher compared to healthy volunteers (Gupta and Wang 2015; Coué and Moro 2015).

Brutsaert et al. (2016) have reported that higher levels of BNP have been associated with decreased risk of T2DM in middle-aged adults and that the interrelation has remained after adjustment for waist circumference, low physical activity, estimated glomerular filtration rate, and high-sensitivity C-reactive protein level. Alternatively, the results of The Multi-Ethnic Study of Atherosclerosis (MESA) have shown that circulating N-terminal pro-B-type NP (NT-proBNP) had a biphasic association with T2DM in which the risk of T2DM incident decreased within so named "physiological range" of evolution in NT-proBNP level (Sanchez et al. 2015; Goetze and Zois 2016). Inversely, a risk of newly established T2DM has increased depending on NT-proBNP concentrations increase in response to some pathophysiological conditions leading to high levels of NT-proBNP including aging, kidney dysfunction, myocardial infarction/acute coronary syndrome, and cardiac dysfunction (Zannad et al. 2015; Brutsaert et al. 2016; Berezin 2016a). Moreover, serum BNP concentrations were able to correspond to homeostasis model assessment of insulin resistance in obese and prediabetics (Fu et al. 2016). Natriuretic Peptides Studies Collaboration (2016) team has gauged the predicted effect of elevated BNP level on HF, stroke, and coronary artery disease using individual participant data from relevant prospective studies. It turned out that BNP level has strongly predicted first-onset HF, augmented CAD, and stroke regardless of presentation of obesity and T2DM. Finally, the peak concentrations of the BNPs are known as the diagnostic and predictive markers of the CV disease, including HF and CAD and T2DM. Obesity was inversely associated with NPs in patients without metabolic syndrome rather than in those who had metabolic syndrome (Zeng et al. 2016). However, the longitudinal studies cannot confirm the increased predictive value of the BNPs in obese individuals without T2DM (McAloon et al. 2016). Thus, isolated obesity remains the crucial condition for decision-making regarding predictive role of NPs and requires more scrutiny in the future. Given all these limitations and controversies mentioned above, BNPs are established biomarkers linked to T2DM regardless of obesity with CV risk and CV outcomes.

6.2.2 *Cardiac Troponins*

Cardiac troponins are frequently considered urgent biomarkers of myocardial injury, which may exhibit high predictive value for CV death in subjects with established CAD, myocardial infarction, acute coronary syndrome, and HF (Pareek et al. 2015; Alonso et al. 2016). Nevertheless, cardiac troponins may appear to be predominant markers independently associated with a higher CV mortality risk in general population including obese and T2DM subpopulations (Scirica et al. 2016). Even mild elevated level of high-sensitivity cardiac troponin T (hs-cTnT) in untreated T2DM patients was recently found as an independent predictor of death beyond traditional CV risk factors, otherwise myocardial infarction and HF (McEvoy et al. 2015;

Hitsumoto and Shirai 2015; Saunders et al. 2011; Looker et al. 2015). Interestingly, elevated level of troponins in diabetics did not always preceded primary cardiac-related diseases, but frequently they have been associated with pulmonary thromboembolism, atrial fibrillation, and thromboembolic complications including stroke (Akoum 2016). Long-term monitoring of hs-cTnT might be useful to stratify the diabetics at higher risk of all-cause mortality, despite their predictive accuracy that may probably be pretty suboptimal, when several comorbidities (worsening of kidney function, pulmonary hypertension, thromboembolism, obesity, and asymptomatic peripheral artery disease) were presented (Resl et al. 2016). Finally, it is supposed that large clinical trials are required to determine whether cardiac troponins correspond to nature evolution of diabetic cardiomyopathy across all its stages.

6.2.3 Copeptin

Copeptin is the C-terminal fragment of arginine vasopressin prohormone. As a nonspecific surrogate biomarker of biomechanical stress, copeptin was found to be associated with a development of gestational diabetes mellitus/T2DM, declined kidney function, and albuminuria in epidemiological studies, as well as with an advanced diabetic-induced nephropathy (Morgenthaler et al. 2006; Morgenthaler 2010; Stoiser et al. 2006; Enhorning et al. 2010, 2015; Dąbrowski et al. 2016). Indeed, circulating level of copeptin was found to be significantly higher in adult, children, and adolescents with diabetes mellitus compared to healthy volunteers (Zhu et al. 2016; Schiel et al. 2016). Recently, plasma copeptin level has exhibited a predicted value for kidney outcomes, CAD, CV mortality, and all-cause mortality in diabetics (Velho et al. 2016; Wannamethee et al. 2016). Previous clinical studies have revealed the ability of elevated level of copeptin to predict CAD, HF, and risk of mortality in T2DM patients regardless of other CV risk factors (Wang et al. 2016; Enhörning et al. 2015). Given the evidence, copeptin could be considered an accurate marker of kidney function, CV risk, and CV mortality beyond traditional risk factors (Zhu et al. 2016). Additionally, the accuracy of CV gauge by copeptin measurement may relate to age and sex of the patients (Wannamethee et al. 2016). Indeed, copeptin has independently predicted an increased risk of incident stroke and CV mortality in men with T2DM but not in men without T2DM. The large clinical trials might be constructed to explain the evidence received recently and much more pretty accurate assay the predictive value of copeptin in diabetic population.

6.3 Biomarkers of Inflammation in Diabetic Population

Recent preclinical and clinical studies have shown a pivotal role of low-intense inflammation in nature evolution of diabetes mellitus and occurrence of the T2DM complications. Various exerted profound effects of pro-inflammatory cytokines through mitochondrial injury, oxidative stress, insulin resistance, and beta cell apoptosis entail glucotoxicity and lipotoxicity, which are crucial factors in the control of metabolic homeostasis. Moreover, diabetes-induced microvascular and macrovascular dysfunction are under tight control of inflammatory response (Jagadapillai et al. 2016).

The wide spectrum of inflammatory biomarkers (C-reactive protein, galectins, tissue necrosis factor-alpha, interleukins, adipocytokines, tissue plasminogen activator, fibrinogen, adhesion molecules, haptoglobins, etc.) has exhibited a strong relationship to T2DM and its complications/outcomes. Some of them, i.e., C-reactive protein (CRP), galectin-3, were found to be predictive biomarkers in T2DM (Lubrano and Balzan 2015; Panteghini 2004; Pepys and Hirschfield 2003). Nevertheless, other inflammatory biomarkers (i.e., growth differentiation factor-15, myeloid-related protein 8/14, pentraxin 3, lectin-like oxidized low-density lipoprotein receptor-1, pregnancy-associated plasma protein-A) have considered surrogated markers of cardiovascular remodeling and atherosclerosis in T2DM patients irrespective of established CAD (Krintus et al. 2014; Marian and Nambi 2004; Otake et al. 2008; Charo and Taubman 2004; Inoue and Sawamura 2007).

6.3.1 High-Sensitivity C-Reactive Protein

Recent epidemiological and clinical studies have revealed that T2DM has associated with a two- to threefold increase in the risk of CV disease independently related to inflammatory response (Fu et al. 2016). The level of hs-CRP in T2DM has well corresponded to insulin resistance, interleukin-6, glycated hemoglobin/glycated albumin in circulation, and non-high-density lipoprotein cholesterol (Reynolds et al. 2016; Fu et al. 2016; Puri et al. 2016). Although current clinical recommendations of different esteemed scientific associations have recommended to use a serial routine measurement of high-sensitivity C-reactive protein (hs-CRP) in T2DM to recap CV risk, a lack of evidence regarding an ability of elevated hs-CRP to distinguish diabetics with and without CV disease remains a serious challenge (Myers et al. 2009; Danesh et al. 2004). Consequently, the significance of hs-CRP as a traditional CV risk factor in diabetics has to probably diversify (Dumic et al. 2006). It seems hs-CRP has engaged in low-grade inflammation in T2DM evolution; this biomarker did not exhibit much more pretty accurate diagnostic and predictive value compared to other metabolic biomarkers to improve CV risk stratification in T2DM. In this context, the precise predictive role of hs-CRP in diabetics requires more investigations.

6.3.2 Galectin-3

Galectin-3 (Gal-3) is a lectin that belongs to soluble beta-galactoside-binding proteins, which are highly expressed on surfaces of the various cells (Dumic et al. 2006). Gal-3 regulates cell-to-cell cooperation, immunity, and extracellular interactions playing an important role in inflammation, coagulation, thrombosis, and malignancy (Creemers and Pinto 2011).

In cross-sectional analyses of 2946 Framingham Heart Study participants, Gal-3 levels were associated with higher body mass index, waist circumference, obesity, and triglycerides (Nayor et al. 2015). Recent studies have shown that the Gal-3 might be an independent marker of vascular remodeling and endothelial dysfunction accompanied with inflammation, proliferation, and atherosclerosis in general population and in diabetics (Papapanagiotou et al. 2015; Ozturk et al. 2015). Among individuals with established CV diseases, increased serum Gal-3 level has now recognized a powerful predictor of HF, T2DM, and CV mortality rate (Pugliese et al. 2014; de Boer et al. 2009; McCullough et al. 2011). In fact, Gal-3 was not able to be superior to hs-CRP, NPs, soluble ST2, or GDF-15 as a predictor of vasculopathy, low kidney function, and CV mortality in patients with T2DM (Shah and Januzzi 2014). Whether novel predictive scores based on Gal-3 would be better than older models for T2DM patients is not fully clear, although there is large body of evidence regarding that the Gal-3 might be a target of the contemporary treatment of the patients with T2DM (Berezin 2015; Coburn and Frishman 2014).

6.3.3 Matricellular Proteins

Matricellular proteins belong to a superfamily of multifunctional growth factors, which are structured components of the extracellular matrix and involved in the regulation of bone development, inflammation, vascular remodeling, and reparation (Alford and Hankenson 2006). The matricellular proteins including osteopontin, osteocalcin, osteoprotegerin, osteonectin, and thrombospondin were found to be the surrogate biomarkers of vascular calcification and endothelial dysfunction that is confirmed by the clear link between glucose metabolism and bone development (Bonnet 2016; Li et al. 2011; Gómez-Ambrosi et al. 2007). Additionally, the predictive role of osteopontin, osteocalcin, and osteoprotegerin in T2DM has been established too, while independence of their ability to reclassify the CV risk in diabetics is under scientific discussion (Frangogiannis 2012; Kong et al. 2014; Kruzliak et al. 2016).

In diabetics, circulating levels of both osteoprotegerin and osteopontin arise compared to healthy volunteers (Schreier et al. 2016). Interestingly, osteopontin and osteoprotegerin are highly upregulated in white adipose tissue and perivascular adipose tissue in obese individuals irrespective to T2DM presentation, and they mediate inflammation and metabolic effects, which entail worsening of endogenous repair system, ectopic calcification, endothelial dysfunction, and atherosclerosis (Berezin 2016c; Berezin et al. 2016; Schreier et al. 2016). Additionally, high glucose, which is suitable for T2DM, can induce the expression of both osteopontin and osteoprotegerin in vasculature and adipose tissue, while patients with obesity and without metabolic syndrome have exhibited increased level of both biomarkers (Talat et al. 2016; Berezin 2016a; Li et al. 2016a, b). However, in diabetics with known CV diseases including HF, osteopontin and osteoprotegerin may be useful in improving the BNP-based model as predictor of decreased global contractility function and endothelial dysfunction (Kruzliak et al. 2016; Schreier et al. 2016). Finally, the interrelationship between CV mortality rate serum levels of matricellular proteins in diabetics is not strongly established and requires more investigations. Other biomarkers of metabolic dysfunction that belong to matricellular proteins require more investigations due to uncertain independent impact on CV mortality and all-cause mortality rates in diabetics.

6.4 Growth Differentiation Factor-15

Growth differentiation factor-15 (GDF-15) belongs to the superfamily of the transforming growth factor beta (TGF-β) (Hughes et al. 2014). GDF-15 is widely presented in the various cells, tissues, and organs including the heart, brain, vasculature, liver, placenta, while the main sources of GDF-15 secretion are macrophages, white adipose tissue, and hepatocytes (Adela and Banerjee 2015). The basic biological functions of the GDF-15 are a regulation of the inflammation, modulation of growth, and cell differentiation, which are under control of ischemia/hypoxia, biomechanical stress, inflammatory cytokines (tumor necrosis factor-alpha, interleukin [IL]-2, IL-4, IL-6), oxidative stress components, growth factors, and angiotensin-II (Pavo et al. 2016). The direct molecular target of GDF-15 is p53 subunit that is induced by oxidative stress and exhibits antiapoptotic effects on the target cells. Therefore, GDF-15 via inhibition of c-Jun N-terminal kinase, Bcl-2-associated death promoter, and epidermal growth factor receptor, may suppress synthesis and secretion of other inflammatory cytokines, i.e., tumor necrosis factor-alpha and IL-6 (Unsicker et al. 2013).

Circulating level of GDF-15 in diabetics is extremely higher than in the healthy volunteers and has been considered a diagnostic marker of asymptomatic atherosclerosis, CAD, HF, cardiomyopathies, pulmonary hypertension, respiratory insufficiency, and kidney failure (Berezin 2016a; Fairlie et al. 1999; Resl et al. 2016). Additionally, in T2DM populations serum level of GDF-15 is positively associated with obesity, waist to height ratio, metabolic parameters (i.e., fasting glucose level, glycated hemoglobin, HOMA index, triglycerides, creatinine), as well as age, blood pressure, and hs-CRP (Berezin 2016b; Adela and Banerjee 2015; Cavusoglu et al. 2015; Li et al. 2014). The predictive value of elevated serum GDF-15 level on CV mortality rate was established in several studies in which were included patients from general population as well as individuals with known asymptomatic atherosclerosis (Dominguez-Rodriguez et al. 2014; Rohatgi et al. 2012). However, GDF-15 is a nonspecific biomarker of metabolic disease due to close relation to malignancy and cancer (Pavo et al. 2016). Although GDF-15 is involved in metabolic homoeostasis and weight loss in obese and T2DM individuals, the role of this biomarker in prediction of disease progression, protective vascular response, and prognosis is uncertain.

6.5 Fibroblast Growth Factor-23

Fibroblast growth factor-23 (FGF-23) is a circulating low-molecular-weight (32-kDa) peptide, which is secreted by the osteocytes in response to hyperphosphatemia and calcitriol therapy. It is known that the FGF-23 acts through binding of α-Klotho known as transmembrane protein. FGF-23 is able to exclusively stimulate FGF receptors on the surface of the target myocytes through activation of phospholipase Cγ/calcineurin/nuclear factor. The final result of the interplay is activation of T cell signaling and proliferative response (Grabner et al. 2015).

FGF-23 is constantly elevated in diabetics especially when chronic kidney disease and phosphate overload are reported (Batra et al. 2016). Recent preclinical and clinical studies have shown the role of FGF-23 as a surrogate biomarker of atherosclerosis in diabetics without known CV disease and CV events (Llauradó et al. 2015). Unfortunately, elevated level of FGF-23 was not able to distinguish diabetics at increased risk of CV mortality depending on presentation of CV disease (Wohlfahrt et al. 2015; Berezin 2015; Hughes et al. 2014; Adela and Banerjee 2015). Probably, FGF-23/α-Klotho ration might be more informative and presents more pretty accurate predictive evidence regarding CV mortality rate in patients with T2DM. Thus, large clinical studies require understanding the role of FGF-23 as a biomarker in diabetics.

6.6 Endothelial Cell-Derived Microparticles

Microparticles (MPs) are defined as a heterogeneous population of vesicles (diameter 100–1000 nm), which originated from the membrane of wide spectrum of cells (endothelial cells, mononuclears, platelets, red blood cells, T cells, etc.). Circulating MPs are powerful autocrine/paracrine regulators of functionality of the target cells. At least growth of tissue, reparation, angiogenesis, inflammation, and regeneration are under control of MPs (Arraud et al. 2014). The cell-to-cell cooperation via MP-related transfer of active molecules/peptides, chromatin material, hormones, and inflammatory and growth factors plays active role in the epigenetic regulation of various pathophysiological processes in diabetes (Mause and Weber 2010).

Recent studies have shown that the increased circulating levels of MPs that originated from endothelial cells, mononuclears, and platelets were found in obese and at the early stage of T2DM. In contrast, among patients with established CV disease including CAD, HF, and asymptomatic atherosclerosis, numerous MPs progressively decreased (Nozaki et al. 2010; Pirro et al. 2008). By now it is known that the impaired interrelation between activated cell-derived MPs and apoptotic cell-derived MPs called "impaired phenotype" might occur prior to metabolic disease and also associate with nature evolution of ones (Berezin et al. 2014; Nozaki et al. 2010). Consequently, the imbalance between numerous different MPs derived from activated and apoptotic cells including endothelial cells has now recognized a strong

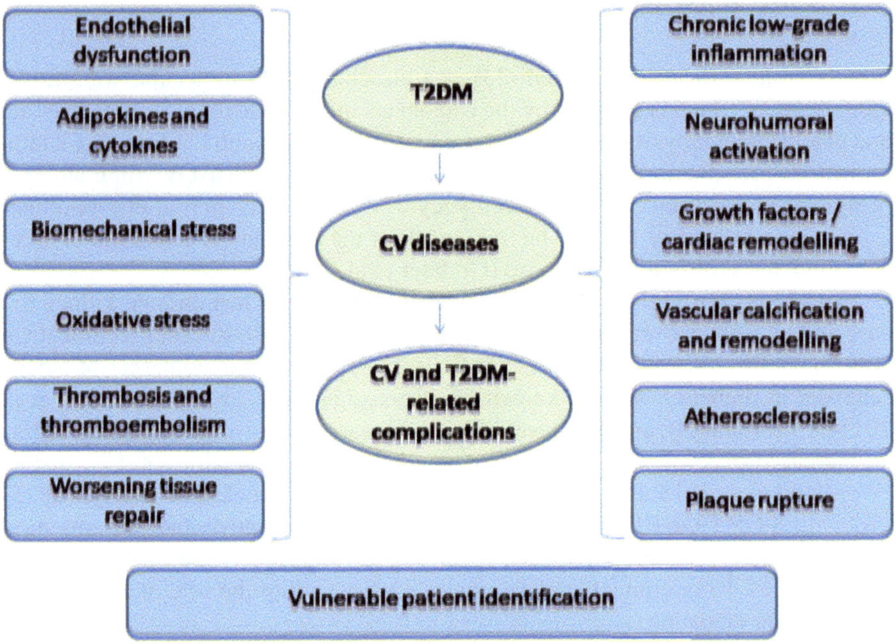

Fig. 6.1 The principal scheme to identification of vulnerable patient with type 2 diabetes mellitus

predictor of diabetic-related outcomes (Berezin and Kremzer 2015; Berezin et al. 2014; Pirro et al. 2008). In this context, the risk stratification of T2DM patients using assay of balance between various immune phenotypes of MPs might be promised for future investigations.

Overall, a broad spectrum of specific and nonspecific cardiac biomarkers might exhibit a prognostic value regarding nature evolution and clinical outcomes of T2DM. Multiple, complementary biomarkers of biomechanical stress and endothelial dysfunction are probably more promising at this way. The principal scheme to identification of vulnerable patient with T2DM, which is given on Fig. 6.1, presents multiple biomarker approach.

6.7 Conclusion

An assessment of diabetic patients is challenging in clinical practice resulting in frequently atypical presentation of signs and symptoms, as well as under recognizing CV risk. Cardiac biomarkers may contribute to improved prediction of CV mortality and CAD incidences in T2DM, but novel clinical data are required to understand what is critically numerous, and combinations of markers are enough to increase risk stratification. Measurement of circulating levels of NPs, Gal-3, and cardiac troponins allows probably to individually stratify the diabetics at risk of CV events. Future directions are associated with discovering of novel biomarkers and their best combinations to provide additional predictive information beyond other traditional CV risk factors.

References

Achelrod D, Gray A, Preiss D, Mihaylova B (2017) Cholesterol- and blood-pressure-lowering drug use for secondary cardiovascular prevention in 2004–2013 Europe. Eur J Prev Cardiol 24(4):426–436

Adela R, Banerjee SK (2015) GDF-15 as a target and biomarker for diabetes and cardiovascular diseases: a translational prospective. J Diabetes Res 2015:490842

Akoum N (2016) New perspectives on atrial fibrillation and stroke. Heart. doi:10.1136/heartjnl-2015-309066. [Epub ahead of print]

Alexander N, Matsushita K, Sang Y, Ballew S, Mahmoodi BK, Astor BC et al (2015) Kidney measures with diabetes and hypertension on cardiovascular disease: the Atherosclerosis Risk in Communities study. Am J Nephrol 41(4–5):409–417

Alford AI, Hankenson KD (2006) Matricellular proteins: extracellular modulators of bone development, remodeling, and regeneration. Bone 38(6):749–757

Alonso N, Lupón J, Barallat J, de Antonio M, Domingo M, Zamora E et al (2016) Impact of diabetes on the predictive value of heart failure biomarkers. Cardiovasc Diabetol 15(1):151

Altabas V (2015) Diabetes, endothelial dysfunction, and vascular repair: what should a Diabetologist keep his eye on? Int J Endocrinol 2015:848272

Arraud N, Linares R, Tan S, Gounou C, Pasquet JM, Mornet S et al (2014) Extracellular vesicles from blood plasma: determination of their morphology, size, phenotype and concentration. J Thromb Haemost 12(5):614–627

Batra J, Buttar RS, Kaur P, Kreimerman J, Melamed ML (2016) FGF-23 and cardiovascular disease: review of literature. Curr Opin Endocrinol Diabetes Obes 23(6):423–429

Berezin AE (2015) Biological markers of cardiovascular diseases. Part 3. Diagnostic and prognostic value of biological markers in stratification of patients with cardiometabolic risk. Lambert Academic Publishing GmbH, Moscow, p 300

Berezin AE (2016a) Biomarkers for cardiovascular risk in patients with diabetes. Heart. doi:10.1136/heartjnl-2016-310197. pii: heartjnl-2016-310197

Berezin AE (2016b) Diabetes mellitus related biomarker: the predictive role of growth-differentiation factor-15. Diabetes Metab Syndr 10(1 Suppl 1):S154–S157

Berezin AE (2016c) Is rationale to decrease serum osteoprotegerin and fetuin-A in type 2 diabetes mellitus patients? Diabetes Metab Syndr 10(3):169–170

Berezin AE, Kremzer AA (2015) Impaired phenotype of circulating endothelial microparticles in chronic heart failure patients: relevance to body mass index. Diabetes Metab Syndr 9(4):230–236

Berezin AE, Kremzer AA, Samura TA, Martovitskaya YV (2014) Circulating endothelial-derived apoptotic microparticles in the patients with ischemic symptomatic chronic heart failure: relevance of pro-inflammatory activation and outcomes. Int Cardiovasc Res J 8(3):116–123

Berezin AE, Kremzer AA, Berezina TA, Martovitskaya YV, Gronenko EA (2016) Data regarding association between serum osteoprotegerin level, numerous of circulating endothelial-derived and mononuclear-derived progenitor cells in patients with metabolic syndrome. Data Brief 8:717–722

Bonnet N (2017) Bone-derived factors: a new gateway to regulate glycemia. Calcif Tissue Int 100(2):174–183

Bozkurt B, Aguilar D, Deswal A, Dunbar SB, Francis GS, Horwich T et al (2016) Contributory risk and management of comorbidities of hypertension, obesity, diabetes mellitus, hyperlipidemia, and metabolic syndrome in chronic heart failure: a scientific statement from the American Heart Association. Circulation 134:e535. [Epub ahead of print]

Brutsaert EF, Biggs ML, Delaney JA, Djoussé L, Gottdiener JS, Ix JH et al (2016) Longitudinal assessment of N-terminal pro-B-type natriuretic peptide and risk of diabetes in older adults: the cardiovascular health study. Metabolism 65(10):1489–1497

Carlsson AC, Sundström J, Carrero JJ, Gustafsson S, Stenemo M, Larsson A et al (2017) Use of a proximity extension assay proteomics chip to discover new biomarkers associated with albuminuria. Eur J Prev Cardiol 24(4):340–348

Cavusoglu E, Marmur JD, Chhabra S, Hojjati MR, Yanamadala S, Chopra V et al (2015) Elevated baseline plasma phospholipid protein (PLTP) levels are an independent predictor of long-term all-cause mortality in patients with diabetes mellitus and known or suspected coronary artery disease. Atherosclerosis 239(2):503–508

Charo IF, Taubman MB (2004) Chemokines in the pathogenesis of vascular disease. Circ Res 95:858–866

Coburn E, Frishman W (2014) Comprehensive review of the prognostic value of galectin-3 in heart failure. Cardiol Rev 22(4):171–175

Coué M, Moro C (2015) Natriuretic peptide control of energy balance and glucose homeostasis. Biochimie. doi:10.1016/j.biochi.2015.05.017. pii: S0300—9084(15)00163-7; [Epub ahead of print]

Creemers EE, Pinto YM (2011) Molecular mechanisms that control interstitial fibrosis in the pressure-overloaded heart. Cardiovasc Res 89(2):265–272

Dąbrowski FA, Jarmużek P, Gondek A, Cudnoch-Jędrzejewska A, Bomba-Opoń D, Wielgoś M (2016) First and third trimester serum concentrations of adropin and copeptin in gestational diabetes mellitus and normal pregnancy. Ginekol Pol 87(9):629–634

Danesh J, Wheeler JG, Hirschfield GM, Eda S, Eiriksdottir G, Rumley A et al (2004) C-reactive protein and other circulating markers of inflammation in the prediction of coronary heart disease. N Engl J Med 350:1387–1397

de Boer RA, Voors AA, Muntendam P, van Gilst WH, van Veldhuisen DJ (2009) Galectin-3: a novel mediator of heart failure development and progression. Eur J Heart Fail 11(9):811–817

Di Pino A, Urbano F, Piro S, Purrello F, Rabuazzo AM (2016) Update on pre-diabetes: focus on diagnostic criteria and cardiovascular risk. World J Diabetes 7(18):423–432

Dixon BS (2016) Is change in albuminuria a surrogate marker for cardiovascular and renal outcomes in type 1 diabetes? Clin J Am Soc Nephrol 11:1921. pii: CJN.09540916. [Epub ahead of print]

Dominguez-Rodriguez A, Abreu-Gonzalez P, Avanzas P (2014) Usefulness of growth differentiation factor-15 levels to predict diabetic cardiomyopathy in asymptomatic patients with type 2 diabetes mellitus. Am J Cardiol 114(6):890–894

Duffy JY, Hameed AB (2015) Cardiovascular disease screening. Semin Perinatol. doi:10.1053/j.semperi.2015.05.004. [Epub ahead of print]

Dumic J, Dabelic S, Flögel M (2006) Galectin-3: an open-ended story. Biochim Biophys Acta 1760(4):616–635

Enhorning S, Wang TJ, Nilsson PM (2010) Plasma copeptin and the risk of diabetes mellitus. Circulation 121:2102–2108

Enhörning S, Hedblad B, Nilsson PM, Engström G, Melander O (2015) Copeptin is an independent predictor of diabetic heart disease and death. Am Heart J 169(4):549–56.e1

Fairlie WD, Moore AG, Bauskin AR, Russell PK, Zhangn HR, Breit SN (1999) MIC-1 is a novel TGF-β superfamily cytokine associated with macrophage activation. J Leukoc Biol 65(1):2–5

Frangogiannis NG (2012) Matricellular proteins in cardiac adaptation and disease. Physiol Rev 92:635–688

Fu S, Ping P, Luo L, Ye P (2016) Deep analyses of the associations of a series of biomarkers with insulin resistance, metabolic syndrome, and diabetes risk in nondiabetic middle-aged and elderly individuals: results from a Chinese community-based study. Clin Interv Aging 11:1531–1538

Goetze JP, Zois NE (2016) Cardiac natriuretic peptides in plasma: from prediction to precision medicine. Lancet Diabetes Endocrinol 4(10):803–805
Gómez-Ambrosi J, Catalán V, Ramírez B, Rodríguez A, Colina I, Silva C et al (2007) Plasma osteopontin levels and expression in adipose tissue are increased in obesity. J Clin Endocrinol Metab 92:3719–3727
Grabner A, Amaral AP, Schramm K, Singh S, Sloan A, Yanucil C et al (2015) Activation of cardiac fibroblast growth factor receptor 4 causes left ventricular hypertrophy. Cell Metab. doi:10.1016/j.cmet.2015.09.002. [Epub ahead of print]
Gupta DK, Wang TJ (2015) Natriuretic peptides and cardiometabolic health. Circ J 79(8):1647–1655
Hitsumoto T, Shirai K (2015) Factors affecting high-sensitivity cardiac troponin T elevation in Japanese metabolic syndrome patients. Diabetes Metab Syndr Obes 9:157–162
Hughes MF, Appelbaum S, Havulinna AS, Jagodzinski A, Zeller T, Kee F, FINRISK and BiomarCaRE investigators et al (2014) ST2 may not be a useful predictor for incident cardiovascular events, heart failure and mortality. Heart 100(21):1715–1721
Imai Y, Dobrian AD, Weaver JR, Butcher MJ, Cole BK, Galkina EV et al (2013) Interaction between cytokines and inflammatory cells in islet dysfunction, insulin resistance and vascular disease. Diabetes Obes Metab 15(Suppl 3):117–129
Inoue N, Sawamura T (2007) Lectin-like oxidized LDL receptor-1 as extracellular chaperone receptor: its versatile functions and human diseases. Methods 43:218–222
Iwanaga Y, Nishi I, Furuichi S, Noguchi T, Sase K, Kihara Y et al (2006) B-type natriuretic peptide strongly reflects diastolic wall stress in patients with chronic heart failure: comparison between systolic and diastolic heart failure. J Am Coll Cardiol 47:742–748
Jagadapillai R, Rane MJ, Lin X, Roberts AM, Hoyle GW, Cai L, Gozal E (2016) Diabetic microvascular disease and pulmonary fibrosis: the contribution of platelets and systemic inflammation. Int J Mol Sci 17(11.) pii: E:1853
Johnson JL (2014) Emerging regulators of vascular smooth muscle cell function in the development and progression of atherosclerosis. Cardiovasc Res 103(4):452–460
Kharroubi AT, Darwish HM (2015) Diabetes mellitus: the epidemic of the century. World J Diabetes 6(6):850–867
Kong P, Cavalera M, Frangogiannis NG (2014) The role of thrombospondin (TSP)-1 in obesity and diabetes. Adipocytes 3(1):81–84
Krintus M, Kozinski M, Kubica J, Sypniewska G (2014) Critical appraisal of inflammatory markers in cardiovascular risk stratification. Crit Rev Clin Lab Sci 51(5):263–279
Kruzliak P, Berezin A, Kremzer A, Samura T, Benacka R, Mozos I et al (2016) Global longitudinal strain and strain rate in type two diabetes patients with chronic heart failure: relevance to osteoprotegerin. Folia Med (Plovdiv) 58(3):164–173
Li Y, Tong X, Rumala C, Clemons K, Wang S (2011) Thrombospondin-1 deficiency reduces obesity-associated inflammation and improves insulin sensitivity in a diet-induced obese mouse model. PLoS One 6:e26656
Li H, Gao F, Xue Y, Qian Y (2014) Value of plasma growth differentiation factor-15 in diagnosis and evaluation of type 2 diabetic nephropathy. Nan Fang Yi Ke Da Xue Bao 34(3):387–390
Li T, Ni L, Liu X, Wang Z, Liu C (2016a) High glucose induces the expression of osteopontin in blood vessels in vitro and in vivo. Biochem Biophys Res Commun 480(2):201–207
Li T, Liu X, Ni L, Wang Z, Wang W, Shi T, Liu X, Liu C (2016b) Perivascular adipose tissue alleviates inflammatory factors and stenosis in diabetic blood vessels. Biochem Biophys Res Commun 480(2):147–152
Llauradó G, Megia A, Cano A, Giménez-Palop O, Simón I, González-Sastre M et al (2015) FGF-23/vitamin D Axis in type 1 diabetes: the potential role of mineral metabolism in arterial stiffness. PLoS One 10(10):e0140222
Looker HC, Colombo M, Agakov F, Zeller T, Groop L, Thorand B, SUMMIT Investigators et al (2015) Protein biomarkers for the prediction of cardiovascular disease in type 2 diabetes. Diabetologia 58(6):1363–1371
Lubrano V, Balzan S (2015) Consolidated and emerging inflammatory markers in coronary artery disease. World J Exp Med 5(1):21–32

Mannarino E, Pirro M (2008) Endothelial injury and repair: a novel theory for atherosclerosis. Angiology 59:69S–72S

Marian AJ, Nambi V (2004) Biomarkers of cardiac disease. Expert Rev Mol Diagn 4:805–820

Martín-Timón I, Sevillano-Collantes C, Segura-Galindo A, del Cañizo-Gómez FJ (2014) Type 2 diabetes and cardiovascular disease: have all risk factors the same strength? World J Diabetes 5(4):444–470

Mause SF, Weber C (2010) Microparticles: protagonists of a novel communication network for intercellular information exchange. Circ Res 107:1047–1057

McAloon CJ, O'Hare P, Osman F, Randeva HS (2016) The interplay between heart failure, metabolism and body composition. Br J Hosp Med (Lond) 77(6):362–364

McCullough PA, Olobatoke A, Vanhecke TE (2011) Galectin-3: a novel blood test for the evaluation and management of patients with heart failure. Rev Cardiovasc Med 12(4):200–210

McEvoy JW, Lazo M, Chen Y, Shen L, Nambi V, Hoogeveen RC et al (2015) Patterns and determinants of temporal change in high-sensitivity cardiac troponin-T: the Atherosclerosis Risk in Communities Cohort study. Int J Cardiol 187:651–657

McKie PM, Rodeheffer RJ, Cataliotti A, Martin FL, Urban LH, Mahoney DW et al (2006) Amino-terminal pro-B-type natriuretic peptide and B-type natriuretic peptide: biomarkers for mortality in a large community-based cohort free of heart failure. Hypertension 47:874–880

McMurray JJ, Adamopoulos S, Anker SD, Auricchio A, Bohm M, Dickstein K et al (2012) ESC guidelines for the diagnosis and treatment of acute and chronic heart failure: the Task Force for the Diagnosis and Treatment of Acute and Chronic Heart Failure 2012 of the European Society of Cardiology. Developed in collaboration with the Heart Failure Association (HFA) of the ESC. Eur J Heart Fail 14:803–869

Mellbin LG, Anselmino M, Rydén L (2010) Diabetes, prediabetes and cardiovascular risk. Eur J Cardiovasc Prev Rehabil 17(supplement 1):S9–S14

Morgenthaler NG (2010) Copeptin: a biomarker of cardiovascular and renal function. Congest Heart Fail 16(Suppl. 1):S37–S44

Morgenthaler NG, Struck J, Alonso C (2006) Assay for the measurement of copeptin, a stable peptide derived from the precursor of vasopressin. Clin Chem 52:112–119

Moro C (2016) Targeting cardiac natriuretic peptides in the therapy of diabetes and obesity. Expert Opin Ther Targets 20(12):1445–1452

Myers GL, Christenson RH, Cushman M, Ballantyne CM, Cooper GR, Pfeiffer CM et al (2009) National Academy of Clinical Biochemistry Laboratory Medicine Practice guidelines: emerging biomarkers for primary prevention of cardiovascular disease. Clin Chem 55:378–384

Natriuretic Peptides Studies Collaboration (2016) Natriuretic peptides and integrated risk assessment for cardiovascular disease: an individual-participant-data meta-analysis. Lancet Diabetes Endocrinol 4(10):840–849

Nayor M, Wang N, Larson MG, Vasan RS, Levy D, Ho JE (2015) Circulating galectin-3 is associated with cardiometabolic disease in the community. J Am Heart Assoc 5(1). doi:10.1161/JAHA.115.002347

Nozaki T, Sugiyama S, Sugamura K, Ohba K, Matsuzawa Y, Konishi M et al (2010) Prognostic value of endothelial microparticles in patients with heart failure. Eur J Heart Fail 12(11):1223–1228

Otake H, Shite J, Shinke T, Watanabe S, Tanino Y, Ogasawara D et al (2008) Relation between plasma adiponectin, high-sensitivity C-reactive protein, and coronary plaque components in patients with acute coronary syndrome. Am J Cardiol 101:1–7

Ozturk D, Celik O, Satilmis S, Aslan S, Erturk M, Cakmak HA et al (2015) Association between serum galectin-3 levels and coronary atherosclerosis and plaque burden/structure in patients with type 2 diabetes mellitus. Coron Artery Dis 26(5):396–401

Panteghini M (2004) Role and importance of biochemical markers in clinical cardiology. Eur Heart J 25:1187–1196

Papapanagiotou A, Siasos G, Kassi E, Gargalionis AN, Papavassiliou AG (2015) Novel inflammatory markers in hyperlipidemia: clinical implications. Curr Med Chem 22(23):2727–2743

Pareek M, Nielsen ML, Leósdóttir M, Nilsson PM, Olsen MH (2015) Baseline cardiac troponin t levels are elevated in subjects with untreated diabetes mellitus: a cross-sectional study. J Hypertens 33(Suppl 1):e54–e55

Pavo N, Wurm R, Neuhold S, Adlbrecht C, Vila G, Strunk G et al (2016) GDF-15 is associated with cancer incidence in patients with type 2 diabetes. Clin Chem 62:1612. pii: clinchem.2016.257212; [Epub ahead of print]

Pepys MB, Hirschfield GM (2003) C-reactive protein: a critical update. J Clin Invest 111:1805–1812

Pirro M, Schillaci G, Bagaglia F, Menecali C, Paltriccia R, Mannarino MR et al (2008) Microparticles derived from endothelial progenitor cells in patients at different cardiovascular risk. Atherosclerosis 197(2):757–767

Pugliese G, Iacobini C, Ricci C, Blasetti Fantauzzi C, Menini S (2014) Galectin-3 in diabetic patients. Clin Chem Lab Med 52(10):1413–1423

Puri R, Nissen SE, Shao M, Elshazly MB, Kataoka Y, Kapadia SR et al (2016) Non-HDL cholesterol and triglycerides: implications for coronary atheroma progression and clinical events. Arterioscler Thromb Vasc Biol 36(11):2220–2228

Raffield LM, Hsu FC, Cox AJ, Carr JJ, Freedman BI, Bowden DW (2015) Predictors of all-cause and cardiovascular disease mortality in type 2 diabetes: diabetes heart study. Diabetol Metab Syndr 7:58

Ramos HR, Birkenfeld AL, de Bold AJ (2015) Interacting disciplines: cardiac natriuretic peptides and obesity: perspectives from an endocrinologist and a cardiologist. Endocr Connect 4(3):R25–R36

Resl M, Clodi M, Vila G, Luger A, Neuhold S, Wurm R et al (2016) Targeted multiple biomarker approach in predicting cardiovascular events in patients with diabetes. Heart 102:1963. doi:10.1136/heartjnl-2015-308949. [Epub ahead of print]

Reynolds AN, Duncan A, Kruimer D, Venn BJ (2016) Glycated albumin is associated with body mass index in euglycemic adults but is not predictive of postprandial blood glucose response. J Clin Lab Anal. doi:10.1002/jcla.22085. [Epub ahead of print]

Rohatgi A, Patel P, Das SR, Ayers CR, Khera A, Martinez-Rumayor A et al (2012) Association of growth differentiation factor-15 with coronary atherosclerosis and mortality in a young, multiethnic population: observations from the Dallas Heart Study. Clin Chem 58(1):172–182

Sanchez OA, Duprez DA, Bahrami H, Peralta CA, Daniels LB, Lima JA et al (2015) Changes in N-terminal pro-B-type natriuretic peptide and incidence of diabetes: the Multi-Ethnic Study of Atherosclerosis (MESA). Diabete Metab 41(5):378–386. doi:10.1016/j.diabet.2015.04.005

Saunders JT, Nambi V, de Lemos JA, Chambless LE, Virani SS, Boerwinkle E et al (2011) Cardiac troponin T measured by a highly sensitive assay predicts coronary heart disease, heart failure, and mortality in the Atherosclerosis Risk in Communities Study. Circulation 123(13):1367–1376

Schiel R, Perenthaler TJ, Steveling A, Stein G (2016) Plasma copeptin in children and adolescents with type 1 diabetes mellitus in comparison to healthy controls. Diabetes Res Clin Pract 118:156–161

Scholte AJ, Schuijf JD, Kharagjitsingh AV, Dibbets-Schneider P, Stokkel MP, van der Wall EE et al (2009) Prevalence and predictors of an abnormal stress myocardial perfusion study in asymptomatic patients with type 2 diabetes mellitus. Eur J Nucl Med Mol Imaging 36:567–575

Schreier M, Schwartze JT, Landgraf K, Scheuermann K, Erbs S, Herberth G et al (2016) Osteopontin is BMI-independently related to early endothelial dysfunction in children. J Clin Endocrinol Metab 101(11):4161–4169

Scirica BM, Bhatt DL, Braunwald E, Raz I, Cavender MA, Im K et al (2016) Prognostic implications of biomarker assessments in patients with type 2 diabetes at high cardiovascular risk: a secondary analysis of a randomized clinical trial. JAMA Cardiol 1(9):989–998

Shah RV, Januzzi JL Jr (2014) Soluble ST2 and galectin-3 in heart failure. Clin Lab Med 34(1):87–97

Sörensen BM, Houben AJ, Berendschot TT, Schouten JS, Kroon AA, van der Kallen CJ et al (2016) Prediabetes and type 2 diabetes are associated with generalized microvascular dysfunction: the Maastricht study. Circulation 134(18):1339–1352

Stoiser B, Mortl D, Hulsmann M (2006) Copeptin, a fragment of the vasopressin precursor, as a novel predictor of outcome in heart failure. Eur J Clin Investig 36:771–778

Talat M, Sherief L, El-Saadany H, Ahmed A, Saleh RM, Sakr M (2016) The role of osteopontin in the pathogenesis and complication of type 1 diabetes mellitus in children and adolescents. J Clin Res Pediatr Endocrinol 8(4):399–404

Torremocha F, Hadjadj S, Carrie F, Rosenberg T, Herpin D, Marechaud R (2001) Prediction of major coronary events by coronary risk profile and silent myocardial ischaemia: prospective follow-up study of primary prevention in 72 diabetic patients. Diabete Metab 27:49–57

Unsicker K, Spittau B, Krieglstein K (2013) The multiple facets of the TGF-β family cytokine growth/differentiation factor-15/macrophage inhibitory cytokine-1. Cytokine Growth Factor Rev 24(4):373–384

Velho G, El Boustany R, Lefèvre G, Mohammedi K, Fumeron F, Potier L et al (2016) Plasma copeptin, kidney outcomes, ischemic heart disease, and all-cause mortality in people with long-standing type 1 diabetes. Diabetes Care 39:2288. [Epub ahead of print]

Wang CB, Zong M, Lu SQ, Tian Z (2016) Plasma copeptin and functional outcome in patients with ischemic stroke and type 2 diabetes. J Diabetes Complicat 30(8):1532–1536

Wannamethee SG, Shaper AG, Whincup PH, Lennon L, Sattar N (2011) Impact of diabetes on cardiovascular disease risk and all-cause mortality in older men: influence of age at onset, diabetes duration, and established and novel risk factors. Arch Int Med 171:404–410

Wannamethee SG, Welsh P, Lennon L, Papacosta O, Whincup PH, Sattar N (2016) Copeptin and the risk of incident stroke, CHD and cardiovascular mortality in older men with and without diabetes: the British regional heart study. Diabetologia 59(9):1904–1912

Wende AR (2015) Post-translational modifications of the cardiac proteome in diabetes and heart failure. Proteomics Clin Appl 10(1):25–38

Wohlfahrt P, Melenovsky V, Kotrc M, Benes J, Jabor A, Franekova J et al (2015) Association of fibroblast growth factor-23 levels and angiotensin-converting enzyme inhibition in chronic systolic heart failure. JACC Heart Fail 3(10):829–839

Yancy CW, Jessup M, Bozkurt B, Butler J, Casey DE Jr, Drazner MH et al (2013) ACCF/AHA guideline for the management of heart failure: a report of the American College of Cardiology Foundation/American Heart Association Task Force on practice guidelines. Circulation 128:e240–e327

Yang ZK, Shen Y, Shen WF, Pu LJ, Meng H, Zhang RY et al (2015) Elevated glycated albumin and reduced endogenous secretory receptor for advanced glycation endproducts levels in serum predict major adverse cardio-cerebral events in patients with type 2 diabetes and stable coronary artery disease. Int J Cardiol 197:241–247

Yen ST, Tan AK, Feisul MI (2016) Awareness and prevalence of diabetes, hypertension, and hypercholesterolemia in Malaysia. J Diabetes. doi:10.1111/1753-0407.12502

Zaccardi F, Khan H, Laukkanen JA (2014) Diabetes mellitus and risk of sudden cardiac death: a systematic review and meta-analysis. Int J Cardiol 177:535–537

Zannad F, Cannon CP, Cushman WC, Bakris GL, Menon V, Perez AT et al (2015) Heart failure and mortality outcomes in patients with type 2 diabetes taking alogliptin versus placebo in EXAMINE: a multicentre, randomised, double-blind trial. Lancet 385(9982):2067–2076

Zeng Q, Dong SY, Wang ML, Li JM, Ren CL, Gao CQ (2016) Obesity and novel cardiovascular markers in a population without diabetes and cardiovascular disease in China. Prev Med 91:62–69

Zhu FX, Wu HL, Tu KS, Chen JX, Zhang M, Shi C (2016) Serum levels of copeptin are associated with type 2 diabetes and diabetic complications in Chinese population. J Diabetes Complicat 30(8):1566–1570

Part II
Kidney Regeneration

Chapter 7
Role of Endothelial Progenitor Cells in Kidney Repair

Jing Zhao and Andrew M.L. Lever

7.1 Introduction

Acute kidney injury (AKI) is a frequent and serious condition associated with a high mortality (Patschan and Muller 2015; Basile et al. 2012). Despite improvements in mortality rates over the last 20 years, the prognosis has not substantially altered (Kribben et al. 2003; Srisawat and Kellum 2011). Currently there are no effective drugs or other therapeutic interventions available. Although AKI is a clinical syndrome with a multitude of causes, they share the same pathological characteristics of tubular cell dysfunction/damage, post-ischemic microvasculopathy, and inflammation (Kang et al. 2002; Wright et al. 2001). There are many different causes for kidney injury, but once renal damage reaches a certain threshold, progression becomes consistent, irreversible, and largely independent of the initial injury and results in chronic kidney disease (CKD) (Nangaku 2004). Recent research suggests that acute kidney injury and chronic kidney disease are part of an integrated clinical syndrome, AKI is a cause of CKD, and preexisting CKD predisposes to AKI (Chawla and Kimmel 2012). Despite significant advances in understanding of the cellular and molecular mechanisms of AKI, the new knowledge has not yet been translated into therapeutic advances. There is still a compelling need for novel and effective approaches to treatment of AKI with the aim of preventing CKD.

The kidney has a remarkable capacity to regenerate and restore its structure and function after acute ischemic and/or toxic injury (Duffield et al. 2005). Injured tubular epithelial cells can be restored within a few days following ischemic injury (Witzgall et al. 1994). Even after prolonged unilateral ureteric obstruction involving

Jing Zhao and Andrew M.L. Lever are supported by the Cambridge Biomedical Research Centre. Their work has been supported by the British Heart Foundation.

J. Zhao (✉) • A.M.L. Lever, MD (✉)
Department of Medicine, Addenbrooke's Hospital, University of Cambridge,
Hills Road, Cambridge CB2 0QQ, UK
e-mail: jg268@medschl.cam.ac.uk; amll1@medschl.cam.ac.uk

P.V. Pham (ed.), *Pancreas, Kidney and Skin Regeneration*, Stem Cells in Clinical Applications, DOI 10.1007/978-3-319-55687-1_7

extensive inflammatory change, tubular necrosis, and apoptosis, the renal cortex can substantially remodel (Cochrane et al. 2005). Diabetic patients with CKD can show reversion of fibrotic lesions in their kidneys 10 years after receiving a pancreas transplant (Fioretto et al. 2006; Fioretto and Mauer 2012). However such self-regeneration is often delayed or inadequate and fails to halt the progressive damage of acute and subsequently chronic kidney disease. The imbalance between renal damage and its renal regeneration contributes to the progressive loss of renal function and development of kidney failure. With no therapy able to protect the kidney or reverse injury in chronic kidney failure, patients often eventually need dialysis or a kidney transplant to stay alive. Advances in the field of regenerative medicine however may provide an alternative solution to enhance endogenous self-repair and prevent progression in chronic kidney failure.

7.2 The Critical Role of Endothelial Cells in Kidney Diseases and Kidney Regeneration

Previously kidney disease pathology was mainly linked with tubular epithelial cell injury; kidney diseases are characterized by loss of renal tubular epithelial cells (Myers and Moran 1986; Weinberg 1991). More recently the important role of the microvascular endothelium in progressive renal disease has been recognized (Kang et al. 2002). It is widely accepted now that vascular endothelium is both a source of and a target for kidney injury (Bonventre and Zuk 2004), often resulting in a vicious cycle that drives progression of kidney injury. Almost all experimental kidney disease models show secondary endothelial injury (Hohenstein et al. 2008). In the kidney, there are two kinds of capillary network, glomerular and peritubular. Respectively they are designed to filter the blood to maintain homeostasis and to ensure tubular interstitial oxygen and nutrient supply (Molema and Aird 2012; Rabelink et al. 2007). The glomerular endothelium plays an essential role in regulating glomerular permeability (Molema and Aird 2012); the progressive loss of peritubular capillaries is closely linked with progression of renal disease and the associated renal scarring and fibrosis (Kang et al. 2001, 2002; Masuda et al. 2001). In addition, deterioration of the renal microvasculature accompanies common systemic diseases affecting the cardiovascular system such as diabetes, hypertension, and atherosclerosis, either as cause or consequence (Chade 2013); they in turn contribute further to kidney disease.

7.2.1 Microvascular Endothelial Cells in Acute Kidney Injury

The most frequent causes for AKI are ischemia, hypoxia, or toxic agents (Patschan and Muller 2015). Ischemic AKI is considered a spectrum of injury which involves renal hemodynamics, tubular and endothelial cell injury, and inflammatory processes, extending from less severe reversible forms to more advanced injury leading to chronic kidney disease (Basile et al. 2012). Maintaining the integrity of the endothelial monolayer is critical since not only does it act as a barrier between the blood and subendothelial matrix proteins it also controls inflammatory cell infiltration and thrombus formation while modulating vascular tone and controlling vascular smooth muscle proliferation (Gimbrone et al. 2000). Altered renal vascular function, especially at the microvasculature level, plays an important role in the initiation and extension phase of AKI and has a detrimental impact on renal function (Molitoris and Sutton 2004; Devarajan 2006; Bonventre and Yang 2011). Glomerular endothelial cell dysfunction precedes podocyte injury in adriamycin-induced nephropathy and initiates the development and progression of glomerulopathy (Sun et al. 2013). Using minimally invasive intravital microscopy, Brodsky et al. demonstrated directly that there was a no-reflow phenomenon in peritubular capillaries in AKI, and invoked endothelial cell dysfunction is the primary cause for this and for the tubular epithelial cell injury seen in AKI (Brodsky et al. 2002a). Systemic or intrarenal administration of fully differentiated endothelial cells into post-ischemic kidneys has produced significant functional protection (Brodsky et al. 2002b), and using surrogate cells expressing endothelial NO synthase achieved similar results in experimental AKI (Brodsky et al. 2002b) hinting at a therapeutic role for endothelial cells in progressive renal disease. However, endothelial cells possess limited regenerative capacity (Caplan and Schwartz 1973; Dimmeler and Zeiher 2004). There is therefore a growing interest in the potential use of endothelial progenitor cells for renal vascular repair.

7.2.2 Microvascular Endothelial Cells in Chronic Kidney Injury

Chronic kidney disease (CKD) develops regardless of etiology when adaptive responses and repair of acute kidney injury fail. Renal fibrosis is the main characteristic of CKD and is the primary cause for CKD progressing to renal failure. The majority of patients with CKD have one or more of type 2 diabetes, hypertension or chronic glomerulonephritis, or other severe comorbidities (Kurokawa et al. 2002). In patients with CKD, ongoing endothelial damage in the capillary system of the renal medulla and accompanying vascular rarefaction are thought to be central drivers of progressive damage. Endothelial dysfunction and atherosclerosis as well as cardiovascular complications are almost universal in CKD (Kurokawa et al. 2002). Endothelial dysfunction constitutes the pathophysiological basis of CKD (Satoh

2012). Persistent damage to endothelial cells not only causes obstruction of the microvasculature but also inflammation by upregulation of adhesion molecules. The loss of the microvasculature directly correlates with the development of glomerular and tubulointerstitial scarring (Kang et al. 2002). Using VEGF to stimulate angiogenesis and/or capillary repair stabilizes renal function and slows progress of fibrosis. This benefit occurs independently of effects on blood pressure or proteinuria, suggesting that angiogenic agents have potential for slowing the progression of renal disease (Kang et al. 2002).

The excessive accumulation of extracellular matrix proteins (ECM) is also an important pathological feature of renal fibrosis (Farris and Colvin 2012). Previously this was thought to be associated with activated fibroblasts in the kidney originating from local, intrinsic renal fibroblasts and epithelial-mesenchymal transition (EMT). The latter is when epithelial cells acquire this phenotype triggered by pathological mediators such as transforming growth factors-β (TGF-β), MCP-1, and matrix metalloproteinases during the inflammatory reaction associated with kidney injury (Mamuya and Duncan 2012; Hills and Squires 2010; Liu 2010). Recent studies have shown that chronic stress and inflammatory factors can also promote endothelial cells to undergo a similar process of endothelial-mesenchymal transition (EndoMT) which contributes to interstitial fibrosis and calcification (Medici et al. 2010; Rieder et al. 2011; Zeisberg et al. 2007).

7.2.3 Endothelial Cell Integrity Is Critical for Kidney Regeneration

Tubular epithelium possesses a remarkable potential to regenerate. This repair capacity is generated primarily by the intrinsic capacity of surviving epithelial cells to dedifferentiate and proliferate (Duffield et al. 2005; Witzgall et al. 1994; Cochrane et al. 2005; Chouchani et al. 2014). More recently, the renal papilla has been recognized as a stem cell niche, and it has been proposed that these cells may play a role in repair as well (Oliver et al. 2004). However this repair often does not occur in a timely fashion, partly due to impairment of peritubular capillary function. These vessels supply oxygen and nutrients essential for epithelial cell regeneration (Miya et al. 2011). Reduction of the microcirculation caused by endothelial cell dysfunction and activation may jeopardize tubular regenerative repair. Microvascular rarefaction is paralleled by significant renal parenchymal fibrosis (Rabelink et al. 2007).

7.3 Definition and Characteristics of Endothelial Progenitor Cells

Asahara and colleagues first identified and named "endothelial progenitor cells" (EPCs) in 1997, after purifying, from adult circulation, CD34+ hematopoietic progenitor cells that express a variety of cell surface markers similar to those expressed by vascular endothelial cells. They showed that these cells could differentiate ex vivo to an endothelial phenotype, capable of incorporating into neovasculature at sites of ischemia and participating in new vessel formation and vascular repair (Asahara et al. 1997). This study challenged the traditional concept of angiogenesis since previously differentiation of mesodermal cells to angioblasts and subsequently to endothelial cells was thought only to occur during embryonic development. Since then these cells have been intensively studied, characterizing them and defining their roles in the repair of damaged vascular endothelium and exploring their potential clinical applications (Rafii and Lyden 2003; Hristov and Weber 2004; Urbich and Dimmeler 2004).

EPCs are a heterogeneous cell population; they can be isolated from the bone marrow, peripheral blood, umbilical cord blood, and spleen (Marsboom and Janssens 2008). EPCs are precursor cells of mesodermal origin and express CD34, CD133, and the vascular endothelial growth factor receptor (VEGFR-2) (Urbich and Dimmeler 2004), cell surface markers shared by hematopoietic stem cell populations. There is a lack of consensus on a definitive phenotype and specific cell surface marker expression, which has caused controversy and confusion about their role in vascular repair. Currently EPCs are defined ultimately by their morphology and function after culture in EC-specific media (Timmermans et al. 2009; Ingram et al. 2005a).

Two distinct populations of putative EPCs are identified in culture: CD14+/CD34+/KDR+, referred to as early-outgrowth EPC, and CD14-/CD34+/KDR+, referred to as late-outgrowth EPC (Prater et al. 2007). Early-outgrowth EPCs (EO-EPCs) are isolated by collecting low-density mononuclear cells (MNC) and plating them in dishes coated with fibronectin in EC-specific medium (Hill et al. 2003; Ito et al. 1999). They can be isolated in culture in relatively large numbers within 7 days. Phenotypically the cells are small, with a rounded or spindle-shaped morphology (Asahara et al. 1997; Hill et al. 2003; Dimmeler and Zeiher 2000; Dimmeler et al. 2001). They express several endothelial markers as well as markers for monocytic lineages such as CD14 and CD45 (Asahara et al. 1997) and secrete angiogenic growth factors (Rehman et al. 2003). Early-outgrowth EPCs are short lived (<2 weeks), have a low proliferative potential, do not differentiate to EC, and cannot form vascular networks (Ziegelhoeffer et al. 2004; Hur et al. 2004). They may restore endothelial function and enhance angiogenesis after tissue ischemia via a paracrine effect from secreted angiogenic growth factors (Rehman et al. 2003). Since numerous blood cell lineages also express integrin receptors specific for fibronectin and attach to plates coated with fibronectin, isolates of early-outgrowth EPC from human peripheral blood are generally a heterogeneous population of hematopoietic cells including monocyte-derived immune cells (Rehman et al. 2003; Zhang et al. 2006, Asakage et al. 2006).

Late-outgrowth EPCs (LO-EPCs) are isolated, based on their colony-forming ability, by plating low-density MNC on dishes coated with collagen I or fibronectin in vitro (Hur et al. 2004; Ingram et al. 2004). LO-EPC colonies are obtained after 2–3 weeks of culture in EC-specific medium. Although both LO-EPC and EO-EPC are generated from the same starting population of MNC, LO-EPCs are derive exclusively from the CD14- fraction while EO-EPCs arise from a CD14+ subpopulation. LO-EPCs are a homogenous cell population with colonies that display a cobblestone appearance. LO-EPC possess a high proliferative potential (Lin et al. 2000; Ingram et al. 2005a, b), differentiate into vascular endothelial cells (Ingram et al. 2004; Lin et al. 2000), and possess de novo vessel-forming ability in vitro and in vivo (Ingram et al. 2004; Zhao et al. 2014; Yoon et al. 2005; Yoder et al. 2007). However LO-EPCs are extremely scarce in the circulation (less than 10 cells/ml) (Prater et al. 2007; Yoder et al. 2007; Angelos et al. 2010). To use them therapeutically, LO-EPC would need to be expanded ex vivo to high concentrations before being delivered back into the circulation. EO-EPC and LO-EPC have a synergistic effect in promoting neovascularization compared to each cell type alone (Yoon et al. 2005; Gulati et al. 2003, 2004).

The term “circulating” EPC is often used in publications since the paper of Asahara et al. in 1997 (Asahara et al. 1997). Circulating EPCs are obtained from peripheral blood, and their definition mainly relies on identification of a particular pattern of cell surface antigen expression, i.e., a CD34+ cell population coexpressing CD133 and KDR. They display a spindle-shaped morphology in culture (Asahara et al. 1997). Circulating EPCs are highly enriched in hematopoietic progenitor activity but do not give rise to endothelial colonies in vitro; therefore their morphology and functional behavior overlap with EO-EPC. Since there are so few LO-EPC in the circulation, the term “circulating EPC” excludes LO-EPC unless specifically indicated.

7.4 Mobilization of EPC

EPCs originating from hematopoietic progenitor cells reside in the bone marrow (BM) compartment. The interaction of progenitor cells with stromal cells in the BM helps to maintain the EPC in an undifferentiated and quiescent state (Heissig et al. 2002). The number of circulating EPC is relatively small under normal conditions. In response to specific signals or injurious stimuli, EPCs within this BM niche proliferate and move to the systemic circulation and migrate to the injury site.

7.4.1 EPC Mobilization from Bone Morrow

Ischemia is one of the most potent signals leading to mobilization of EPC from the BM. EPCs mobilized to ischemic sites were demonstrated both clinically and experimentally in myocardial ischemia, myocardial infarction, and ischemic stroke (Takahashi et al. 1999; Patschan et al. 2007; Adams et al. 2004: Yip et al. 2008; Shintani et al. 2001). Ischemia after acute kidney injury rapidly mobilizes EPC, and transplantation of EPC provided partial renoprotection (Kale et al. 2003; Patschan et al. 2006). Injury-induced EPC mobilization is mediated by generation of hypoxia-inducible factor-1 regulated release of vascular endothelial growth factor (VEGF), stromal cell-derived factor-1 (SDF-1), erythropoietin (EPO) as well as placental growth factor (PlGF), and granulocyte- and granulocyte-macrophage colony-stimulating factors (G-CSF and GM-CSF) in the injury site (Gill et al. 2001; Ceradini et al. 2004; Bahlmann et al. 2003, 2004; Hattori et al. 2002). Expression of corresponding cognate receptors on EPC allows them to be recognized, recruited, and retained in the ischemic tissues. Increasing the circulating level of VEGF using either expression vector plasmids or recombinant protein enhanced the levels of circulating EPCs in experimental models and in clinical pilot trials (Kalka et al. 2000; Moore et al. 2001). The chemokine SDF-1 and its receptor CXCR4 are believed to play a role in the recruitment and retention of circulating EPC to ischemic tissues (Lapidot et al. 2005; Askari et al. 2003; Kollet et al. 2003). SDF-1 is constitutively expressed but its levels are rapidly upregulated by a wide range of injurious stimuli such as inflammatory mediators, changes in the extracellular matrix, altered mechanical forces, and hypoxia (Ceradini et al. 2004). Renal SDF-1 signals mobilized CXCR4-positive EPC to home to the kidney after ischemic injury (Togel et al. 2005). Activated platelets secrete high levels of SDF-1 so that aggregated platelets on exposed extracellular matrix proteins in injured vessels provide an additional signal for mobilization and recruitment of circulating EPC to a damaged area (Ceradini et al. 2004). The picture is not completely clear however since, using the atherosclerotic renal artery stenosis (ARAS) experimental model, Chade et al. showed only a minor contributory role of the SDF-1/CXCR5 axis in renal homing (Chade et al. 2009). Uric acid, which is overproduced by ischemic tissue, is capable of mobilizing EPC to the ischemic kidney (Patschan et al. 2007). EPO is another important physiological determinant of EPC mobilization and has a similar potency to VEGF in experimental models in vivo (Heeschen et al. 2003) and in peripheral blood of patients with ischemic heart disease (Bahlmann et al. 2004).

GM-CSF and G-CSF are used routinely in bone marrow transplantation. Although they mobilize EPC from BM, they also promote inflammation by increasing circulating leukocytes (Morimoto et al. 1990). The safety of G-CSF application in patients with acute or chronic myocardial infarction was questioned after aggravation of restenosis was observed (Kang et al. 2004). In keeping with this, the use of cyclophosphamide and granulocyte colony-stimulating factor (G-CSF)-induced mobilization of endogenous hematopoietic stem cells (HSC) in the experimental renal ischemia model resulted in a worsening of renal failure, possibly due to its pro-inflammatory effect and the concurrent induced granulocytosis (Togel et al. 2004). Therefore the strategy of using cytokines to mobilize EPC from the bone marrow to achieve "sufficient" numbers in the circulation needs caution.

EPC mobilization can also be achieved by physical exercise (Aicher et al. 2005) and pharmacological stimulation; various drugs such as statins (Walter et al. 2002), glitazones (Sorrentino et al. 2007), or insulin (Humpert et al. 2008) have been shown to stimulate circulating EPC. Some of the drugs used to treat kidney diseases have an effect on EPC mobilization including angiotensin-converting enzyme inhibitors ramipril (Min et al. 2004; Wang et al. 2006) and enalapril (Wang et al. 2006) as well as angiotensin receptor blockers like olmesartan, irbesartan, losartan, or telmisartan (Bahlmann et al. 2010). Interestingly, while short-term statin therapy has a positive effect on circulating EPC numbers, long-term continuous treatment with high amounts of statins inversely correlates with the EPC number (Hristov et al. 2007).

7.4.2 Mechanisms of Mobilization from the Bone Marrow

To release BM EPC from their close interaction with BM stromal cells and/or extracellular matrix within BM, the microenvironment needs to be disrupted; this can be affected by matrix metalloproteinase-9 (MMP-9). Increasing the local concentration of MMP-9 in BM cleaves membrane bound Kit ligand (mKitL) and releases soluble Kit ligand (KitL; also known as stem cell factor) and transfers EPC from the quiescent to the proliferative niche (Heissig et al. 2002). VEGF-, SDF-1-, and PlGF-induced EPC mobilization are dependent on MMP-9 (Heissig et al. 2002; Goligorsky et al. 2009) although currently it is still uncertain whether this applies to G-CSF-induced EPC mobilization (Heissig et al. 2002; Robinson et al. 2003). Interactions between hematopoietic progenitor cells and stromal cells are mediated in part through $\alpha 4\beta 1$ (VLA4)/vascular cell adhesion molecule-1 (VCAM-1). G-CSF-induced EPC mobilization could be mediated by downregulation of the VLA4/VCAM-1 interaction since inducible ablation of VLA4 and conditional ablation of VCAM-1 are both associated with enhancement of G-CSF-induced EPC mobilization (Scott et al. 2003; Papayannopoulou 2004; Hu et al. 2003).

7.4.3 EPC and the Spleen

The spleen is rich in EPC and serves as an important reservoir during EPC trafficking (Zampetaki et al. 2008). The spleen provides a specific microenvironment where sinusoidal endothelial cells capture EPC and secrete factors to maintain and regulate their function (Kiel et al. 2005; Zhao et al. 2010). EPC homing to the spleen is mediated by the SDF-1/CXCR4 axis (Zhao et al. 2010). EPCs isolated from the spleen demonstrated endothelial cell characteristics and are able to perform angiogenic functions in vitro and in vivo (Zhao et al. 2007; Wassmann et al. 2006; Werner et al. 2003, 2005). So far there is no evidence that EPC from this niche can mobilize to the circulation in response to injury (Zampetaki et al. 2008). Patschan et al. showed that the spleen acts as a temporary reservoir for mobilized EPC and that ischemic preconditioning mobilized EPC from the splenic pool to the renal ischemic site (Patschan et al. 2006). Splenectomy increases the number of circulating EPC and the homing of infused cells to an injury site, suggesting that the dynamic interaction between EPC and the spleen microenvironment contributes significantly to the efficiency of EPC homing (Patschan et al. 2006). Modulation of the interaction of circulating EPC with the spleen may provide an attractive strategy for inhibiting EPC sequestration in the spleen and increasing the number of peripheral blood EPC.

7.5 EPC Homing

Injury and pharmacological drugs are able to mobilize EPC and increase EPC concentration in the circulation. However these mechanisms are easily overwhelmed, and therefore supplementation with ex vivo expanded EPC may be required for optimal vascular repair. When cells are infused via the vascular route, homing and engraftment are a prerequisite for EPC to exhibit their angiogenic activity in the target tissue. Understanding the mechanisms of EPC homing to an injury site is essential to improve the efficacy of cell-based therapies and the development of new specific therapeutic strategies.

7.5.1 Circulating EPC or Early-Outgrowth EPC Homing

The majority of previous studies on EPC homing used early-outgrowth EPC. The recruitment of circulating EPC or infused early-outgrowth EPC to injury sites strongly resembles that of an inflammatory response. Circulating EPCs interact with the damaged endothelial monolayer and exposed extracellular matrix proteins in a similar way to leukocytes interacting with activated endothelial cells, and they share some common features such as a coordinated sequence of multistep adhesive events including an initial phase of rolling and final firm adhesion (Chavakis et al. 2008).

Adhesion molecules involved in leukocyte homing are also utilized for early-outgrowth EPC homing to an injury site and influence both EPC rolling and firm adhesion (Chavakis et al. 2008). The initial phase of rolling is mediated by E-selectin and P-selectin (Chavakis et al. 2008; Foubert et al. 2007). Activation of P-selectin has been shown to augment EPC-mediated neovascularization (Werner et al. 2003). Inhibition of P-selectin ligand, PSGL-1, impaired the pro-angiogenic and adhesive effects of EPC (Foubert et al. 2007). The initial phase of rolling is reversible. To stop EPC dissociating from the endothelial surface, the low-affinity rolling interactions must be replaced by high-affinity adhesion. Firm adhesion of EPC to endothelium is mediated by integrins (Gong et al. 2011; Caiado and Dias 2012). Integrins are glycosylated heterodimeric proteins expressed on the cell surface, consisting of non-covalently linked α- and β-subunits. They mediate adhesion of cells to extracellular matrix proteins (cell–matrix adhesion) and to other cells (cell–cell adhesion) (Hynes 2002). Integrins are mandatory for the leukocyte homing to sites of inflammation. Circulating EPCs (early-outgrowth EPC) have a similar integrin expression profile to leukocytes, with high levels of integrin β2, αM, αX, αL, and α4 (Chavakis et al. 2005), which are the receptors for ICAM-1/2 (Caiado and Dias 2012). Upregulation of ICAM-1 in the ischemic muscle was shown to be associated with enhanced EPC recruitment to ischemic limbs (Yoon et al. 2006). Overexpression of integrin-linked kinase (ILK), a hypoxia-responsive gene inducing both ICAM-1 and SDF-1 in endothelial cells, increased recruitment of EPC to ischemic tissue (Lee et al. 2006), confirming an important role of integrin β2. Integrin α4 also showed a crucial role in mobilizing and homing of EPC to ischemic vasculature (Jin et al. 2006; Qin et al. 2006).

Neovascularization requires transmigration. It has been reported that the invasive capacity of EPC was mediated by cathepsin L (Urbich et al. 2005). This protease is highly expressed in EPCs and is essential for matrix degradation and invasion. Cathepsin L knockout mice demonstrated an impaired recovery following hind-limb ischemia, and cathepsin L knockout EPCs neither homed to sites of ischemia nor augmented neovascularization (Urbich et al. 2005). MMP-2 was also found to influence the invasive properties of endothelial progenitors. EPCs from MMP2 -/- mice exhibit reduced ECM degradation and respond poorly to hind-limb ischemia (Cheng et al. 2007).

7.5.2 Late-Outgrowth EPC Homing

Currently the mechanism of late-outgrowth EPC homing to the injury site has not been studied as extensively as that of early-outgrowth EPC. The interaction of LO-EPC with injury sites involves both cell–cell and cell–matrix contact. Infused cells through vascular route must undergo a sequence of distinct steps homing to the injury site before performing vascular repair. Using an in vitro flow system to simulate the physical conditions of blood circulation, our data suggested that for LO-EPC, initial rolling and subsequent stable adherence to EC is not dependent on adhesion molecules, since upregulating the cell surface adhesion molecules E-selectin, ICAM-1, and VCAM-1 in the endothelial cells did not increased the interaction of LO-EPC with EC (Zhao et al. 2016), in contrast to early-outgrowth EPC and leukocytes (Gong et al. 2011; Ley et al. 2011). The integrin expression profile of LO-EPC could be responsible for this. LO-EPCs show low expression of integrins αLβ2, αMβ2, αXβ2, and αDβ2 (Zhao et al. 2014) that are responsible for binding to vascular ligands such as ICAM-1, ICAM-2, and VCAM-1 (Simon and Green 2005; Sahin and Buitenhuis 2012). Flowing LO-EPC readily adhered to and spread at discontinuities in an EC monolayer paracellularly between adjacent cells. Adhesion junction protein VE-cadherin may play a role in this stable adhesion, since blocking VE-cadherin in EC led to a significant decrease in the interaction (Zhao et al. 2016). Using a unilateral ureteral obstruction-induced experimental renal fibrosis model, Yamaguchi et al. showed that the progressive rarefaction of peritubular capillaries was associated with a significant increase in VE-cadherin protein expression (Yamaguchi et al. 2012), suggesting that VE-cadherin upregulation in EC represents a compensatory response to reduce the permeability associated with renal damage. Increased VE-cadherin expression may however also enhance the interaction with LO-EPC. The interaction of LO-EPC with ECM was integrin dependent. Integrins $\alpha5\beta1$, $\alpha V\beta1$, and $\alpha V\beta3$ are responsible for this interaction with ECM making EPC highly adherent to fibronectin and vitronectin but less so to collagen IV, collagen I, and laminin (Zhao et al. 2016; Angelos et al. 2010). The maximum adhesion of EPC to fibronectin was influenced by the perfused cell density and shear stress (Angelos et al. 2010).

7.6 Endothelial Progenitor Cells in Kidney Disease

EPC number and functionality reflect the capacity for endogenous vascular repair. Progressive kidney disease increases the need for tissue repair, but paradoxically increased severity of kidney disease may impair endogenous repair. Understanding the influence of disease-related risk factors and the injury microenvironment on the intrinsic EPC pool or manipulating systemically infused EPC may provide an effective strategy to maximize the endogenous repair and prevent progress of chronic kidney failure.

7.6.1 Influence of Disease-Related Risk Factors on EPC

The number and function of circulating EPC are influenced by physiological and pathological factors. Aging is associated with decreased numbers and impaired function of circulating EPC (Edelberg et al. 2002; Rauscher et al. 2003; Scheubel et al. 2003; Shantsila et al. 2007; Heiss et al. 2005). An inverse correlation exists between circulating EPC numbers and risk factors for cardiovascular disease, such as atherosclerosis, diabetes, stroke, hypercholesterolemia, and hypertension (Choi et al. 2004; Heiss et al. 2005; Urbich and Dimmeler 2005; Schmidt-Lucke et al. 2005; Zampetaki et al. 2008; Xiao et al. 2007; Vasa et al. 2001b). Kidney disease and cardiovascular disorders are often linked together. Cardiovascular disease is frequently associated with CKD (Levey et al. 1998), and chronic renal failure is associated with an increased cardiovascular risk with coronary artery disease being the major cause of morbidity and mortality in these patients (Locatelli et al. 2000, 2004). Thus several factors present in patients with CKD have a negative impact on EPC number and function, such as hypertension, glucose intolerance, dyslipidaemia, inflammation, oxidative stress, low levels of erythropoietin (EPO), as well as uremic toxins (reviewed in (Bahlmann et al. 2010)). Previous studies showed a decreased number and impaired angiogenic function of EPC in patients with chronic renal failure (Choi et al. 2004; Krenning et al. 2009). These changes were already detectable at stage 1 CKD and correlated with progression of renal disease (Krenning et al. 2009). Patients with end-stage renal failure (ESRF) have markedly decreased EPC number and function (Lee et al. 2015). Initiation of hemodialysis or successful kidney transplantation improves EPC number and function (Herbrig et al. 2006). Long-term hemodialysis increases EPC number but impairs function (Herbrig et al. 2004), and dialysis treatment only partially ameliorates EPC dysfunction in patients with CKD (Krenning et al. 2009). EPC function is substantially improved in patients with ESRF after transplantation (Lee et al. 2006), and kidney graft function directly correlates with EPC number (Herbrig et al. 2006; Steiner et al. 2006).

Impaired circulating EPC number and function may hamper physiological vascular repair and contribute to the increased risk for cardiovascular disease observed in CKD and progressive kidney disease. How cardiovascular risk factors influence circulating EPC is not known but they may influence the mobilization of circulating EPC or cause depletion or exhaustion of the endogenous EPC pool because of the continued requirement for repair of damaged endothelia. The former was supported by observations that the initiation of statin therapy increases the levels of circulating EPC (Vasa et al. 2001a; Dimmeler et al. 2001), the latter by the observation that circulating endothelial cells (dislodged from injured vessels) correlated with the progression of CKD (Zhang et al. 2014).

The influence of risk factors on late-outgrowth EPC is controversial. We showed that the number of LO-EPC colonies was significantly higher in patients with ESRF and the angiogenic function of LO-EPC from patients with ESRF was comparable to that of healthy controls, suggesting their potential role in autologous endothelial repair (Zhao et al. 2014). This same discrepancy between LO-EPC and EO-EPC was also shown in patients with coronary artery disease who had reduced numbers of EO-EPC (Kollet et al. 2003), but increased numbers of LO-EPC (Stroncek et al. 2009).

7.6.2 Influence of Injury Microenvironment of EPC

Vascular injury is associated with prolonged local inflammation, a reduced oxygen supply, and increased free radical production. Such a harsh microenvironment may result in alteration of EPC phenotype, hindering their angiogenic function and compromising their therapeutic benefits. There are conflicting reports on the effect of ischemia on EPC. We showed that ex vivo expanded LO-EPC withstand simulated ischemia-reperfusion injury-induced apoptosis and maintain their capacity to proliferate, migrate, and form vascular networks to a much greater degree than mature EC (Zhao et al. 2012). The enhanced resistance to oxidative stress correlated with high endogenous expression and activity of the antioxidant manganese superoxide dismutase (MnSOD) (He et al. 2004; Dernbach et al. 2004). In contrast, Ingram et al. claimed that LO-EPCs derived from both cord blood and adult blood are more sensitive to oxidative stress compared to mature EC (Ingram et al. 2007). The difference in findings could be due to different reagents used to provoke oxidative stress in EPC causing effects on different pathways.

LO-EPC responses to inflammatory signals are similar to that of mature endothelial cells. TNF-α activates LO-EPC generating a pro-inflammatory phenotype by upregulating expression of the surface adhesion molecule ICAM-1 (Duffield et al. 2005; Fioretto and Mauer 2012; Myers and Moran 1986). To enhance the ability of LO-EPC to mediate vascular repair and maintain their angiogenic phenotype and function in the inflammatory environment, we transduced LO-EPC with the anti-apoptotic gene A20. Lentiviral vector-mediated overexpression of A20 in LO-EPC protected them against TNF-α-mediated apoptosis, reducing intracellular adhesion molecule-1 (ICAM-1) expression and inflammatory cytokine secretion (Zhao et al. 2012), suggesting that gene-modified LO-EPCs are functionally more robust than native LO-EPC and may provide a better therapeutic option for repair.

7.7 Endothelial Progenitor Cells in Kidney Repair

Since EPCs were demonstrated to contribute to re-endothelialization (Dimmeler and Zeiher 2000; Dimmeler et al. 2001; Rehman et al. 2003) and neovascularization (Dimmeler et al. 2001; Ziegelhoeffer et al. 2004; Hur et al. 2004), they have been moved into the clinic and tested in patients with acute and chronic ischemic heart disease. It has proven safe to deliver EPC prepared from the bone marrow (Abdel-Latif et al. 2007; Schachinger et al. 2006). The results of the largest phase II/III trial, REPAIR-AMI trial, even suggested they might improve clinical outcome (Schachinger et al. 2006). However, the application of EPC in kidney repair is just beginning; most studies were experimental, and so far relatively there are few data. Replacement of the lost tubular cells is critical for recovery of kidney function. This regenerative capacity manifests itself by regeneration of tubular cells from dedifferentiated cells of tubular origin that have survived the initial ischemic insult and that can divide and are able to redifferentiate into a mature tubular cell phenotype (Witzgall et al. 1994; Humphreys et al. 2008; Kusaba et al. 2014). The contribution of bone marrow-derived hematopoietic stem cells (BMSC) in regeneration of tubular epithelial cells is controversial. Several studies claim a role for adult BMSCs in regeneration of tubules of the injured kidney (Poulsom et al. 2001; Kale et al. 2003; Lin et al. 2003). However one study pointed out that BMSC could indirectly affect tubule regeneration by contribution to repopulation of the injured vascular endothelium (Duffield et al. 2005). The return of blood flow and inhibition of inflammation are dependent upon endothelial cell integrity and are essential for functional renal recovery. EPC-mediated repair of endothelial injury could thus be based on either mobilizing EPC from the bone marrow and/or delivery of EPC after culture in vitro.

7.7.1 EPC Repair of Damaged Microvascular Endothelial Cells

Recent progress on endothelial progenitor cells has opened up new possibilities for enhancing the endogenous capacity for regeneration and repair of damaged endothelial cells in kidney injury. Endothelial damage and dysfunction are characterized by activation of inflammatory genes, cell swelling, and disruption of endothelial junctions and exposed basement membrane (Sutton et al. 2002). Both mobilizing EPC from BM and transplanted EPC have been shown to contribute to glomerular capillary repair and transferred partial renoprotection after acute renal ischemia injury (Patschan et al. 2006). EPCs mobilized into the damaged glomeruli and were involved in capillary repair of damaged glomeruli in habu snake venom (HSV) induced glomerulonephritis (Abe-Yoshio et al. 2008). Intrarenal infusion of autologous early and late outgrowth EPC preserved microvascular architecture and function and decreased remodeling in chronic renal artery stenosis (Chade et al. 2009). Intrarenal arterial injection of EPC reduces endothelial injury and mesangial cell activation by incorporation into the glomerular endothelial lining and production of angiogenic factors in experimental ant-Thy1 glomerulonephritis (Uchimura et al. 2005).

The effect of infused late-outgrowth EPC on renal vascular repair was demonstrated in their role in IgA nephropathy (IgAN), the most common form of glomerular disease throughout the world in which 25 % patients will develop end-stage renal disease (ESRD) within 5–25 years of diagnosis (Guo et al. 2014). Glomerulosclerosis and interstitial fibrosis are irreversible pathological changes during the progression of IgAN to ESRD. So far there is no effective therapeutic treatment to prevent this. The exact cause for it is unclear; microvascular endothelial cell dysfunction and damage may contribute to the upregulation of fibrosis-related genes and accumulation of extracellular matrix (ECM) (Guo et al. 2014; Choi et al. 2000, Ohashi et al. 2000). In experimental IgAN intravenously injected LO-EPCs were detected in glomerular and interstitial areas and resulted in an increased peritubular capillary density, a reduced expansion of extracellular matrix in glomerular, slower renal pathogenesis progression, and improved renal function (Guo et al. 2014).

7.7.2 EPC Contribution to Renal Tubular Epithelium Regeneration

Peritubular capillaries surrounding the tubule act as a source of growth factors, oxygen, and nutrients required for tubular recovery and therefore play an important role in regulation of this process after renal injury (Miya et al. 2011). However, following acute ischemic injury, renal vascular endothelial cells lack a regenerative potential comparable to epithelial cells, and the peritubular vasculature displays decreased capillary density (Basile 2007; Horbelt et al. 2007; Maeshima et al. 2015; Becherucci et al. 2009). Loss of peritubular endothelial integrity is directly associated with impaired tubular regeneration. Recruitment of BM-derived EPC during tubular regeneration is essential for tubular cell regeneration. It reported that renal ischemia mobilizes EPC and induces the accumulation of EPC in the renal medulla and that transplantation of EPC-enriched cells from the medullary parenchyma affords partial renoprotection after renal ischemia (Patschan et al. 2006, Becherucci et al. 2009). EPCs were found to have incorporated in peritubular capillaries in human kidney and bone marrow transplantation. Paracrine mechanisms of therapeutic effect of EPC are also reported. Microvesicles derived from EPC protect the kidney from ischemic acute injury by delivering their RNA content as a microRNA cargo that converts surviving resident renal cells into a more regenerative state (Cantaluppi et al. 2012).

The regenerative process of the kidney after injury was also influenced by the immune system (Rabelink et al. 2007; Maeshima et al. 2015). Peritubular capillary endothelium is involved in this inflammatory process by promoting the accumulation of leukocytes through upregulation of ICAM-1 expression in response to kidney damage. Upregulation of the adhesion molecules and attraction of leukocytes to the site of injury will cause margination and adhesion of leukocytes in the capillary, thus reducing blood flow (Rabelink et al. 2007). EPCs restore peritubular endothelial integrity, and termination of renal inflammation will enable tubular regeneration by inhibiting tubular cell necrosis or by driving a phenotype switch from pro-inflammatory to anti-inflammatory immune cells (Maeshima et al. 2015).

7.8 Conclusion and Perspectives

Kidney disease is a progressive disease involving a complex interplay between renal hemodynamics, tubular and endothelial cell injury, and inflammatory processes. The balance between loss of renal function and self-repair contributes to the disease progression. Whereas the capacity of epithelial regeneration has long been recognized, endothelial cells have a limited repair capacity. Understanding of the role of endothelial cells in kidney diseases and the angiogenic potential of endothelial progenitor cells provides us with a new approach to tackle the kidney disease. Using EPC for repair and maintenance of the glomerular and peritubular capillaries could be a therapeutic target for prevention of progressive kidney disease. Of all the endothelial progenitors so far identified, LO-EPCs are promising candidates for cell therapy that could make a significant contribution in this field.

References

Abdel-Latif A, Bolli R, Tleyjeh IM, Montori VM, Perin EC, Hornung CA, Zuba-Surma EK, Al-Mallah M, Dawn B (2007) Adult bone marrow-derived cells for cardiac repair: a systematic review and meta-analysis. Arch Intern Med 167:989–997

Abe-Yoshio Y, Abe K, Miyazaki M, Furusu A, Nishino T, Harada T, Koji T, Kohno S (2008) Involvement of bone marrow-derived endothelial progenitor cells in glomerular capillary repair in habu snake venom-induced glomerulonephritis. Virchows Arch 453:97–106

Adams V, Lenk K, Linke A, Lenz D, Erbs S, Sandri M, Tarnok A, Gielen S, Emmrich F, Schuler G, Hambrecht R (2004) Increase of circulating endothelial progenitor cells in patients with coronary artery disease after exercise-induced ischemia. Arterioscler Thromb Vasc Biol 24:684–690

Aicher A, Zeiher AM, Dimmeler S (2005) Mobilizing endothelial progenitor cells. Hypertension 45:321–325

Angelos MG, Brown MA, Satterwhite LL, Levering VW, Shaked NT, Truskey GA (2010) Dynamic adhesion of umbilical cord blood endothelial progenitor cells under laminar shear stress. Biophys J 99:3545–3554

Asahara T, Murohara T, Sullivan A, Silver M, van der Zee R, Li T, Witzenbichler B, Schatteman G, Isner JM (1997) Isolation of putative progenitor endothelial cells for angiogenesis. Science 275:964–967

Asakage M, Tsuno NH, Kitayama J, Kawai K, Okaji Y, Yazawa K, Kaisaki S, Osada T, Watanabe T, Takahashi K, Nagawa H (2006) Early-outgrowth of endothelial progenitor cells can function as antigen-presenting cells. Cancer Immunol Immunother 55:708–716

Askari AT, Unzek S, Popovic ZB, Goldman CK, Forudi F, Kiedrowski M, Rovner A, Ellis SG, Thomas JD, Dicorleto PE, Topol EJ, Penn MS (2003) Effect of stromal-cell-derived factor 1 on stem-cell homing and tissue regeneration in ischaemic cardiomyopathy. Lancet 362:697–703

Bahlmann FH, Degroot K, Duckert T, Niemczyk E, Bahlmann E, Boehm SM, Haller H, Fliser D (2003) Endothelial progenitor cell proliferation and differentiation is regulated by erythropoietin. Kidney Int 64:1648–1652

Bahlmann FH, De Groot K, Spandau JM, Landry AL, Hertel B, Duckert T, Boehm SM, Menne J, Haller H, Fliser D (2004) Erythropoietin regulates endothelial progenitor cells. Blood 103:921–926

Bahlmann FH, Speer T, Fliser D (2010) Endothelial progenitor cells in chronic kidney disease. Nephrol Dial Transplant 25:341–346

Basile DP (2007) The endothelial cell in ischemic acute kidney injury: implications for acute and chronic function. Kidney Int 72:151–156

Basile DP, Anderson MD, Sutton TA (2012) Pathophysiology of acute kidney injury. Compr Physiol 2:1303–1353

Becherucci F, Mazzinghi B, Ronconi E, Peired A, Lazzeri E, Sagrinati C, Romagnani P, Lasagni L (2009) The role of endothelial progenitor cells in acute kidney injury. Blood Purif 27:261–270

Bonventre JV, Yang L (2011) Cellular pathophysiology of ischemic acute kidney injury. J Clin Invest 121:4210–4221

Bonventre JV, Zuk A (2004) Ischemic acute renal failure: an inflammatory disease? Kidney Int 66:480–485

Brodsky SV, Yamamoto T, Tada T, Kim B, Chen J, Kajiya F, Goligorsky MS (2002a) Endothelial dysfunction in ischemic acute renal failure: rescue by transplanted endothelial cells. Am J Physiol Ren Physiol 282:F1140–F1149

Brodsky SV, Yamamoto T, Tada T, Kim B, Chen J, Kajiya F, Goligorsky MS (2002b) Endothelial dysfunction in ischemic acute renal failure: rescue by transplanted endothelial cells. Am J Physiol Renal Physiol 282:F1140–F1149

Caiado F, Dias S (2012) Endothelial progenitor cells and integrins: adhesive needs. Fibrogenesis Tissue Repair 5:4

Cantaluppi V, Gatti S, Medica D, Figliolini F, Bruno S, Deregibus MC, Sordi A, Biancone L, Tetta C, Camussi G (2012) Microvesicles derived from endothelial progenitor cells protect the kidney from ischemia-reperfusion injury by microRNA-dependent reprogramming of resident renal cells. Kidney Int 82:412–427

Caplan BA, Schwartz CJ (1973) Increased endothelial cell turnover in areas of in vivo Evans blue uptake in the pig aorta. Atherosclerosis 17:401–417

Ceradini DJ, Kulkarni AR, Callaghan MJ, Tepper OM, Bastidas N, Kleinman ME, Capla JM, Galiano RD, Levine JP, Gurtner GC (2004) Progenitor cell trafficking is regulated by hypoxic gradients through HIF-1 induction of SDF-1. Nat Med 10:858–864

Chade AR (2013) Renal vascular structure and rarefaction. Compr Physiol 3:817–831

Chade AR, Zhu X, Lavi R, Krier JD, Pislaru S, Simari RD, Napoli C, Lerman A, Lerman LO (2009) Endothelial progenitor cells restore renal function in chronic experimental renovascular disease. Circulation 119:547–557

Chavakis E, Aicher A, Heeschen C, Sasaki K-I, Kaiser R, El Makhfi N, Urbich C, Peters T, Scharffetter-kochanek K, Zeiher AM, Chavakis T, Dimmeler S (2005) Role of beta 2-integrins for homing and neovascularization capacity of endothelial progenitor cells. J Exp Med 201:63–72

Chavakis E, Urbich C, Dimmeler S (2008) Homing and engraftment of progenitor cells: a prerequisite for cell therapy. J Mol Cell Cardiol 45:514–522

Chawla LS, Kimmel PL (2012) Acute kidney injury and chronic kidney disease: an integrated clinical syndrome. Kidney Int 82:516–524

Cheng XW, Kuzuya M, Nakamura K, Maeda K, Tsuzuki M, Kim W, Sasaki T, Liu Z, Inoue N, Kondo T, Jin H, Numaguchi Y, Okumura K, Yokota M, Iguchi A, Murohara T (2007) Mechanisms underlying the impairment of ischemia-induced neovascularization in matrix metalloproteinase 2-deficient mice. Circ Res 100:904–913

Choi YJ, Chakraborty S, Nguyen V, Nguyen C, Kim BK, Shim SI, Suki WN, Truong LD (2000) Peritubular capillary loss is associated with chronic tubulointerstitial injury in human kidney: altered expression of vascular endothelial growth factor. Hum Pathol 31:1491–1497

Choi J-H, Kim KL, Huh W, Kim B, Byun J, Suh W, Sung J, Jeon E-S, Oh H-Y, Kim D-K (2004) Decreased number and impaired angiogenic function of endothelial progenitor cells in patients with chronic renal failure. Arterioscler Thromb Vasc Biol 24:1246–1252

Chouchani ET, Pell VR, Gaude E, Aksentijevic D, Sundier SY, Robb EL, Logan A, Nadtochiy SM, Ord ENJ, Smith AC, Eyassu F, Shirley R, Hu C-H, Dare AJ, James AM, Rogatti S, Hartley RC, Eaton S, Costa ASH, Brookes PS, Davidson SM, Duchen MR, Saeb-Parsy K, Shattock MJ, Robinson AJ, Work LM, Frezza C, Krieg T, Murphy MP (2014) Ischaemic accumulation of succinate controls reperfusion injury through mitochondrial ROS. Nature 515:431–435

Cochrane AL, Kett MM, Samuel CS, Campanale NV, Anderson WP, Hume DA, Little MH, Bertram JF, Ricardo SD (2005) Renal structural and functional repair in a mouse model of reversal of ureteral obstruction. J Am Soc Nephrol 16:3623–3630

Dernbach E, Urbich C, Brandes RP, Hofmann WK, Zeiher AM, Dimmeler S (2004) Antioxidative stress-associated genes in circulating progenitor cells: evidence for enhanced resistance against oxidative stress. Blood 104:3591–3597

Devarajan P (2006) Update on mechanisms of ischemic acute kidney injury. J Am Soc Nephrol 17:1503–1520

Dimmeler S, Zeiher AM (2000) Endothelial cell apoptosis in angiogenesis and vessel regression. Circ Res 87:434–439

Dimmeler S, Zeiher AM (2004) Vascular repair by circulating endothelial progenitor cells: the missing link in atherosclerosis? J Mol Med (Berl) 82:671–677

Dimmeler S, Aicher A, Vasa M, Mildner-Rihm C, Adler K, Tiemann M, Rutten H, Fichtlscherer S, Martin H, Zeiher AM (2001) HMG-CoA reductase inhibitors (statins) increase endothelial progenitor cells via the PI 3-kinase/Akt pathway. J Clin Invest 108:391–397

Duffield JS, Park KM, Hsiao LL, Kelley VR, Scadden DT, Ichimura T, Bonventre JV (2005) Restoration of tubular epithelial cells during repair of the postischemic kidney occurs independently of bone marrow-derived stem cells. J Clin Invest 115:1743–1755

Edelberg JM, Tang L, Hattori K, Lyden D, Rafii S (2002) Young adult bone marrow-derived endothelial precursor cells restore aging-impaired cardiac angiogenic function. Circ Res 90:E89–E93

Farris AB, Colvin RB (2012) Renal interstitial fibrosis: mechanisms and evaluation. Curr Opin Nephrol Hypertens 21:289–300

Fioretto P, Mauer M (2012) Reversal of diabetic nephropathy: lessons from pancreas transplantation. J Nephrol 25:13–18

Fioretto P, Sutherland DE, Najafian B, Mauer M (2006) Remodeling of renal interstitial and tubular lesions in pancreas transplant recipients. Kidney Int 69:907–912

Foubert P, Silvestre JS, Souttou B, Barateau V, Martin C, Ebrahimian TG, Lere-Dean C, Contreres JO, Sulpice E, Levy BI, Plouet J, Tobelem G, le Ricousse-Roussanne S (2007) PSGL-1-mediated activation of EphB4 increases the proangiogenic potential of endothelial progenitor cells. J Clin Invest 117:1527–1537

Gill M, Dias S, Hattori K, Rivera ML, Hicklin D, Witte L, Girardi L, Yurt R, Himel H, Rafii S (2001) Vascular trauma induces rapid but transient mobilization of VEGFR2(+)AC133(+) endothelial precursor cells. Circ Res 88:167–174

Gimbrone MA Jr, Topper JN, Nagel T, Anderson KR, Garcia-Cardena G (2000) Endothelial dysfunction, hemodynamic forces, and atherogenesis. Ann N Y Acad Sci 902:230–239. discussion 239-40

Goligorsky MS, Kuo MC, Patschan D, Verhaar MC (2009) Review article: endothelial progenitor cells in renal disease. Nephrology (Carlton) 14:291–297

Gong X-B, Li Y-Q, Gao Q-C, Cheng B-B, Shen B-R, Yan Z-Q, Jiang Z-L (2011) Adhesion behavior of endothelial progenitor cells to endothelial cells in simple shear flow. Acta Mech Sinica 27:1071–1080

Gulati R, Jevremovic D, Peterson TE, Witt TA, Kleppe LS, Mueske CS, Lerman A, Vile RG, Simari RD (2003) Autologous culture-modified mononuclear cells confer vascular protection after arterial injury. Circulation 108:1520–1526

Gulati R, Jevremovic D, Witt TA, Kleppe LS, Vile RG, Lerman A, Simari RD (2004) Modulation of the vascular response to injury by autologous blood-derived outgrowth endothelial cells. Am J Physiol Heart Circ Physiol 287:H512–H517

Guo W, Feng J-M, Yao L, Sun L, Zhu G-Q (2014) Transplantation of endothelial progenitor cells in treating rats with IgA nephropathy. BMC Nephrol 15:110–110

Hattori K, Heissig B, Wu Y, Dias S, Tejada R, Ferris B, Hicklin DJ, Zhu Z, Bohlen P, Witte L, Hendrikx J, Hackett NR, Crystal RG, Moore MA, Werb Z, Lyden D, Rafii S (2002) Placental growth factor reconstitutes hematopoiesis by recruiting VEGFR1(+) stem cells from bone-marrow microenvironment. Nat Med 8:841–849

He T, Peterson TE, Holmuhamedov EL, Terzic A, Caplice NM, Oberley LW, Katusic ZS (2004) Human endothelial progenitor cells tolerate oxidative stress due to intrinsically high expression of manganese superoxide dismutase. Arterioscler Thromb Vasc Biol 24:2021–2027

Heeschen C, Aicher A, Lehmann R, Fichtlscherer S, Vasa M, Urbich C, Mildner-Rihm C, Martin H, Zeiher AM, Dimmeler S (2003) Erythropoietin is a potent physiologic stimulus for endothelial progenitor cell mobilization. Blood 102:1340–1346

Heiss C, Keymel S, Niesler U, Ziemann J, Kelm M, Kalka C (2005) Impaired progenitor cell activity in age-related endothelial dysfunction. J Am Coll Cardiol 45:1441–1448

Heissig B, Hattori K, Dias S, Friedrich M, Ferris B, Hackett NR, Crystal RG, Besmer P, Lyden D, Moore MA, Werb Z, Rafii S (2002) Recruitment of stem and progenitor cells from the bone marrow niche requires MMP-9 mediated release of kit-ligand. Cell 109:625–637

Herbrig K, Pistrosch F, Oelschlaegel U, Wichmann G, Wagner A, Foerster S, Richter S, Gross P, Passauer J (2004) Increased total number but impaired migratory activity and adhesion of endothelial progenitor cells in patients on long-term hemodialysis. Am J Kidney Dis 44:840–849

Herbrig K, Gebler K, Oelschlaegel U, Pistrosch F, Foerster S, Wagner A, Gross P, Passauer J (2006) Kidney transplantation substantially improves endothelial progenitor cell dysfunction in patients with end-stage renal disease. Am J Transplant 6:2922–2928

Hill JM, Zalos G, Halcox JP, Schenke WH, Waclawiw MA, Quyyumi AA, Finkel T (2003) Circulating endothelial progenitor cells, vascular function, and cardiovascular risk. N Engl J Med 348:593–600

Hills CE, Squires PE (2010) TGF-beta1-induced epithelial-to-mesenchymal transition and therapeutic intervention in diabetic nephropathy. Am J Nephrol 31:68–74

Hohenstein B, Braun A, Amann KU, Johnson RJ, HUGO CPM (2008) A murine model of site-specific renal microvascular endothelial injury and thrombotic microangiopathy. Nephrol Dial Transplant 23:1144–1156

Horbelt M, Lee SY, Mang HE, Knipe NL, Sado Y, Kribben A, Sutton TA (2007) Acute and chronic microvascular alterations in a mouse model of ischemic acute kidney injury. Am J Physiol Renal Physiol 293:F688–F695

Hristov M, Weber C (2004) Endothelial progenitor cells: characterization, pathophysiology, and possible clinical relevance. J Cell Mol Med 8:498–508

Hristov M, Fach C, Becker C, Heussen N, Liehn EA, Blindt R, Hanrath P, Weber C (2007) Reduced numbers of circulating endothelial progenitor cells in patients with coronary artery disease associated with long-term statin treatment. Atherosclerosis 192:413–420

Hu Y, Davison F, Zhang Z, Xu Q (2003) Endothelial replacement and angiogenesis in arteriosclerotic lesions of allografts are contributed by circulating progenitor cells. Circulation 108:3122–3127

Humpert PM, Djuric Z, Zeuge U, Oikonomou D, Seregin Y, Laine K, Eckstein V, Nawroth PP, Bierhaus A (2008) Insulin stimulates the clonogenic potential of angiogenic endothelial progenitor cells by IGF-1 receptor-dependent signaling. Mol Med 14:301–308

Humphreys BD, Valerius MT, Kobayashi A, Mugford JW, Soeung S, Duffield JS, McMahon AP, Bonventre JV (2008) Intrinsic epithelial cells repair the kidney after injury. Cell Stem Cell 2:284–291

Hur J, Yoon CH, Kim HS, Choi JH, Kang HJ, Hwang KK, Oh BH, Lee MM, Park YB (2004) Characterization of two types of endothelial progenitor cells and their different contributions to neovasculogenesis. Arterioscler Thromb Vasc Biol 24:288–293

Hynes RO (2002) Integrins: bidirectional, allosteric signaling machines. Cell 110:673–687

Ingram DA, Mead LE, Tanaka H, Meade V, Fenoglio A, Mortell K, Pollok K, Ferkowicz MJ, Gilley D, Yoder MC (2004) Identification of a novel hierarchy of endothelial progenitor cells using human peripheral and umbilical cord blood. Blood 104:2752–2760

Ingram DA, Caplice NM, Yoder MC (2005a) Unresolved questions, changing definitions, and novel paradigms for defining endothelial progenitor cells. Blood 106:1525–1531

Ingram DA, Mead LE, Moore DB, Woodard W, Fenoglio A, Yoder MC (2005b) Vessel wall-derived endothelial cells rapidly proliferate because they contain a complete hierarchy of endothelial progenitor cells. Blood 105:2783–2786

Ingram DA, Krier TR, Mead LE, McGuire C, Prater DN, Bhavsar J, Saadatzadeh MR, Bijangi-Vishehsaraei K, Li F, Yoder MC, Haneline LS (2007) Clonogenic endothelial progenitor cells are sensitive to oxidative stress. Stem Cells 25:297–304

Ito H, Rovira II, Bloom ML, Takeda K, Ferrans VJ, Quyyumi AA, Finkel T (1999) Endothelial progenitor cells as putative targets for angiostatin. Cancer Res 59:5875–5877

Jin H, Aiyer A, Su J, Borgstrom P, Stupack D, Friedlander M, Varner J (2006) A homing mechanism for bone marrow-derived progenitor cell recruitment to the neovasculature. J Clin Invest 116:652–662

Kale S, Karihaloo A, Clark PR, Kashgarian M, Krause DS, Cantley LG (2003) Bone marrow stem cells contribute to repair of the ischemically injured renal tubule. J Clin Invest 112:42–49

Kalka C, Masuda H, Takahashi T, Gordon R, Tepper O, Gravereaux E, Pieczek A, Iwaguro H, Hayashi SI, Isner JM, Asahara T (2000) Vascular endothelial growth factor(165) gene transfer augments circulating endothelial progenitor cells in human subjects. Circ Res 86:1198–1202

Kang DH, Hughes J, Mazzali M, Schreiner GF, Johnson RJ (2001) Impaired angiogenesis in the remnant kidney model: II. Vascular endothelial growth factor administration reduces renal fibrosis and stabilizes renal function. J Am Soc Nephrol 12:1448–1457

Kang D-H, Kanellis J, Hugo C, Truong L, Anderson S, Kerjaschki D, Schreiner GF, Johnson RJ (2002) Role of the microvascular endothelium in progressive renal disease. J Am Soc Nephrol 13:806–816

Kang HJ, Kim HS, Zhang SY, Park KW, Cho HJ, Koo BK, Kim YJ, Soo Lee D, Sohn DW, Han KS, Oh BH, Lee MM, Park YB (2004) Effects of intracoronary infusion of peripheral blood stem-cells mobilised with granulocyte-colony stimulating factor on left ventricular systolic function and restenosis after coronary stenting in myocardial infarction: the MAGIC cell randomised clinical trial. Lancet 363:751–756

Kiel MJ, Yilmaz OH, Iwashita T, Terhorst C, Morrison SJ (2005) SLAM family receptors distinguish hematopoietic stem and progenitor cells and reveal endothelial niches for stem cells. Cell 121:1109–1121

Kollet O, Shivtiel S, Chen YQ, Suriawinata J, Thung SN, Dabeva MD, Kahn J, Spiegel A, Dar A, Samira S, Goichberg P, Kalinkovich A, Arenzana-Seisdedos F, Nagler A, Hardan I, Revel M, Shafritz DA, Lapidot T (2003) HGF, SDF-1, and MMP-9 are involved in stress-induced human CD34+ stem cell recruitment to the liver. J Clin Invest 112:160–169

Krenning G, Dankers PYW, Drouven JW, Waanders F, Franssen CFM, van Luyn MJA, Harmsen MC, Popa ER (2009) Endothelial progenitor cell dysfunction in patients with progressive chronic kidney disease. Am J Physiol Ren Physiol 296:F1314–F1322

Kribben A, Herget-Rosenthal S, Pietruck F, Philipp T (2003) Acute renal failure – an review. Dtsch Med Wochenschr 128:1231–1236

Kurokawa K, Nangaku M, Saito A, Inagi R, Miyata T (2002) Current issues and future perspectives of chronic renal failure. J Am Soc Nephrol 13(Suppl 1):S3–S6

Kusaba T, Lalli M, Kramann R, Kobayashi A, Humphreys BD (2014) Differentiated kidney epithelial cells repair injured proximal tubule. Proc Natl Acad Sci U S A 111:1527–1532

Lapidot T, Dar A, Kollet O (2005) How do stem cells find their way home? Blood 106:1901–1910

Lee SP, Youn SW, Cho HJ, Li L, Kim TY, Yook HS, Chung JW, Hur J, Yoon CH, Park KW, Oh BH, Park YB, Kim HS (2006) Integrin-linked kinase, a hypoxia-responsive molecule, controls post-natal vasculogenesis by recruitment of endothelial progenitor cells to ischemic tissue. Circulation 114:150–159

Lee HJ, Kim W, Kim WS, Woo JS, Kim YG, Moon JY, Lee SH, Ihm CG, Lee TW, Jeong KH (2015) Circulating endothelial progenitor cell levels predict cardiovascular events in end-stage renal disease patients on maintenance hemodialysis. Nephron 130:151–158

Levey AS, Beto JA, Coronado BE, Eknoyan G, Foley RN, Kasiske BL, Klag MJ, Mailloux LU, Manske CL, Meyer KB, Parfrey PS, Pfeffer MA, Wenger NK, Wilson PW, Wright JT Jr (1998) Controlling the epidemic of cardiovascular disease in chronic renal disease: what do we know? What do we need to learn? Where do we go from here? National Kidney Foundation Task Force on Cardiovascular Disease. Am J Kidney Dis 32:853–906

Ley K, Miller YI, Hedrick CC (2011) Monocyte and macrophage dynamics during atherogenesis. Arterioscler Thromb Vasc Biol 31:1506–1516

Lin Y, Weisdorf DJ, Solovey A, Hebbel RP (2000) Origins of circulating endothelial cells and endothelial outgrowth from blood. J Clin Invest 105:71–77

Lin F, Cordes K, Li L, Hood L, Couser WG, Shankland SJ, Igarashi P (2003) Hematopoietic stem cells contribute to the regeneration of renal tubules after renal ischemia-reperfusion injury in mice. J Am Soc Nephrol 14:1188–1199

Liu Y (2010) New insights into epithelial-mesenchymal transition in kidney fibrosis. J Am Soc Nephrol 21:212–222

Locatelli F, Marcelli D, Conte F, D'Amico M, del Vecchio L, Limido A, Malberti F, Spotti D (2000) Cardiovascular disease in chronic renal failure: the challenge continues. Registro Lombardo Dialisi e Trapianto. Nephrol Dial Transplant 15(Suppl 5):69–80

Locatelli F, Covic A, Chazot C, Leunissen K, Luno J, Yaqoob M (2004) Hypertension and cardiovascular risk assessment in dialysis patients. Nephrol Dial Transplant 19:1058–1068

Maeshima A, Takahashi S, Nakasatomi M, Nojima Y (2015) Diverse cell populations involved in regeneration of renal tubular epithelium following acute kidney injury. Stem Cells Int 2015:964849

Mamuya FA, Duncan MK (2012) aV integrins and TGF-beta-induced EMT: a circle of regulation. J Cell Mol Med 16:445–455

Marsboom G, Janssens S (2008) Endothelial progenitor cells: new perspectives and applications in cardiovascular therapies. Expert Rev Cardiovasc Ther 6:687–701

Masuda Y, Shimizu A, Mori T, Ishiwata T, Kitamura H, Ohashi R, Ishizaki M, Asano G, Sugisaki Y, Yamanaka N (2001) Vascular endothelial growth factor enhances glomerular capillary repair and accelerates resolution of experimentally induced glomerulonephritis. Am J Pathol 159:599–608

Medici D, Shore EM, Lounev VY, Kaplan FS, Kalluri R, Olsen BR (2010) Conversion of vascular endothelial cells into multipotent stem-like cells. Nat Med 16:1400–1406

Min TQ, Zhu CJ, Xiang WX, Hui ZJ, Peng SY (2004) Improvement in endothelial progenitor cells from peripheral blood by ramipril therapy in patients with stable coronary artery disease. Cardiovasc Drugs Ther 18:203–209

Miya M, Maeshima A, Mishima K, Sakurai N, Ikeuchi H, Kuroiwa T, Hiromura K, Yokoo H, Nojima Y (2011) Enhancement of in vitro human tubulogenesis by endothelial cell-derived factors: implications for in vivo tubular regeneration after injury. Am J Physiol Ren Physiol 301:F387–F395

Molema G, Aird WC (2012) Vascular heterogeneity in the kidney. Semin Nephrol 32:145–155

Molitoris BA, Sutton TA (2004) Endothelial injury and dysfunction: role in the extension phase of acute renal failure. Kidney Int 66:496–499

Moore MA, Hattori K, Heissig B, Shieh JH, Dias S, Crystal RG, Rafii S (2001) Mobilization of endothelial and hematopoietic stem and progenitor cells by adenovector-mediated elevation of serum levels of SDF-1, VEGF, and angiopoietin-1. Ann N Y Acad Sci 938:36–45. discussion 45-7

Morimoto A, Sakata Y, Watanabe T, Murakami N (1990) Leucocytosis induced in rabbits by intravenous or central injection of granulocyte colony stimulating factor. J Physiol 426:117–126

Myers BD, Moran SM (1986) Hemodynamically mediated acute renal failure. N Engl J Med 314:97–105

Nangaku M (2004) Mechanisms of tubulointerstitial injury in the kidney: final common pathways to end-stage renal failure. Intern Med 43:9–17

Ohashi R, Kitamura H, Yamanaka N (2000) Peritubular capillary injury during the progression of experimental glomerulonephritis in rats. J Am Soc Nephrol 11:47–56

Oliver JA, Maarouf O, Cheema FH, Martens TP, Al-Awqati Q (2004) The renal papilla is a niche for adult kidney stem cells. J Clin Invest 114:795–804

Papayannopoulou T (2004) Current mechanistic scenarios in hematopoietic stem/progenitor cell mobilization. Blood 103:1580–1585

Patschan D, Muller GA (2015) Acute kidney injury. J Inj Violence Res 7:19–26

Patschan D, Krupincza K, Patschan S, Zhang Z, Hamby C, Goligorsky MS (2006) Dynamics of mobilization and homing of endothelial progenitor cells after acute renal ischemia: modulation by ischemic preconditioning. Am J Physiol Ren Physiol 291:F176–F185

Patschan D, Patschan S, Gobe GG, Chintala S, Goligorsky MS (2007) Uric acid heralds ischemic tissue injury to mobilize endothelial progenitor cells. J Am Soc Nephrol 18:1516–1524
Poulsom R, Forbes SJ, Hodivala-Dilke K, Ryan E, Wyles S, Navaratnarasah S, Jeffery R, Hunt T, Alison M, Cook T, Pusey C, Wright NA (2001) Bone marrow contributes to renal parenchymal turnover and regeneration. J Pathol 195:229–235
Prater DN, Case J, Ingram DA, Yoder MC (2007) Working hypothesis to redefine endothelial progenitor cells. Leukemia 21:1141–1149
Qin G, Ii M, Silver M, Wecker A, Bord E, Ma H, Gavin M, Goukassian DA, Yoon YS, Papayannopoulou T, Asahara T, Kearney M, Thorne T, Curry C, Eaton L, Heyd L, Dinesh D, Kishore R, Zhu Y, Losordo DW (2006) Functional disruption of alpha4 integrin mobilizes bone marrow-derived endothelial progenitors and augments ischemic neovascularization. J Exp Med 203:153–163
Rabelink TJ, Wijewickrama DC, de Koning EJ (2007) Peritubular endothelium: the Achilles heel of the kidney? Kidney Int 72:926–930
Rafii S, Lyden D (2003) Therapeutic stem and progenitor cell transplantation for organ vascularization and regeneration. Nat Med 9:702–712
Rauscher FM, Goldschmidt-Clermont PJ, Davis BH, Wang T, Gregg D, Ramaswami P, Pippen AM, Annex BH, Dong C, Taylor DA (2003) Aging, progenitor cell exhaustion, and atherosclerosis. Circulation 108:457–463
Rehman J, Li J, Orschell CM, March KL (2003) Peripheral blood "endothelial progenitor cells" are derived from monocyte/macrophages and secrete angiogenic growth factors. Circulation 107:1164–1169
Rieder F, Kessler SP, West GA, Bhilocha S, de la Motte C, Sadler TM, Gopalan B, Stylianou E, Fiocchi C (2011) Inflammation-induced endothelial-to-mesenchymal transition: a novel mechanism of intestinal fibrosis. Am J Pathol 179:2660–2673
Robinson SN, Pisarev VM, Chavez JM, Singh RK, Talmadge JE (2003) Use of matrix metalloproteinase (MMP)-9 knockout mice demonstrates that MMP-9 activity is not absolutely required for G-CSF or Flt-3 ligand-induced hematopoietic progenitor cell mobilization or engraftment. Stem Cells 21:417–427
Sahin AO, Buitenhuis M (2012) Molecular mechanisms underlying adhesion and migration of hematopoietic stem cells. Cell Adhes Migr 6:39–48
Satoh M (2012) Endothelial dysfunction as an underlying pathophysiological condition of chronic kidney disease. Clin Exp Nephrol 16:518–521
Schachinger V, Erbs S, Elsasser A, Haberbosch W, Hambrecht R, Holschermann H, Yu J, Corti R, Mathey DG, Hamm CW, Suselbeck T, Werner N, Haase J, Neuzner J, Germing A, Mark B, Assmus B, Tonn T, Dimmeler S, Zeiher AM (2006) Improved clinical outcome after intracoronary administration of bone-marrow-derived progenitor cells in acute myocardial infarction: final 1-year results of the REPAIR-AMI trial. Eur Heart J 27:2775–2783
Scheubel RJ, Zorn H, Silber RE, Kuss O, Morawietz H, Holtz J, Simm A (2003) Age-dependent depression in circulating endothelial progenitor cells in patients undergoing coronary artery bypass grafting. J Am Coll Cardiol 42:2073–2080
Schmidt-Lucke C, Rössig L, Fichtlscherer S, Vasa M, Britten M, Kämper U, Dimmeler S, Zeiher AM (2005) Reduced number of circulating endothelial progenitor cells predicts future cardiovascular events. Circulation 111:2981–2987
Scott LM, Priestley GV, Papayannopoulou T (2003) Deletion of alpha4 integrins from adult hematopoietic cells reveals roles in homeostasis, regeneration, and homing. Mol Cell Biol 23:9349–9360
Shantsila E, Watson T, Lip GY (2007) Endothelial progenitor cells in cardiovascular disorders. J Am Coll Cardiol 49:741–752
Shintani S, Murohara T, Ikeda H, Ueno T, Honma T, Katoh A, Sasaki K-I, Shimada T, Oike Y, Imaizumi T (2001) Mobilization of endothelial progenitor cells in patients with acute myocardial infarction. Circulation 103:2776–2779
Simon SI, Green CE (2005) Molecular mechanics and dynamics of leukocyte recruitment during inflammation. Annu Rev Biomed Eng 7:151–185

Sorrentino SA, Bahlmann FH, Besler C, Muller M, Schulz S, Kirchhoff N, Doerries C, Horvath T, Limbourg A, Limbourg F, Fliser D, Haller H, Drexler H, Landmesser U (2007) Oxidant stress impairs in vivo reendothelialization capacity of endothelial progenitor cells from patients with type 2 diabetes mellitus: restoration by the peroxisome proliferator-activated receptor-gamma agonist rosiglitazone. Circulation 116:163–173

Srisawat N, Kellum JA (2011) Acute kidney injury: definition, epidemiology, and outcome. Curr Opin Crit Care 17:548–555

Steiner S, Winkelmayer WC, Kleinert J, Grisar J, Seidinger D, Kopp CW, Watschinger B, Minar E, Horl WH, Fodinger M, Sunder-Plassmann G (2006) Endothelial progenitor cells in kidney transplant recipients. Transplantation 81:599–606

Stroncek JD, Grant BS, Brown MA, Povsic TJ, Truskey GA, Reichert WM (2009) Comparison of endothelial cell phenotypic markers of late-outgrowth endothelial progenitor cells isolated from patients with coronary artery disease and healthy volunteers. Tissue Eng A 15:3473–3486

Sun YB, Qu X, Zhang X, Caruana G, Bertram JF, Li J (2013) Glomerular endothelial cell injury and damage precedes that of podocytes in adriamycin-induced nephropathy. PLoS One 8:e55027

Sutton TA, Fisher CJ, Molitoris BA (2002) Microvascular endothelial injury and dysfunction during ischemic acute renal failure. Kidney Int 62:1539–1549

Takahashi T, Kalka C, Masuda H, Chen D, Silver M, Kearney M, Magner M, Isner JM, Asahara T (1999) Ischemia- and cytokine-induced mobilization of bone marrow-derived endothelial progenitor cells for neovascularization. Nat Med 5:434–438

Timmermans F, Plum J, Yoder MC, Ingram DA, Vandekerckhove B, Case J (2009) Endothelial progenitor cells: identity defined? J Cell Mol Med 13:87–102

Togel F, Isaac J, Westenfelder C (2004) Hematopoietic stem cell mobilization-associated granulocytosis severely worsens acute renal failure. J Am Soc Nephrol 15:1261–1267

Togel F, Isaac J, Hu Z, Weiss K, Westenfelder C (2005) Renal SDF-1 signals mobilization and homing of CXCR4-positive cells to the kidney after ischemic injury. Kidney Int 67:1772–1784

Uchimura H, Marumo T, Takase O, Kawachi H, Shimizu F, Hayashi M, Saruta T, Hishikawa K, Fujita T (2005) Intrarenal injection of bone marrow-derived angiogenic cells reduces endothelial injury and mesangial cell activation in experimental glomerulonephritis. J Am Soc Nephrol 16:997–1004

Urbich C, Dimmeler S (2004) Endothelial progenitor cells: characterization and role in vascular biology. Circ Res 95:343–353

Urbich C, Dimmeler S (2005) Risk factors for coronary artery disease, circulating endothelial progenitor cells, and the role of HMG-CoA reductase inhibitors. Kidney Int 67:1672–1676

Urbich C, Heeschen C, Aicher A, Sasaki K, Bruhl T, Farhadi MR, Vajkoczy P, Hofmann WK, Peters C, Pennacchio LA, Abolmaali ND, Chavakis E, Reinheckel T, Zeiher AM, Dimmeler S (2005) Cathepsin L is required for endothelial progenitor cell-induced neovascularization. Nat Med 11:206–213

Vasa M, Fichtlscherer S, Adler K, Aicher A, Martin H, Zeiher AM, Dimmeler S (2001a) Increase in circulating endothelial progenitor cells by statin therapy in patients with stable coronary artery disease. Circulation 103:2885–2890

Vasa M, Fichtlscherer S, Aicher A, Adler K, Urbich C, Martin H, Zeiher AM, Dimmeler S (2001b) Number and migratory activity of circulating endothelial progenitor cells inversely correlate with risk factors for coronary artery disease. Circ Res 89:e1–e7

Walter DH, Rittig K, Bahlmann FH, Kirchmair R, Silver M, Murayama T, Nishimura H, Losordo DW, Asahara T, Isner JM (2002) Statin therapy accelerates reendothelialization: a novel effect involving mobilization and incorporation of bone marrow-derived endothelial progenitor cells. Circulation 105:3017–3024

Wang CH, Verma S, Hsieh IC, Chen YJ, Kuo LT, Yang NI, Wang SY, Wu MY, Hsu CM, Cheng CW, Cherng WJ (2006) Enalapril increases ischemia-induced endothelial progenitor cell mobilization through manipulation of the CD26 system. J Mol Cell Cardiol 41:34–43

Wassmann S, Werner N, Czech T, Nickenig G (2006) Improvement of endothelial function by systemic transfusion of vascular progenitor cells. Circ Res 99:e74–e83
Weinberg JM (1991) The cell biology of ischemic renal injury. Kidney Int 39:476–500
Werner N, Junk S, Laufs U, Link A, Walenta K, Bãhm M, Nickenig G (2003) Intravenous transfusion of endothelial progenitor cells reduces Neointima formation after vascular injury. Circ Res 93:e17–e24
Werner L, Deutsch V, Barshack I, Miller H, Keren G, George J (2005) Transfer of endothelial progenitor cells improves myocardial performance in rats with dilated cardiomyopathy induced following experimental myocarditis. J Mol Cell Cardiol 39:691–697
Witzgall R, Brown D, Schwarz C, Bonventre JV (1994) Localization of proliferating cell nuclear antigen, vimentin, c-Fos, and clusterin in the postischemic kidney. Evidence for a heterogenous genetic response among nephron segments, and a large pool of mitotically active and dedifferentiated cells. J Clin Invest 93:2175–2188
Wright JR, Duggal A, Thomas R, Reeve R, Roberts IS, Kalra PA (2001) Clinicopathological correlation in biopsy-proven atherosclerotic nephropathy: implications for renal functional outcome in atherosclerotic renovascular disease. Nephrol Dial Transplant 16:765–770
Xiao Q, Kiechl S, Patel S, Oberhollenzer F, Weger S, Mayr A, Metzler B, Reindl M, Hu Y, Willeit J, Xu Q (2007) Endothelial progenitor cells, cardiovascular risk factors, cytokine levels and atherosclerosis – results from a large population-based study. PLoS One 2:e975
Yamaguchi I, Tchao BN, Burger ML, Yamada M, Hyodo T, Giampietro C, Eddy AA (2012) Vascular endothelial cadherin modulates renal interstitial fibrosis. Nephron Exp Nephrol 120:e20–e31
Yip H-K, Chang L-T, Chang W-N, Lu C-H, Liou C-W, Lan M-Y, Liu JS, Youssef AA, Chang H-W (2008) Level and value of circulating endothelial progenitor cells in patients after acute ischemic stroke. Stroke 39:69–74
Yoder MC, Mead LE, Prater D, Krier TR, Mroueh KN, Li F, Krasich R, Temm CJ, Prchal JT, Ingram DA (2007) Redefining endothelial progenitor cells via clonal analysis and hematopoietic stem/progenitor cell principals. Blood 109:1801–1809
Yoon CH, Hur J, Park KW, Kim JH, Lee CS, Oh IY, Kim TY, Cho HJ, Kang HJ, Chae IH, Yang HK, Oh BH, Park YB, Kim HS (2005) Synergistic neovascularization by mixed transplantation of early endothelial progenitor cells and late outgrowth endothelial cells: the role of angiogenic cytokines and matrix metalloproteinases. Circulation 112:1618–1627
Yoon CH, Hur J, Oh IY, Park KW, Kim TY, Shin JH, Kim JH, Lee CS, Chung JK, Park YB, Kim HS (2006) Intercellular adhesion molecule-1 is upregulated in ischemic muscle, which mediates trafficking of endothelial progenitor cells. Arterioscler Thromb Vasc Biol 26:1066–1072
Zampetaki A, Kirton JP, Xu Q (2008) Vascular repair by endothelial progenitor cells. Cardiovasc Res 78:413–421
Zeisberg EM, Tarnavski O, Zeisberg M, Dorfman AL, McMullen JR, Gustafsson E, Chandraker A, Yuan X, Pu WT, Roberts AB, Neilson EG, Sayegh MH, Izumo S, Kalluri R (2007) Endothelial-to-mesenchymal transition contributes to cardiac fibrosis. Nat Med 13:952–961
Zhang SJ, Zhang H, Wei YJ, Su WJ, Liao ZK, Hou M, Zhou JY, Hu SS (2006) Adult endothelial progenitor cells from human peripheral blood maintain monocyte/macrophage function throughout in vitro culture. Cell Res 16:577–584
Zhang K, Yin F, Lin L (2014) Circulating endothelial cells and chronic kidney disease. Biomed Res Int 2014:7
Zhao X, Huang L, Yin Y, Fang Y, Zhou Y (2007) Autologous endothelial progenitor cells transplantation promoting endothelial recovery in mice. Transpl Int 20:712–721
Zhao X, Wu N, Huang L (2010) Endothelial progenitor cells and spleen: new insights in regeneration medicine. Cytotherapy 12:7–16
Zhao J, Bolton EM, Ormiston ML, Bradley JA, Morrell NW, Lever AM (2012) Late outgrowth endothelial progenitor cells engineered for improved survival and maintenance of function in transplant-related injury. Transpl Int 25:229–241

Zhao J, Bolton EM, Randle L, Bradley JA, Lever AM (2014) Functional characterization of late outgrowth endothelial progenitor cells in patients with end-stage renal failure. Transpl Int 27:437–451

Zhao J, Mitrofan CG, Appleby S, Morrell NW, Lever AM (2016) Disrupted endothelial cell layer and exposed extracellular matrix protein present sufficient adhesive signals to capture late outgrowth endothelial progenitor cells. Stem Cells Int 2016: Article ID 1406304, 13 pages

Ziegelhoeffer T, Fernandez B, Kostin S, Heil M, Voswinckel R, Helisch A, Schaper W (2004) Bone marrow-derived cells do not incorporate into the adult growing vasculature. Circ Res 94:230–238

Chapter 8
Mesenchymal Stem Cell Transplantation for Kidney Diseases

Phuc Van Pham

Abbreviations

ADPKD	Autosomal dominant polycystic kidney disease
ADSC	Adipose-derived stem cells
ALB	Albumin
ANA	Antinuclear antibodies
BILAG	British Isles Lupus Assessment Group
BM-MSC	Bone marrow-derived mesenchymal stem cells
BPAR	Biopsy-proven acute rejection
BUN	Blood urea nitrogen
CMV	Cytomegalovirus
CNI	Calcineurin inhibitor
DGF	Delayed graft function
DSA	Donor-specific antibody
ELISPOT	Enzyme-linked ImmunoSpot
GFR	Glomerular filtration rate
HLA	Human leukocyte antigen
i.a.	Intra-arterial
i.v.	Intravenous
MDRD	Modification of diet in renal disease
MSC	Mesenchymal stem cells
NAG	N-acetyl-p-D-glucosaminidase enzyme
NGAL	Neutrophil gelatinase-associated lipocalin
RBC	Red blood cells
SAE	Severe adverse effects
sCr	Serum creatinine
SLE	Systemic lupus erythematosus

P.V. Pham, MSc, PhD (✉)
Laboratory of Stem Cell Research and Application, University of Science, Vietnam National University, Ho Chi Minh City, Vietnam
e-mail: pvphuc@hcmuns.edu.vn

P.V. Pham (ed.), *Pancreas, Kidney and Skin Regeneration*, Stem Cells in Clinical Applications, DOI 10.1007/978-3-319-55687-1_8

SLEDAI	Systemic lupus erythematosus disease activity index
UAE	Urinary albumin excretion
UCB-MSC	Umbilical cord blood-derived mesenchymal stem cells
UCMSC	Umbilical cord-derived mesenchymal stem cells

8.1 Mesenchymal Stem Cell Sources

8.1.1 Mesenchymal Stem Cells

Mesenchymal stem cells (MSCs) are predominant stem cells in human body. MSCs were firstly isolated from bone marrow in 1974 by Friedenstein et al. (1974). MSCs and/or cells with similar phenotype have been detected and isolated from various tissue sources, including adipose tissue (Zuk et al. 2001, 2002), peripheral blood (Fernandez et al. 1997; Huss et al. 2000; Purton et al. 1998), umbilical cord blood (Erices et al. 2000; Lee et al. 2004; Mareschi et al. 2001), banked umbilical cord blood (Phuc et al. 2011, 2012), umbilical cord (Kestendjieva et al. 2008; Romanov et al. 2003), umbilical cord membrane (Kita et al. 2010), umbilical cord vein (Covas et al. 2003), Wharton's jelly of the umbilical cord (Hou et al. 2009), placenta (Rylova et al. 2015), amniotic fluid (Peng et al. 2007; Savickiene et al. 2015), amniotic membrane (Pirjali et al. 2013; Shaer et al. 2014), dental pulp (Jo et al. 2007; Pierdomenico et al. 2005), menstrual blood (Gargett and Masuda 2010; Musina et al. 2008), breast milk (Patki et al. 2010; Sani et al. 2015), and urine (Fu et al. 2014; Qin et al. 2014)… (Fig. 8.1).

Since MSCs can be isolated from various tissues and under various conditions, the minimal criteria for the MSC phenotype were established, per suggestion by Dominici et al. (2006). The recommendation was accepted by the International Society of Cellular Therapy (ISCT) (Dominici et al. 2006). According to Dominici et al., there are three minimal criteria for cells to be termed MSCs:

- Firstly, the candidate cells must be plastic adherent when maintained in standard culture conditions.
- Secondly, the candidates must express CD105, CD73, and CD90 and lack expression of CD45, CD34, CD14 or CD11b, CD79alpha or CD19, and HLA-DR surface molecules.
- Thirdly, the candidates must differentiate into osteoblasts, adipocytes, and chondroblasts in vitro.

Although there are a lot of sources of MSCs, there are some important sources used in clinical applications.

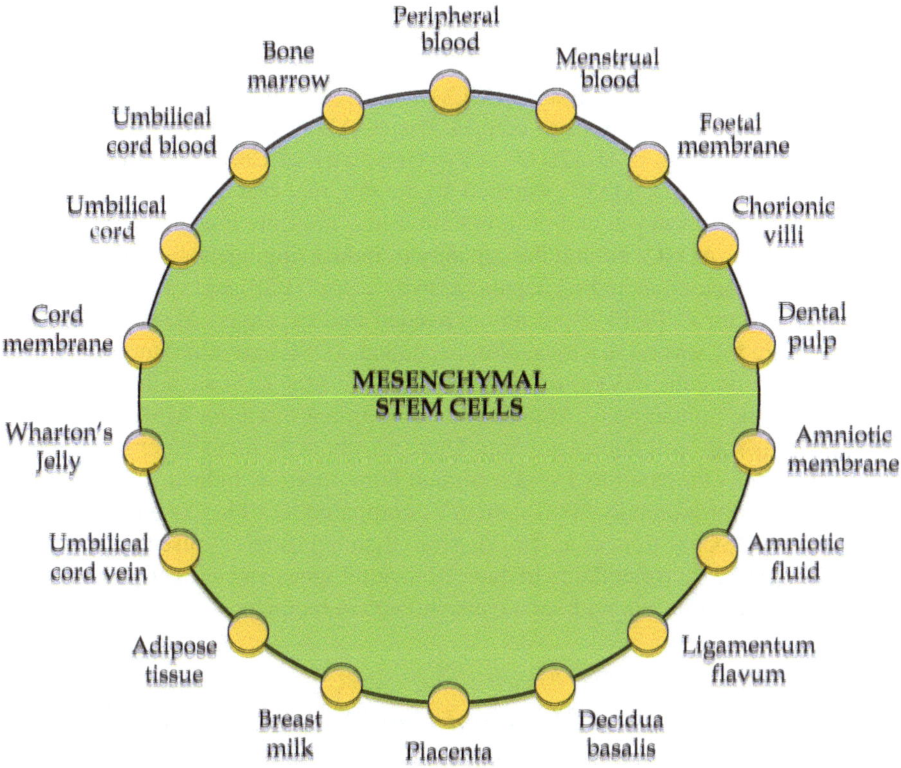

Fig. 8.1 Various sources for MSC isolation in humans. In human, almost all tissues contained MSCs from liquid tissues, such as bone marrow, peripheral blood, milk, etc., to solid tissues such as dental pulp, amniotic membrane, placenta, adipose tissue

8.1.2 Bone Marrow-Derived Mesenchymal Stem Cells

Bone marrow (BM) was the first source of MSCs discovered in 1974 by Friedenstein et al. BM-MSCs were found to be adherent cells with fibroblast-like shape. These cells rapidly proliferate in vitro with differentiation potential into multiple mesoderm cells, including osteoblasts, chondrocytes, and adipocytes. From 10 mL of bone marrow, MSCs can be isolated and cultured to enough cells for transplant in some diseases. To date, MSCs from bone marrow have been clinically used in some diseases, such as graft versus host disease (GVHD), Crohn's, osteoarthritis, heart ischemia, hind-limb ischemia, and liver cirrhosis. MSCs from bone marrow has also been used to produce the first stem cell drug (commercial name: Prochymal) in the world, manufactured by Osiris Therapeutics.

8.1.3 Adipose Tissue-Derived Stem Cells

Adipose tissue-derived stem cells (ADSCs) exhibit MSC phenotypes and are sometimes referred to as MSCs from adipose tissue. ADSCs were first isolated from adipose tissue by Zuk et al. (2002) at the University of California, Los Angeles, USA (Zuk et al. 2002). ADSCs satisfied all the minimal criteria for MSCs, including long-term self-renewal and differentiation into multiple lineages of mesoderm cells. Moreover, ADSCs exhibit the capacity to induce angiogenesis via production of angiogenic factors, including hepatic growth factor (HGF) and vascular endothelial growth factor (VEGF). Both non-expanded and expanded ADSCs have been used in clinic for several diseases. Non-expanded ADSCs are a mixture of stromal vascular fraction cells (SVFs) of which 1–10% are MSCs. Since adipose tissue is abundant in the human body, ADSCs are a convenient tissue for MSCs, especially in autologous stem cell transplantation. There are about 50 times more stem cells in 1 g of fat when compared to 1 g of aspirated bone marrow. Indeed, 1 g of adipose tissue yields approximately 5000 stem cells compared to 100–1000 cells/mL from BM-derived MSCs (Strem et al. 2005). More than ten different diseases have been treated by ADSC transplantation to date. Both autologous and allogeneic transplantation of ADSCs are safe, without adverse effects in recipients.

8.1.4 Umbilical Cord Blood and Umbilical Cord-Derived Mesenchymal Stem Cells

Both umbilical cord blood (UCB) and umbilical cord (UC) are considered as medical wastes after the birthing process. However, MSCs from these sources can be feasibly and conveniently isolated and expanded over a long period of time. In allogeneic transplantation for therapeutic applications, UCB-MSCs and UC-MSCs have been used in clinic since 2010. In some regards, MSCs from UCB and UC can be superior to ADSCs and BM-MSCs (Van Pham et al. 2016).

8.2 Role of Mesenchymal Stem Cell Roles in Kidney and Kidney-Related Diseases

8.2.1 The Homing and Transdifferentiation of MSCs

MSCs can migrate into damaged tissues and differentiate into desired target cells (Fig. 8.2). This homing process is related to the interaction between chemokines (produced by injured tissues) and certain receptors (on MSCs). For example, CD44 expressed on MSCs can interact with hyaluronic acid expressed at the injured kidney (Herrera et al. 2007). Indeed, blockade of CD44 by specific antibodies can

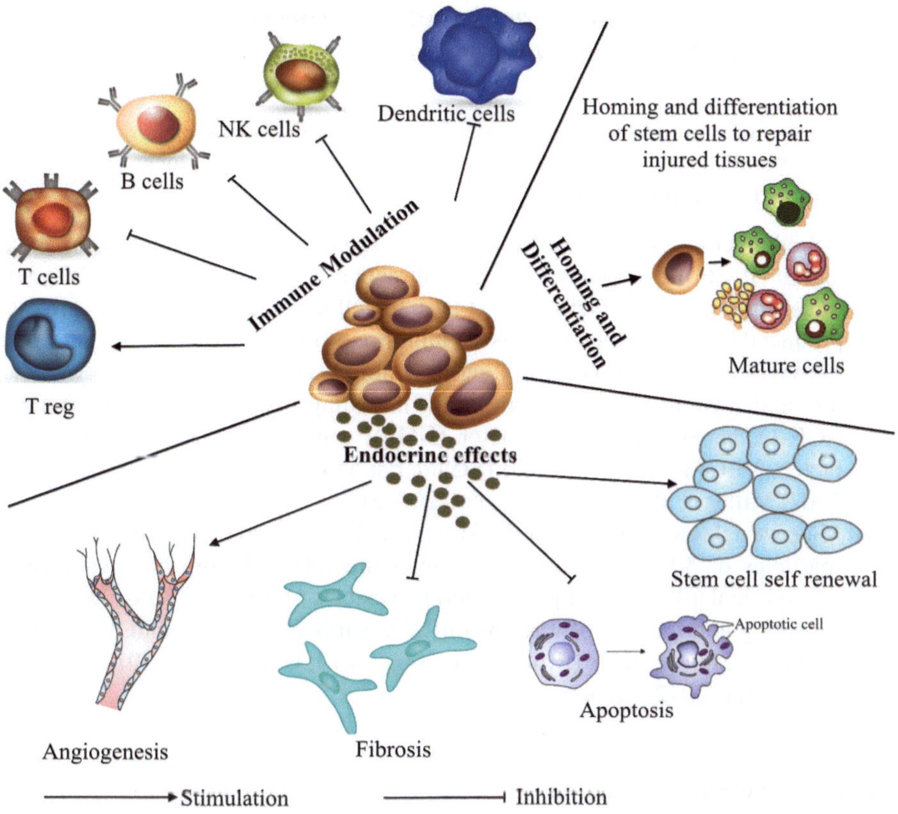

Fig. 8.2 Some mechanisms of MSCs in kidney and kidney related disease treatment. (1) MSCs can modulate the host's immune system; (2) MSCs can home and differentiate into specific cells that replace the injured cells at some tissues; (3) MSCs release a lot of cytokines and growth factors that can inhibit the fibrosis as well as apoptosis at injured tissues; and trigger the self renewal process of stem cells; and stimulate the angiogenesis

block the homing of MSCs into the injured kidney (Herrera et al. 2007). Another mechanism relating to homing is the stromal derived factor-1–CXCR4/CXCR7 axis. Indeed, injured kidney can produce numerous cytokines, chemokines, secreted proteins, and growth factors, especially chemokine stromal cell-derived factor-1 (SDF-1, also known as CXCL12). SDF-1 will recruit MSCs via receptors (CXCR4/CXCR7) on MSCs. CXCR4 is highly expressed on MSCs (Son et al. 2006; Wynn et al. 2004). In models of injured kidney, Liu et al. systemically administered MSCs to mice and showed that grafted MSCs homed to the ischemic kidney, improved renal function, and reduced cell apoptosis. The study implicated the SDF-1–CXCR4/CXCR7 axis in the kidney repair (Liu et al. 2012).

Some studies showed that MSCs can differentiate into mesangial cells (Herrera et al. 2007; Morigi et al. 2004; Yokoo et al. 2006). In swine renovascular disease, MSCs were detected in all regions of the kidney but mostly at the renal interstitium after 4 weeks of intrarenal infusion (Ebrahimi et al. 2013; Eirin et al. 2012).

8.2.2 *Effects of Soluble Factors and Microvesicles from Mesenchymal Stem Cells*

Besides the differentiation effects of MSCs, the cells can also have biological effects on grafted bodies based on active biological factors that they produce. These secretors can be divided into two forms:

(i) Soluble factors: several soluble factors are produced by MSCs, including interleukin (IL)-6, IL-8, monocyte chemoattractant protein-1 (MCP-1/CCL2), and transforming growth factor (TGF)-*β*; extracellular matrix remodelers are also produced by MSCs, including tissue inhibitor of metalloproteinase 2 (TIMP-2), fibronectin, periostin, collagen, and metalloproteinase inhibitors; growth factors and regulators are also produced, including insulin-like growth factor 1 (IGF-1), hepatocyte growth factor (HGF), and vascular endothelial growth factor (VEGF) (Chen et al. 2011; Galderisi and Giordano 2014; Imberti et al. 2007; Miller et al. 1994; Zhang et al. 2004a).

(ii) Microvesicles are divisible in shedding vesicles released by membrane budding (particles of 50–200 nm), and exosomes are released from intracytoplasmic multivesicular bodies (bilipid membrane vesicles of 50 nm or less). These bodies can contain a pool of mRNA, miRNA, surface receptors, and biologically active lipids or proteins (Bruno et al. 2009, 2012; Collino et al. 2010; Thebaud and Stewart 2012; Zhou et al. 2013).

8.2.2.1 Immune Modulation

MSCs have a remarkable capacity to regulate immune responses. They can regulate immune responses both in vitro and in vivo via affecting the four main kinds of immune cells of the immune system, namely, T lymphocytes (Aggarwal and Pittenger 2005; Di Nicola et al. 2002; English et al. 2009), B lymphocytes (Asari et al. 2009; Augello et al. 2005; Corcione et al. 2006), natural killer (NK) cells (Sotiropoulou et al. 2006; Spaggiari et al. 2006), and dendritic cells (DCs) (Chen et al. 2007; Zhang et al. 2004b).

The effects of MSCs are related to certain bioactive molecules that they produce (Table 8.1). Almost all of these molecules are anti-inflammatory agents (e.g., IL-10, prostaglandin E2, and interleukin-1 receptor antagonist) or antiproliferative agents (e.g., TGF-β1), hepatocyte growth factor (HGF), and human leukocyte antigen G isoform (HLA-G5)).

8.2.2.2 Stimulation of Angiogenesis, Inhibition of Reactive Oxygen Species, Fibrosis, and Apoptosis

MSCs can protect the damaged kidney via the release of growth factors, proangiogenic factors and anti-inflammatory cytokines. Indeed, in vitro, MSCs are shown to produce large amounts of the proangiogenic factor VEGF (which facilitates

Table 8.1 Important bioactive molecules secreted by MSCs and their functions

Bioactive molecules	Functions
Prostaglandin-E2 (PGE2)	Antiproliferative mediators (Bouffi et al. 2010)
	Anti-inflammatory (Foraker et al. 2011)
Interleukin-10 (IL-10)	Anti-inflammatory (Nemeth et al. 2009)
Transforming growth factor β-1 (TGFβ1), hepatocyte growth factor (HGF)	Suppress T-lymphocyte proliferation (Di Nicola et al. 2002)
Interleukin-1 receptor antagonist	Anti-inflammatory (Ortiz et al. 2007)
Human leukocyte antigen G isoform (HLA-G5)	Antiproliferative for naïve T cells (Selmani et al. 2008)
LL-3	Antimicrobial peptide and reduce inflammation (Krasnodembskaya et al. 2010)
Angiopoietin-1	Restore epithelial protein permeability (Fang et al. 2010)
MMP3, MMP9	Mediating neovascularization (Kim et al. 2007)
Keratinocyte growth factor	Alveolar epithelial fluid transport (Lee et al. 2009)
Vascular endothelial growth factor (VEGF), basic fibroblast growth factor (bFGF), placental growth factor (PlGF), and monocyte chemoattractant protein-1 (MCP-1)	Enhance proliferation of endothelial cells and smooth muscle cells (Kinnaird et al. 2004a, b)

glomerular and tubular recovery) (Eirin et al. 2012; Kunter et al. 2006) and IGF-1 (which protects renal function and tubular structure from murine cisplatin-induced kidney injury) (Imberti et al. 2007).

Microvesicles can also protect acute kidney injury (AKI) in mice models. Ju et al. injected microvesicles obtained from cultures of human UC-MSCs to treat rats of AKI by mediating RNA transfer and synthesis of human HGF inside vesicles into rats (Ju et al. 2015). Exosomes from UC-MSCs were shown to repair cisplatin-induced AKI in rats and NRK-52E cell injury by ameliorating oxidative stress and cell apoptosis, promoting cell proliferation in vivo and in vitro (Zhou et al. 2013).

8.3 Preclinical Trials of Mesenchymal Stem Cell Transplantation for Kidney Diseases

In preclinical studies of kidney disease models, a major limitation is that all models pertain to chronic kidney diseases (CKD). CKD in animals can be induced by diabetes, hypertension, and chronic allograft nephropathy. Some preclinical trials are summarized in Table 8.1. In animal models, autologous, allogeneic, and xenogenic transplantations of MSCs have been evaluated for treatment of kidney diseases (Table 8.2).

Table 8.2 Preclinical studies using mesenchymal stem cells for the treatment of chronic kidney disease

Disease	Source	Delivery method	Mechanism of action	Side effects	Reference
Diabetic nephropathy	Mice BM-MSC	i.v.	Engraftment Direct effect	None	Ezquer et al. (2008)
Diabetic nephropathy	Human BM-MSC	Intracardiac	Engraftment Direct effect	None	Lee et al. (2006)
Partial nephrectomy	Rat BM-MSC	i.v.	Paracrine effect	None	Choi et al. (2009)
Chronic allograft nephropathy	Rat BM-MSC	i.v.	Immunomodulatory effect	None	Franquesa et al. (2012)
Renal revascularization	Allogeneic swine ADSC	Intrarenal	Engraftment Direct effect Paracrine	None	Ebrahimi et al. (2013), Eirin et al. (2012)
Renal artery stenosis	Autologous swine ADSC	Intrarenal	Engraftment Direct effect Paracrine	None	Zhu et al. (2013)

In mice, kidney disease can be induced by type 1 diabetes mellitus models whereby mice are injected with streptozotocin, which causes mice to develop hyperglycemia and glycosuria. Mice treated with MSC transplantation have shown histologically normal glomeruli and decreased albuminuria (Ezquer et al. 2008). Moreover, Lee et al. have treated immunodeficient mice with type 2 diabetes (produced with multiple low doses of streptozotocin) with intracardiac infusions of human BM-MSCs (Lee et al. 2006). The results from Lee's study showed that a few injected human MSCs were capable of differentiating into glomerular endothelial cells at the kidneys of MSC-treated diabetic mice (Lee et al. 2006).

In rat models, MSC transplantation at 1 day after nephrectomy could lead to recovered renal function and attenuated renal injury (Choi et al. 2009). Treated animals significantly attenuated progression of proteinuria (Choi et al. 2009). MSC transplantation was also beneficial in that it led to decrease of interstitial fibrosis, tubular atrophy, T-cell and macrophage infiltration, and expression of inflammatory cytokines (Franquesa et al. 2012).

Atherosclerotic renovascular disease has been tested in larger animal models, such as pigs. In these models, MSCs were isolated from adipose tissue and intrarenally injected into the swines (Zhu et al. 2013). The result showed that MSCs could protect the stenotic kidney, attenuate renal inflammation, improve renal function, as well as reduce inflammation, oxidative stress, and apoptosis (Ebrahimi et al. 2013; Eirin et al. 2012; Zhu et al. 2013).

According to a review by Papazova et al. (2015), 71 studies have evaluated stem cell transplantation for the treatment of kidney diseases. Most studies (58%) have used MSCs to evaluate the therapeutic efficacy on CKD, three studies have evaluated both preventive as well as rescue cell-based interventions, 38% have evaluated prevention, and 58% have evaluated rescue only interventions. Papazova showed that most of the studies (68%) examined single administration, 23% used multiple administrations (two to eight administrations), and 9% of the studies investigated both. In animals, there were five transplantation methods, which included renal artery, intra-arterial nonrenal, intravenous, parenchymal or subcapsular, and intraperitoneal, with the most studies using intravenous cell administration (68%). Generally, treatment of CKD with cells or cell products significantly improved functional and histological parameters and decreased plasma creatinine and plasma urea (Papazova et al. 2015).

8.4 Clinical Trials of Mesenchymal Stem Cell Transplantation for Kidney Diseases

From promising results in animal models of kidney diseases, clinical trials were subsequently carried out since 2010. According to clinicaltrials.gov, there have been 30 clinical trials pertaining to kidney or kidney-related diseases (Table 8.3). These 30 trials were classified into six kinds of kidney diseases and kidney-related diseases, including:

- Acute kidney injury
- Chronic kidney injury
- Focal segmental glomerulosclerosis
- Diabetic kidney disease
- Autoimmune disease
- Kidney transplantation

8.4.1 Acute Kidney Injury

AKI refers to a sudden episode of kidney failure or kidney damage that happens within a few hours or a few days. AKI causes the accumulation of waste products in the blood. The function of the kidney is lost; therefore it cannot maintain the right balance of fluid in the human body. The accumulation of waste inside the blood can subsequently affect other organs, including the brain, heart, and lungs. AKI is characterized by the following symptoms: too little urine leaving the body; swelling in legs, ankles, and eye area; fatigue; shortness of breath; confusion; nausea; seizures; coma (in severe cases); and chest pain/pressure. At the present time, AKI is treated by dialysis to replace kidney function until the kidney recovers. Some patients completely recover, some develop chronic kidney injury (CKI), and about 50% of AKI patients die.

To date, there have been three clinical trials recorded in the clinicaltrials.gov database. The first clinical trial (NCT00733876) was conducted in 2013. It was a phase I clinical trial with 16 patients. In this study, CKI patients were transplanted

Table 8.3 Clinical trials using MSC transplantation for kidney and kidney-related diseases

NCT number	Title/phase	Country	Conditions	Enrollment (planned)	Type of MSC	Delivery methods
Acute kidney injury						
NCT01275612	Mesenchymal stem cells in cisplatin-induced acute renal failure in patients with solid organ cancers Phase I	Italy	Cisplatin-induced AKI	9	Allogeneic BM-MSC	Single i.v. infusion
NCT01602328	A study to evaluate the safety and efficacy of AC607 for the treatment of kidney injury in cardiac surgery subjects Phase II	USA	Postcardiac surgery AKI	156	Allogeneic AC607 BM-MSC	Single i.v. infusion
NCT00733876	Allogeneic multipotent stromal cell treatment for acute kidney injury following cardiac surgery Phase I	USA	Postoperative AKI (patients who require on-pump cardiac surgery)	15	Allogeneic BM-MSC	Intra-aortic infusion
Chronic kidney disease						
NCT02166489	Mesenchymal stem cells transplantation in patients with chronic renal failure due to polycystic kidney disease Phase I	Iran	Chronic renal failure due to autosomal dominant polycystic kidney disease (ADPKD)	6	Autologous BM-MSC	Single i.v. infusion
NCT02266394	Hypoxia and inflammatory injury in human renovascular hypertension Phase I	USA	Renal artery stenosis, ischemic nephropathy, renovascular disease, chronic kidney disease in human renovascular hypertension	42	Autologous ADSC	Single i.a. infusion

NCT01840540	MSC for occlusive disease of the kidney Phase I	USA	Atherosclerotic renal artery stenosis, ischemic nephropathy, renovascular hypertension	6	Autologous ADSC	Single i.a. infusion
NCT02195323	Autologous bone marrow-derived mesenchymal stromal cells (BM-MSC) in patients with chronic kidney disease (CKD) Phase I	Iran	Chronic kidney disease	7	Autologous BM-MSC	Single i.v. infusion
Focal segmental glomerular sclerosis						
NCT02382874	Allogeneic adMSC transplantation in idiopathic nephrotic syndrome (focal segmental glomerulosclerosis) Phase I	Iran	Focal segmental glomerulosclerosis	5	Allogeneic ADSC	Single i.v. injection
Diabetic kidney disease						
NCT02585622	Novel stromal cell therapy for diabetic kidney disease (NEPHSTROM) Phase I/II	UK	Diabetic kidney disease	48	Allogeneic BM-MSC	i.v. infusion 3 doses
Autoimmune disease						
NCT00698191	Mesenchymal stem cells transplantation for refractory systemic lupus erythematosus Phase I/II	China	Refractory systemic lupus erythematosus	20	Allogeneic BM-MSC	i.v. infusion
NCT01539902	Phase 2 study of human umbilical cord-derived mesenchymal stem cell for the treatment of lupus nephritis Phase II	China	Lupus nephritis	25	Allogeneic UCMSC	i.v. infusion

(continued)

Table 8.3 (continued)

NCT number	Title/phase	Country	Conditions	Enrollment (planned)	Type of MSC	Delivery methods
NCT02633163	A controlled trial of allogeneic mesenchymal stem cells for the treatment of refractory lupus Phase II	USA	Systemic lupus erythematosus	81	Allogeneic UCMSC	Single i.v. infusion
NCT01741857	Umbilical cord-derived mesenchymal stem cells transplantation for active and refractory systemic lupus erythematosus Phase I/II	China	Systemic lupus erythematosus	40	Allogeneic UCMSC	N/A
NCT00659217	Effect of mesenchymal stem cell transplantation for lupus nephritis Phase I/II	China	Lupus nephritis	20	Autologous MSC	Infusion
Kidney transplantation						
NCT00659620	Mesenchymal stem cell transplantation in the treatment of chronic allograft nephropathy Phase I/II	China	Kidney transplant, chronic allograft nephropathy	20	Allogeneic MSC	Infusion
NCT02409940	To elucidate the effect of mesenchymal stem cells on the T-cell repertoire of the kidney transplant patients Phase I	India	Renal transplant rejection	30	Allogeneic/ autologous MSC	Infusion
NCT02561767	Effect of BM-MSC in DCD kidney transplantation Phase I/II	China	Kidney transplantation, acute kidney tubular necrosis	120	Allogeneic BM-MSC	i.v. infusion

NCT01429038	Mesenchymal stem cells after renal or liver transplantation Phase I/II	Belgium	Kidney failure	40	Allogeneic BM-MSC	Single infusion
NCT00658073	Induction therapy with autologous mesenchymal stem cells for kidney allografts	China	Renal transplant rejection	165	Autologous BM-MSC	Two infusions, one at releasing renal artery clamp and one 2 weeks after transplantation
NCT02563366	Effect of BM-MSC on early graft function recovery after DCD kidney transplant Phase I/II	China	Kidney transplantation, acute kidney tubular necrosis	120	Allogeneic BM-MSC	Four i.v. administration doses
NCT02490020	A perspective multicenter controlled study on application of mesenchymal stem cell (MSC) to prevent rejection after renal transplantation by donation after cardiac death Phase I	China	Disorder related to renal transplantation, renal transplant rejection	260	BM-MSC	i.v. infusion
NCT00752479	Mesenchymal stem cells under basiliximab/low-dose RATG to induce renal transplant tolerance	Italy	Kidney transplant	4	Syngeneic BM-MSC	Infusion
NCT02563340	Effect of BM-MSC on chronic AMR after kidney transplantation Phase I/II	China	Kidney transplant	60	Allogeneic BM-MSC	Four i.v. infusion
NCT02492490	Effect of SVF-derived MSC in DCD renal transplantation Phase I/II	China	Kidney transplant	120	Autologous ADSC	Four i.v. infusions

(continued)

Table 8.3 (continued)

NCT number	Title/phase	Country	Conditions	Enrollment (planned)	Type of MSC	Delivery methods
NCT00734396	Mesenchymal stem cells and subclinical rejection Phase I/II	Netherlands	Kidney transplant	15	Autologous BM-MSC	Two i.v. infusions
NCT02492308	Induction with SVF-derived MSC in living-related kidney transplantation Phase I/II	China	Living-relative kidney transplantation	120	Autologous BM-MSC	Four i.v. infusions
NCT02387151	Allogeneic mesenchymal stromal cell therapy in renal transplant recipients Phase I	Netherlands	Rejection, graft loss	10	Allogeneic BM-MSC	Two i.v. infusions
NCT02012153	Mesenchymal stromal cells in kidney transplant recipients Phase I	Italy	Kidney transplant rejection	6	Autologous BM-MSC	Single i.v. infusion
NCT02565459	MSC and kidney transplant tolerance Phase I	Italy	Kidney transplant	22	Allogeneic BM-MSC	Single i.v. infusion
NCT02057965	Mesenchymal stromal cell therapy in renal recipients Phase II	Netherlands	Renal transplant rejection, fibrosis	70	Autologous BM-MSC	Three i.v. infusions

via distal thoracic aorta with allogeneic MSCs derived from bone marrow. Distal thoracic aorta transfusion is used to avoid cell entrapment in the lungs. The results from the study showed that after 6 months, there were no any adverse effects in any of the grafted patients. Subsequently, phase II clinical trials have been performed. In one clinical trial (NCT01275612), patients were systemically transfused with ex vivo expanded allogeneic MSCs. In another trial (NCT01602328), patients were transplanted with allogeneic BM-MSCs after undergoing cardiac surgery. Results and analyses from both phase II clinical trials have not yet been completed.

8.4.2 Chronic Kidney Disease

CKD is becoming increasingly prevalent, affecting about 8–16% of the general population. Of note, more than 30% of CKD patients are over 70 years of age (Bruck et al. 2015). The gradual increase of CKD is related to the rise in atherosclerosis and type 2 diabetes. CKD is defined by damage of the kidneys and decrease of kidney function. CKD patients develop complications such as high blood pressure, anemia, weak bones, poor nutritional health, and nerve damage.

Currently, there are four clinical trials using stem cells for CKD treatment. All of these studies are phase I studies which evaluate the safety of MSC transplantation. Two of the studies investigated the use of autologous BM-MSCs (NCT02166489 and NCT02195323), and two studies investigated ADSCs (NCT02266394 and NCT01840540). No preliminary results have been reported to date.

8.4.3 Diabetic Kidney Disease

Diabetic kidney disease (DKD), also known as diabetic nephropathy, is a complication that occurs in some diabetic patients. In DKD patients the filters of the kidneys (glomeruli) are damaged. Therefore, abnormal amounts of protein (especially albumin from the blood) will deposit in the urine. In healthy people, only a tiny amount of albumin would be found in the urine. Thus, the level of album in urine is one diagnostic marker to indicate DKD. Based on the level of albumin, DKD can be classified into two groups:

- *Microalbuminuria*: the amount of albumin detected in the urine ranges from 30 to 300 mg per day.
- *Proteinuria*: the amount of albumin detected in the urine is >300 mg per day.

DKD is a clinical syndrome associated with kidney damage which progresses to CKD. The 5-year mortality rate is 39% – a rate comparable to many cancers. The current treatments for DKD entail early detection, glycemic control, and tight blood pressure management with preferential use of renin-angiotensin system blockade (Gallagher and Suckling 2016). These treatments can treat the symptoms but also reduce the side effects of DKD.

The first clinical trial using stem cell therapy for DKD (NCT02585622) was performed in 2015. It was a controlled phase I/II clinical trial. The study was conducted after success in a preclinical trial (in mice) using BM-MSCs (Ezquer et al. 2012). Accordingly, the clinical trial also used allogeneic BM-MSCs to transplant in DKD patients; the primary goal was to evaluate the safety, feasibility, and tolerability of the allogeneic BM-MSCs, and the secondary goal was to evaluate the preliminary efficacy.

8.4.4 *Focal Segmental Glomerulosclerosis*

Focal segmental glomerulosclerosis (FSGS) is one of the most common causes of primary glomerular diseases in adults. In essence, FSGS is a progressive form of kidney disease, accounting for 2.3% of patients with end-stage renal disease (ESRD). FSGS is characterized by generalized edema, massive proteinuria, hypoalbuminemia, and hyperlipidemia. However, in the collapsing form of FSGS, patients experience severe hypertension, more massive proteinuria, poor response to corticosteroids, and a much faster rate of progression to ESRD. The traditional treatment for FSGS includes corticosteroids and calcineurin inhibitors. However, low response has been recorded in almost all patients with FSGS. Hence, alternative therapy has been evaluated for this disease. Some preclinical trials in animal models of FSGS have shown that MSC transplantation can give rise to some benefits and can lead to overall improvement of disease (Ma et al. 2013). Promise from such preclinical studies has led to the design of clinical trials. The first clinical trial (NCT02382874) administered allogeneic ADSCs as an intravenous infusion to treat five refractory FSGS patients. To date, there have been no published results from this study.

8.4.5 *Autoimmune Disease*

Although there are many autoimmune diseases, the autoimmune condition which has the greatest effect on the kidneys is systemic lupus erythematosus (SLE). SLE can affect many organs and tissues in the human body. However, the most significant manifestation of SLE is nephritis. SLE patients can be treated by high doses of corticosteroids, cyclophosphamides, and other immunosuppressive and biological agents. Although, patient outcome improves greatly with these treatments, the strong side effects (e.g., infection, ovarian failure, and secondary malignancy) can lead to patient death (Crow 2009; Houssiau and Ginzler 2008).

MSC transplantation has produced remarkable results in all clinical trials of SLE therapy. Almost all the trials have shown that MSC transplantation is effective for conditions across the SLE spectrum. The first report by Sun et al. (2009) showed that some SLE patients who were unresponsive to monthly i.v. cyclophosphamide and oral prednisone ($\geq$20 mg/day) (Sun et al. 2009) could be treated by MSC transplantation.

All patients significantly improved at 1, 6, and 12 months after transplantation, especially in their urinary protein. More importantly, no complications or serious side effects were detected at the 12–18 month follow-up (Sun et al. 2009). In another clinical trial, Liang et al. (2010) showed that allogeneic BM-MSC transplantation effectively improved clinical and serological features in grafted patients. There was a significant decrease in proteinuria at 24 h after transplantation, and in anti-dsDNA antibodies at 1 month and 3 months posttransplantation (Liang et al. 2010).

In another phase II study, Wang et al. (2013) showed that allogeneic UC-MSCs could be successfully used to treat SLE. With 4 years of follow-up (mean 27 months), patients exhibited better clinical results with an overall survival rate of 94% (Wang et al. 2013). Recently a report came out from the first multicenter clinical trial to use allogeneic UC-MSCs to treat SLE. The study results showed that 32.5% of patients achieved a major clinical response and another 27.5% of patients achieved a partial clinical response during a 12-month follow-up (Wang et al. 2014). Moreover other clinical trials (NCT01539902 and NCT00659217) are in progress.

8.4.6 Kidney Transplantation

Kidney transplant is the best chance to increase survival and improve overall health-related quality in ESRD patients. In almost all kidney transplantations, potent immunosuppressive drugs are used which suppress the host's immune system. These immunosuppressive drugs can increase the survival percentage of the graft but not by much. Moreover, many of their side effects have been recorded in patients.

With their capacity for immune modulation, MSCs have been used to modulate immune responses to the graft. To date, there have been 16 trials (both ongoing and completed) which have been registered on clinicaltrials.gov. All these trials were conducted to evaluate the safety and efficacy of MSC transplantation following kidney transplantation (Pileggi et al. 2013).

In almost all of the trials, autologous BM-MSCs were administered via systemic transfusion. The first clinical trial (NCT00752479), by Perico et al., used autologous BM-MSCs to treat two patients with ESRD and undergoing kidney transplant (Perico et al. 2011; Pileggi et al. 2013). BM-MSCs were transfused into patients after 1 week following surgery. The results showed that the grafted kidney displayed a good graft function at 1-year follow-up. Particularly, regulatory T cells were increased while the number of memory CD8+ T cells decreased.

In another study, autologous BM-MCSs were also used in kidney transplantation. These cells were transfused into two patients (who required kidney replacement) at 1 day before kidney transplantation. The results showed that one patient developed acute cellular rejection (ACR) 2 weeks later due to higher HLA haplotype mismatch and was treated with steroid pulses. However, both patients had excellent graft function at the 1-year follow-up (Casiraghi et al. 2016; Perico et al. 2013).

In 2012, Tan et al. conducted the largest clinical trial (NCT00658073) which compared the use of BM-MSC transplantation versus anti-CD25 antibody in ESRD patients undergoing kidney transplant (Tan et al. 2012). At the 1-year follow-up, results showed that replacement of CD25 blockade did not affect graft survival, but MSC transplantation helped to recover the kidney function faster. Moreover, BM-MSC transplantation led to less severe ACR compared to the anti-CD25 antibody. Graft function at 1 year was comparable in all groups.

Besides BM-MSCs, there have been two clinical trials which have used ADSCs from donors to transfuse to kidney recipients. The first clinical trial examined ADSCs in combination with hematopoietic stem cell transplantation (versus HSCT alone) in 100 patients indicated for allograft of kidney for ESRD (Vanikar et al. 2011). The results showed that in the group receiving ADSC transplantation (but not the control group), graft survival was improved, and chimerism levels were sustained at the 18-month follow-up period (Vanikar et al. 2011). In a larger clinical trial testing ADSC in 916 patients, the patients were transfused with ADSCs from kidney donors to induce hyporesponsiveness. Preliminary analyses of the results obtained from the study show that ADSC transplantation is effective in minimizing immunosuppression in kidney transplant, resulting in good graft function and patient/graft survival even out to 4 years (Vanikar and Trivedi 2012).

8.5 Conclusion

Kidney diseases have gradually increased in current years. These diseases can be developed as a result of diabetes mellitus. Preclinical trials and clinical trials using MSCs for kidney disease therapy have only been conducted in the last 5 years but already show great promise. Although the number of clinical trials is low, the preliminary results show that MSC transplantation is a promising approach to treat kidney diseases, especially CKD, DKD, autoimmune related diseases, and kidney transplantation. The role of MSCs have also been explored by direct (homing and transdifferentiation of MSCs toward renal cells) and indirect (production of soluble factors, exosome, and microvesicles) effects of MSCs. From such results, it is clear that MSC transplantation shows promise to become a new therapeutic treatment for kidney disease in the near future.

References

Aggarwal S, Pittenger MF (2005) Human mesenchymal stem cells modulate allogeneic immune cell responses. Blood 105:1815–1822

Asari S, Itakura S, Ferreri K, Liu CP, Kuroda Y, Kandeel F, Mullen Y (2009) Mesenchymal stem cells suppress B-cell terminal differentiation. Exp Hematol 37:604–615

Augello A, Tasso R, Negrini SM, Amateis A, Indiveri F, Cancedda R, Pennesi G (2005) Bone marrow mesenchymal progenitor cells inhibit lymphocyte proliferation by activation of the programmed death 1 pathway. Eur J Immunol 35:1482–1490

Bouffi C, Bony C, Courties G, Jorgensen C, Noel D (2010) IL-6-dependent PGE2 secretion by mesenchymal stem cells inhibits local inflammation in experimental arthritis. PLoS One 5:e14247

Bruck K, Stel VS, Fraser S, De Goeij MC, Caskey F, Abu-Hanna A, Jager KJ (2015) Translational research in nephrology: chronic kidney disease prevention and public health. Clin Kidney J 8:647–655

Bruno S, Grange C, Deregibus MC, Calogero RA, Saviozzi S, Collino F, Morando L, Busca A, Falda M, Bussolati B et al (2009) Mesenchymal stem cell-derived microvesicles protect against acute tubular injury. J Am Soc Nephrol: JASN 20:1053–1067

Bruno S, Grange C, Collino F, Deregibus MC, Cantaluppi V, Biancone L, Tetta C, Camussi G (2012) Microvesicles derived from mesenchymal stem cells enhance survival in a lethal model of acute kidney injury. PLoS One 7:e33115

Casiraghi F, Perico N, Cortinovis M, Remuzzi G (2016) Mesenchymal stromal cells in renal transplantation: opportunities and challenges. Nat Rev Nephrol 12:241–253

Chen L, Zhang W, Yue H, Han Q, Chen B, Shi M, Li J, Li B, You S, Shi Y et al (2007) Effects of human mesenchymal stem cells on the differentiation of dendritic cells from CD34+ cells. Stem Cells Dev 16:719–731

Chen Y, Qian H, Zhu W, Zhang X, Yan Y, Ye S, Peng X, Li W, Xu W (2011) Hepatocyte growth factor modification promotes the amelioration effects of human umbilical cord mesenchymal stem cells on rat acute kidney injury. Stem Cells Dev 20:103–113

Choi S, Park M, Kim J, Hwang S, Park S, Lee Y (2009) The role of mesenchymal stem cells in the functional improvement of chronic renal failure. Stem Cells Dev 18:521–529

Collino F, Deregibus MC, Bruno S, Sterpone L, Aghemo G, Viltono L, Tetta C, Camussi G (2010) Microvesicles derived from adult human bone marrow and tissue specific mesenchymal stem cells shuttle selected pattern of miRNAs. PLoS One 5:e11803

Corcione A, Benvenuto F, Ferretti E, Giunti D, Cappiello V, Cazzanti F, Risso M, Gualandi F, Mancardi GL, Pistoia V et al (2006) Human mesenchymal stem cells modulate B-cell functions. Blood 107:367–372

Covas DT, Siufi JL, Silva AR, Orellana MD (2003) Isolation and culture of umbilical vein mesenchymal stem cells. Braz J Med Biol Res 36:1179–1183

Crow MK (2009) Developments in the clinical understanding of lupus. Arthritis Res Ther 11:245

Di Nicola M, Carlo-Stella C, Magni M, Milanesi M, Longoni PD, Matteucci P, Grisanti S, Gianni AM (2002) Human bone marrow stromal cells suppress T-lymphocyte proliferation induced by cellular or nonspecific mitogenic stimuli. Blood 99:3838–3843

Dominici M, Le Blanc K, Mueller I, Slaper-Cortenbach I, Marini F, Krause D, Deans R, Keating A, Prockop D, Horwitz E (2006) Minimal criteria for defining multipotent mesenchymal stromal cells. The International Society for Cellular Therapy position statement. Cytotherapy 8:315–317

Ebrahimi B, Eirin A, Li Z, Zhu XY, Zhang X, Lerman A, Textor SC, Lerman LO (2013) Mesenchymal stem cells improve medullary inflammation and fibrosis after revascularization of swine atherosclerotic renal artery stenosis. PLoS One 8:e67474

Eirin A, Zhu XY, Krier JD, Tang H, Jordan KL, Grande JP, Lerman A, Textor SC, Lerman LO (2012) Adipose tissue-derived mesenchymal stem cells improve revascularization outcomes to restore renal function in swine atherosclerotic renal artery stenosis. Stem Cells (Dayton, Ohio) 30:1030–1041

English K, Ryan JM, Tobin L, Murphy MJ, Barry FP, Mahon BP (2009) Cell contact, prostaglandin E(2) and transforming growth factor beta 1 play non-redundant roles in human mesenchymal stem cell induction of CD4+CD25(high) forkhead box P3+ regulatory T cells. Clin Exp Immunol 156:149–160

Erices A, Conget P, Minguell JJ (2000) Mesenchymal progenitor cells in human umbilical cord blood. Br J Haematol 109:235–242

Ezquer FE, Ezquer ME, Parrau DB, Carpio D, Yanez AJ, Conget PA (2008) Systemic administration of multipotent mesenchymal stromal cells reverts hyperglycemia and prevents nephropathy in type 1 diabetic mice. Biol Blood Marrow Transplant: J Am Soc Blood Marrow Transplant 14:631–640

Ezquer ME, Ezquer FE, Arango-Rodriguez ML, Conget PA (2012) MSC transplantation: a promising therapeutic strategy to manage the onset and progression of diabetic nephropathy. Biol Res 45:289–296

Fang X, Neyrinck AP, Matthay MA, Lee JW (2010) Allogeneic human mesenchymal stem cells restore epithelial protein permeability in cultured human alveolar type II cells by secretion of angiopoietin-1. J Biol Chem 285:26211–26222

Fernandez M, Simon V, Herrera G, Cao C, Del Favero H, Minguell JJ (1997) Detection of stromal cells in peripheral blood progenitor cell collections from breast cancer patients. Bone Marrow Transplant 20:265–271

Foraker JE, Oh JY, Ylostalo JH, Lee RH, Watanabe J, Prockop DJ (2011) Cross-talk between human mesenchymal stem/progenitor cells (MSCs) and rat hippocampal slices in LPS-stimulated cocultures: the MSCs are activated to secrete prostaglandin E2. J Neurochem 119:1052–1063

Franquesa M, Herrero E, Torras J, Ripoll E, Flaquer M, Goma M, Lloberas N, Anegon I, Cruzado JM, Grinyo JM et al (2012) Mesenchymal stem cell therapy prevents interstitial fibrosis and tubular atrophy in a rat kidney allograft model. Stem Cells Dev 21:3125–3135

Friedenstein AJ, Deriglasova UF, Kulagina NN, Panasuk AF, Rudakowa SF, Luria EA, Ruadkow IA (1974) Precursors for fibroblasts in different populations of hematopoietic cells as detected by the in vitro colony assay method. Exp Hematol 2:83–92

Fu Y, Guan J, Guo S, Guo F, Niu X, Liu Q, Zhang C, Nie H, Wang Y (2014) Human urine-derived stem cells in combination with polycaprolactone/gelatin nanofibrous membranes enhance wound healing by promoting angiogenesis. J Transl Med 12:274

Galderisi U, Giordano A (2014) The gap between the physiological and therapeutic roles of mesenchymal stem cells. Med Res Rev 34:1100–1126

Gallagher H, Suckling RJ (2016) Diabetic nephropathy: where are we on the journey from pathophysiology to treatment? Diabetes Obes Metab 18:641–647

Gargett CE, Masuda H (2010) Adult stem cells in the endometrium. Mol Hum Reprod 16: 818–834

Herrera MB, Bussolati B, Bruno S, Morando L, Mauriello-Romanazzi G, Sanavio F, Stamenkovic I, Biancone L, Camussi G (2007) Exogenous mesenchymal stem cells localize to the kidney by means of CD44 following acute tubular injury. Kidney Int 72:430–441

Hou T, Xu J, Wu X, Xie Z, Luo F, Zhang Z, Zeng L (2009) Umbilical cord Wharton's jelly: a new potential cell source of mesenchymal stromal cells for bone tissue engineering. Tissue Eng Part A 15:2325–2334

Houssiau FA, Ginzler EM (2008) Current treatment of lupus nephritis. Lupus 17:426–430

Huss R, Lange C, Weissinger EM, Kolb HJ, Thalmeier K (2000) Evidence of peripheral blood-derived, plastic-adherent CD34(−/low) hematopoietic stem cell clones with mesenchymal stem cell characteristics. Stem Cells 18:252–260

Imberti B, Morigi M, Tomasoni S, Rota C, Corna D, Longaretti L, Rottoli D, Valsecchi F, Benigni A, Wang J et al (2007) Insulin-like growth factor-1 sustains stem cell mediated renal repair. J Am Soc Nephrol: JASN 18:2921–2928

Jo YY, Lee HJ, Kook SY, Choung HW, Park JY, Chung JH, Choung YH, Kim ES, Yang HC, Choung PH (2007) Isolation and characterization of postnatal stem cells from human dental tissues. Tissue Eng 13:767–773

Ju G-Q, Cheng J, Zhong L, Wu S, Zou X-Y, Zhang G-Y, Gu D, Miao S, Zhu Y-J, Sun J (2015) Microvesicles derived from human umbilical cord mesenchymal stem cells facilitate tubular epithelial cell dedifferentiation and growth via hepatocyte growth factor induction. PLoS One 10:e0121534

Kestendjieva S, Kyurkchiev D, Tsvetkova G, Mehandjiev T, Dimitrov A, Nikolov A, Kyurkchiev S (2008) Characterization of mesenchymal stem cells isolated from the human umbilical cord. Cell Biol Int 32:724–732

Kim Y, Kim H, Cho H, Bae Y, Suh K, Jung J (2007) Direct comparison of human mesenchymal stem cells derived from adipose tissues and bone marrow in mediating neovascularization in response to vascular ischemia. Cell Physiol Biochem 20:867–876

Kinnaird T, Stabile E, Burnett MS, Lee CW, Barr S, Fuchs S, Epstein SE (2004a) Marrow-derived stromal cells express genes encoding a broad spectrum of arteriogenic cytokines and promote in vitro and in vivo arteriogenesis through paracrine mechanisms. Circ Res 94:678–685

Kinnaird T, Stabile E, Burnett MS, Shou M, Lee CW, Barr S, Fuchs S, Epstein SE (2004b) Local delivery of marrow-derived stromal cells augments collateral perfusion through paracrine mechanisms. Circulation 109:1543–1549

Kita K, Gauglitz GG, Phan TT, Herndon DN, Jeschke MG (2010) Isolation and characterization of mesenchymal stem cells from the sub-amniotic human umbilical cord lining membrane. Stem Cells Dev 19:491–502

Krasnodembskaya A, Song Y, Fang X, Gupta N, Serikov V, Lee JW, Matthay MA (2010) Antibacterial effect of human mesenchymal stem cells is mediated in part from secretion of the antimicrobial peptide LL-37. Stem Cells 28:2229–2238

Kunter U, Rong S, Djuric Z, Boor P, Muller-Newen G, Yu D, Floege J (2006) Transplanted mesenchymal stem cells accelerate glomerular healing in experimental glomerulonephritis. J Am Soc Nephrol: JASN 17:2202–2212

Lee OK, Kuo TK, Chen WM, Lee KD, Hsieh SL, Chen TH (2004) Isolation of multipotent mesenchymal stem cells from umbilical cord blood. Blood 103:1669–1675

Lee RH, Seo MJ, Reger RL, Spees JL, Pulin AA, Olson SD, Prockop DJ (2006) Multipotent stromal cells from human marrow home to and promote repair of pancreatic islets and renal glomeruli in diabetic NOD/scid mice. Proc Natl Acad Sci U S A 103:17438–17443

Lee JW, Fang X, Gupta N, Serikov V, Matthay MA (2009) Allogeneic human mesenchymal stem cells for treatment of *E. coli* endotoxin-induced acute lung injury in the ex vivo perfused human lung. Proc Natl Acad Sci U S A 106:16357–16362

Liang J, Zhang H, Hua B, Wang H, Lu L, Shi S, Hou Y, Zeng X, Gilkeson GS, Sun L (2010) Allogenic mesenchymal stem cells transplantation in refractory systemic lupus erythematosus: a pilot clinical study. Ann Rheum Dis 69:1423–1429

Liu H, Liu S, Li Y, Wang X, Xue W, Ge G, Luo X (2012) The role of SDF-1-CXCR4/CXCR7 axis in the therapeutic effects of hypoxia-preconditioned mesenchymal stem cells for renal ischemia/reperfusion injury. PLoS One 7:e34608

Ma H, Wu Y, Xu Y, Sun L, Zhang X (2013) Human umbilical mesenchymal stem cells attenuate the progression of focal segmental glomerulosclerosis. Am J Med Sci 346:486–493

Mareschi K, Biasin E, Piacibello W, Aglietta M, Madon E, Fagioli F (2001) Isolation of human mesenchymal stem cells: bone marrow versus umbilical cord blood. Haematologica 86: 1099–1100

Miller SB, Martin DR, Kissane J, Hammerman MR (1994) Hepatocyte growth factor accelerates recovery from acute ischemic renal injury in rats. Am J Physiol 266:F129–F134

Morigi M, Imberti B, Zoja C, Corna D, Tomasoni S, Abbate M, Rottoli D, Angioletti S, Benigni A, Perico N et al (2004) Mesenchymal stem cells are renotropic, helping to repair the kidney and improve function in acute renal failure. J Am Soc Nephrol: JASN 15:1794–1804

Musina RA, Belyavski AV, Tarusova OV, Solovyova EV, Sukhikh GT (2008) Endometrial mesenchymal stem cells isolated from the menstrual blood. Bull Exp Biol Med 145:539–543

Nemeth K, Leelahavanichkul A, Yuen PS, Mayer B, Parmelee A, Doi K, Robey PG, Leelahavanichkul K, Koller BH, Brown JM et al (2009) Bone marrow stromal cells attenuate sepsis via prostaglandin E(2)-dependent reprogramming of host macrophages to increase their interleukin-10 production. Nat Med 15:42–49

Ortiz LA, Dutreil M, Fattman C, Pandey AC, Torres G, Go K, Phinney DG (2007) Interleukin 1 receptor antagonist mediates the antiinflammatory and antifibrotic effect of mesenchymal stem cells during lung injury. Proc Natl Acad Sci U S A 104:11002–11007

Papazova DA, Oosterhuis NR, Gremmels H, van Koppen A, Joles JA, Verhaar MC (2015) Cell-based therapies for experimental chronic kidney disease: a systematic review and meta-analysis. Dis Model Mech 8:281–293

Patki S, Kadam S, Chandra V, Bhonde R (2010) Human breast milk is a rich source of multipotent mesenchymal stem cells. Hum Cell 23:35–40

Peng HH, Wang TH, Chao AS, Chang SD (2007) Isolation and differentiation of human mesenchymal stem cells obtained from second trimester amniotic fluid; experiments at Chang Gung Memorial Hospital. Chang Gung Med J 30:402–407

Perico N, Casiraghi F, Introna M, Gotti E, Todeschini M, Cavinato RA, Capelli C, Rambaldi A, Cassis P, Rizzo P et al (2011) Autologous mesenchymal stromal cells and kidney transplantation: a pilot study of safety and clinical feasibility. Clin J Am Soc Nephrol: CJASN 6:412–422

Perico N, Casiraghi F, Gotti E, Introna M, Todeschini M, Cavinato RA, Capelli C, Rambaldi A, Cassis P, Rizzo P et al (2013) Mesenchymal stromal cells and kidney transplantation: pretransplant infusion protects from graft dysfunction while fostering immunoregulation. Transplant Int: Off J Eur Soc Organ Transplant 26:867–878

Phuc PV, Nhung TH, Loan DT, Chung DC, Ngoc PK (2011) Differentiating of banked human umbilical cord blood-derived mesenchymal stem cells into insulin-secreting cells. In Vitro Cell Dev Biol Anim 47:54–63

Phuc PV, Ngoc VB, Lam DH, Tam NT, Viet PQ, Ngoc PK (2012) Isolation of three important types of stem cells from the same samples of banked umbilical cord blood. Cell Tissue Bank 13:341–351

Pierdomenico L, Bonsi L, Calvitti M, Rondelli D, Arpinati M, Chirumbolo G, Becchetti E, Marchionni C, Alviano F, Fossati V et al (2005) Multipotent mesenchymal stem cells with immunosuppressive activity can be easily isolated from dental pulp. Transplantation 80:836–842

Pileggi A, Xu X, Tan J, Ricordi C (2013) Mesenchymal stromal (stem) cells to improve solid organ transplant outcome: lessons from the initial clinical trials. Curr Opin Organ Transplant 18:672–681

Pirjali T, Azarpira N, Ayatollahi M, Aghdaie MH, Geramizadeh B, Talai T (2013) Isolation and characterization of human mesenchymal stem cells derived from human umbilical cord Wharton's jelly and amniotic membrane. Int J Organ Transplant Med 4:111–116

Purton LE, Mielcarek M, Torok-Storb B (1998) Monocytes are the likely candidate 'stromal' cell in G-CSF-mobilized peripheral blood. Bone Marrow Transplant 21:1075–1076

Qin D, Long T, Deng J, Zhang Y (2014) Urine-derived stem cells for potential use in bladder repair. Stem Cell Res Ther 5:69

Romanov YA, Svintsitskaya VA, Smirnov VN (2003) Searching for alternative sources of postnatal human mesenchymal stem cells: candidate MSC-like cells from umbilical cord. Stem Cells 21:105–110

Rylova YV, Milovanova NV, Gordeeva MN, Savilova AM (2015) Characteristics of multipotent mesenchymal stromal cells from human terminal placenta. Bull Exp Biol Med 159:253–257

Sani M, Hosseini SM, Salmannejad M, Aleahmad F, Ebrahimi S, Jahanshahi S, Talaei-Khozani T (2015) Origins of the breast milk-derived cells; an endeavor to find the cell sources. Cell Biol Int 39:611–618

Savickiene J, Treigyte G, Baronaite S, Valiuliene G, Kaupinis A, Valius M, Arlauskiene A, Navakauskiene R (2015) Human amniotic fluid mesenchymal stem cells from second- and third-trimester amniocentesis: differentiation potential, molecular signature, and proteome analysis. Stem Cells Int 2015:319238

Selmani Z, Naji A, Zidi I, Favier B, Gaiffe E, Obert L, Borg C, Saas P, Tiberghien P, Rouas-Freiss N et al (2008) Human leukocyte antigen-G5 secretion by human mesenchymal stem cells is required to suppress T lymphocyte and natural killer function and to induce CD4+ CD25highFOXP3+ regulatory T cells. Stem Cells 26:212–222

Shaer A, Azarpira N, Aghdaie MH, Esfandiari E (2014) Isolation and characterization of human mesenchymal stromal cells derived from placental decidua basalis; umbilical cord Wharton's jelly and amniotic membrane. Pak J Med Sci 30:1022–1026

Son BR, Marquez-Curtis LA, Kucia M, Wysoczynski M, Turner AR, Ratajczak J, Ratajczak MZ, Janowska-Wieczorek A (2006) Migration of bone marrow and cord blood mesenchymal stem cells in vitro is regulated by stromal-derived factor-1-CXCR4 and hepatocyte growth factor-c-met axes and involves matrix metalloproteinases. Stem Cells (Dayton, Ohio) 24:1254–1264

Sotiropoulou PA, Perez SA, Gritzapis AD, Baxevanis CN, Papamichail M (2006) Interactions between human mesenchymal stem cells and natural killer cells. Stem Cells 24:74–85

Spaggiari GM, Capobianco A, Becchetti S, Mingari MC, Moretta L (2006) Mesenchymal stem cell-natural killer cell interactions: evidence that activated NK cells are capable of killing MSCs, whereas MSCs can inhibit IL-2-induced NK-cell proliferation. Blood 107:1484–1490

Strem BM, Hicok KC, Zhu M, Wulur I, Alfonso Z, Schreiber RE, Fraser JK, Hedrick MH (2005) Multipotential differentiation of adipose tissue-derived stem cells. Keio J Med 54:132–141

Sun L, Akiyama K, Zhang H, Yamaza T, Hou Y, Zhao S, Xu T, Le A, Shi S (2009) Mesenchymal stem cell transplantation reverses multiorgan dysfunction in systemic lupus erythematosus mice and humans. Stem Cells 27:1421–1432

Tan J, Wu W, Xu X, Liao L, Zheng F, Messinger S, Sun X, Chen J, Yang S, Cai J et al (2012) Induction therapy with autologous mesenchymal stem cells in living-related kidney transplants: a randomized controlled trial. JAMA 307:1169–1177

Thebaud B, Stewart DJ (2012) Exosomes: cell garbage can, therapeutic carrier, or trojan horse? Circulation 126:2553–2555

Van Pham P, Vu NB, Phan NK (2016) Umbilical cord-derived stem cells (MODULATISTTM) show strong immunomodulation capacity compared to adipose tissue-derived or bone marrow-derived mesenchymal stem cells. Biomed Res Ther 3:687–696

Vanikar AV, Trivedi HL (2012) Stem cell transplantation in living donor renal transplantation for minimization of immunosuppression. Transplantation 94:845–850

Vanikar AV, Trivedi HL, Feroze A, Kanodia KV, Dave SD, Shah PR (2011) Effect of co-transplantation of mesenchymal stem cells and hematopoietic stem cells as compared to hematopoietic stem cell transplantation alone in renal transplantation to achieve donor hypo-responsiveness. Int Urol Nephrol 43:225–232

Wang D, Zhang H, Liang J, Li X, Feng X, Wang H, Hua B, Liu B, Lu L, Gilkeson GS et al (2013) Allogeneic mesenchymal stem cell transplantation in severe and refractory systemic lupus erythematosus: 4 years of experience. Cell Transplant 22:2267–2277

Wang D, Li J, Zhang Y, Zhang M, Chen J, Li X, Hu X, Jiang S, Shi S, Sun L (2014) Umbilical cord mesenchymal stem cell transplantation in active and refractory systemic lupus erythematosus: a multicenter clinical study. Arthritis Res Ther 16:R79

Wynn RF, Hart CA, Corradi-Perini C, O'Neill L, Evans CA, Wraith JE, Fairbairn LJ, Bellantuono I (2004) A small proportion of mesenchymal stem cells strongly expresses functionally active CXCR4 receptor capable of promoting migration to bone marrow. Blood 104:2643–2645

Yokoo T, Fukui A, Ohashi T, Miyazaki Y, Utsunomiya Y, Kawamura T, Hosoya T, Okabe M, Kobayashi E (2006) Xenobiotic kidney organogenesis from human mesenchymal stem cells using a growing rodent embryo. J Am Soc Nephrol: JASN 17:1026–1034

Zhang J, Li Y, Chen J, Yang M, Katakowski M, Lu M, Chopp M (2004a) Expression of insulin-like growth factor 1 and receptor in ischemic rats treated with human marrow stromal cells. Brain Res 1030:19–27

Zhang W, Ge W, Li C, You S, Liao L, Han Q, Deng W, Zhao RC (2004b) Effects of mesenchymal stem cells on differentiation, maturation, and function of human monocyte-derived dendritic cells. Stem Cells Dev 13:263–271

Zhou Y, Xu H, Xu W, Wang B, Wu H, Tao Y, Zhang B, Wang M, Mao F, Yan Y et al (2013) Exosomes released by human umbilical cord mesenchymal stem cells protect against cisplatin-induced renal oxidative stress and apoptosis in vivo and in vitro. Stem Cell Res Ther 4:34

Zhu XY, Urbieta-Caceres V, Krier JD, Textor SC, Lerman A, Lerman LO (2013) Mesenchymal stem cells and endothelial progenitor cells decrease renal injury in experimental swine renal artery stenosis through different mechanisms. Stem Cells (Dayton, Ohio) 31:117–125

Zuk PA, Zhu M, Mizuno H, Huang J, Futrell JW, Katz AJ, Benhaim P, Lorenz HP, Hedrick MH (2001) Multilineage cells from human adipose tissue: implications for cell-based therapies. Tissue Eng 7:211–228

Zuk PA, Zhu M, Ashjian P, De Ugarte DA, Huang JI, Mizuno H, Alfonso ZC, Fraser JK, Benhaim P, Hedrick MH (2002) Human adipose tissue is a source of multipotent stem cells. Mol Biol Cell 13:4279–4295

Part III
Skin Regeneration

Chapter 9
Use of Stem Cells in Acute and Complex Wounds

Yusef Yousuf, Saeid Amini-Nik, and Marc G. Jeschke

9.1 Introduction

Millions of people in the United States and around the world will suffer from acute and chronic wounds. It is estimated that 300,000 people are hospitalized each year in the United States due to acute wounds along with 11 million people affected (Hostetler et al. 2006). Chronic wounds affect approximately 6.5 million people, and it costs an excess of 25 billion annually to treat these wounds, a number that represents 2% of annual health care spending (Sen et al. 2009). These alarming numbers are mainly the result of an aging population and a rise in the incidence of diabetes. For example, roughly five million people suffering from diabetes will develop chronic wounds that will fail to heal (National Diabetes Statistical Report 2014). In recent decades, clinicians and scientists have expanded upon their knowledge of the mechanisms of wound repair and wound care (Dreifke et al. 2015). As a result, the goal of wound therapy has

Y. Yousuf
Institute of Medical Science, University of Toronto, Toronto, Canada

Sunnybrook Research Institute, Toronto, Canada

S. Amini-Nik
Sunnybrook Research Institute, Toronto, Canada

Laboratory Medicine and Pathobiology, University of Toronto, Toronto, Canada

Department of Surgery, Division of Plastic Surgery, University of Toronto, Toronto, Canada

M.G. Jeschke, MD, PhD (✉)
Institute of Medical Science, University of Toronto, Toronto, Canada

Sunnybrook Research Institute, Toronto, Canada

Department of Surgery, Division of Plastic Surgery, University of Toronto, Toronto, Canada

Department of Immunology, University of Toronto, Toronto, Canada

Ross-Tilley Burn Centre, Sunnybrook Health Sciences Centre, Toronto, Canada
e-mail: marc.jeschke@sunnybrook.ca

P.V. Pham (ed.), *Pancreas, Kidney and Skin Regeneration*, Stem Cells in Clinical Applications, DOI 10.1007/978-3-319-55687-1_9

shifted from managing symptoms to a more practical approach that will ultimately promote optimal wound healing and improve patients' quality of life. Despite these advancements wound healing remains a challenge. Stem cells are essential for tissue homeostasis and repair, and their versatility holds tremendous potential for tissue regeneration in a number of different clinical applications, including acute and chronic wounds. The unique self-renewal and differentiation capacity of stem cells make them attractive alternative to traditional wound treatments (Cha and Falanga 2007). Stem cell therapy aims to enhance cutaneous regeneration by completely restoring the structure and function of tissue so it is indistinguishable from its native state. In this chapter, we will focus on current and emerging stem cell-based treatments in the management of acute and chronic skin wounds. We will address what type of stem cells are used, how these stem cells are administered to damaged tissue, and finally the challenges and future directions that stem cell therapy faces.

9.2 Wound Healing

The skin is composed of two main layers: the epidermis which forms a barrier to the external environment and the dermis which is made of connective tissue that provides the skin with its mechanical properties. The mammalian epidermis is made up of stratified keratinized epithelium with hair follicles and glands scattered throughout (Heng 2011). In contrast, the dermis is subdivided into the papillary region and reticular dermis. The papillary region is superficial to the dermis, while the reticular dermis is a deeper and thicker layer; both regions consist of collagen, elastic fibers, and extrafibrillar matrix (Watt and Fujiwara 2011). Following injury to the skin and if operative, there is a coordinated and complex response that aims to achieve tissue integrity and homeostasis. There are a number of intracellular and extracellular events that become activated, and the intensity of the response depends on the severity of the injury, the size of wound, and the type of wound (Gurtner et al. 2008; Bielefeld et al. 2013).

9.2.1 Stages of Wound Healing

The healing process can be summarized in three phases that coincide: hemostasis/inflammation, proliferation, and tissue maturation and remodeling. The hemostasis phase is responsible for preventing blood loss and infection at the wound site. There is vascular constriction and platelet aggregation that form a fibrin clot to temporarily seal the wound (Gurtner et al. 2008; Schultz et al. 2011; Guo and Dipietro 2010). The formation of the fibrin clot is crucial to the infiltration of inflammatory cells such as neutrophils, macrophages, and lymphocytes (Gurtner et al. 2008; Schultz et al. 2011; Guo and Dipietro 2010). Neutrophils remove foreign objects and bacteria, while macrophages phagocytose foreign particles and dead neutrophils

(Mahdavian Delavary et al. 2011). The infiltration of these cells also results in the release of cytokines and growth factors such as transforming growth factor (TGF-β), platelet-derived growth factor (PDGF), tumor necrosis factor (TNF)-α, interleukin (IL), and fibroblast growth factor (FGF) (Gurtner et al. 2008). The release of these cytokines and growth factors stimulates the migration and proliferation of fibroblasts and keratinocytes into the wound, a process that is crucial to initiating the subsequent healing process (Singer and Clark 1999).

The proliferative phase occurs approximately 3 days after injury in humans and is marked by angiogenesis, formation of granulation tissue, and reepithelization of the epidermis (Singer and Clark 1999). The dermis is repaired by fibroblasts that produce fibronectin, promote collagen deposition, and secrete growth factors (Bielefeld et al. 2013). Matrix metalloproteinases (MMPs) cleave the ECM allowing for the passage of fibroblasts and keratinocytes into the wound bed (Singer and Clark 1999). As a consequence, a temporary matrix is formed that epithelial cells can migrate on and proliferate leading to the restoration of a functional epidermal layer (Clark 1990). A number of different growth factors such as keratinocyte growth factor (KGF), epidermal growth factor (EGF), fibroblast growth factor (FGF), TGF-β, heparin-binding EGF, FGF-2, and vascular endothelial growth factor (VEGF) help facilitate the proliferative phase (Gurtner et al. 2008; Bielefeld et al. 2013; Werner and Grose 2003). This includes regulating the proliferation and mobility of keratinocytes (Gurtner et al. 2008) and supporting the formation of new capillaries and blood vessels (Werner and Grose 2003). Lastly, the proliferative stage is augmented by stem cell populations within the hair follicles and epidermis that contribute to the reepithelization of the epidermis in a wound environment (Vagnozzi et al. 2015). Stem cells from the hair follicle and interfollicular epidermis migrate to the wound site in mouse full-thickness wounds (Mascre et al. 2012; Tumbar et al. 2004). Interestingly, lineage tracing studies have demonstrated that hair follicle stem cells (marked by K15) transiently contribute to wound reepithelization and are subsequently lost from the epidermis weeks later (Ito et al. 2005). These stem cells produce transit amplifying cells which can later differentiate into the stratified epithelial layer. This suggests that stem cells from the hair follicle are not necessary for the long-term maintenance of the epidermis after injury.

Once there is adequate deposition of collagen and closure of the epithelial gap, the maturation and remodeling phase begins. The properties of the tissue begin to change as type III collagen and fibronectin are slowly replaced by more organized type I collagen (Gurtner et al. 2008). Fibroblasts also transition to a more contractile myofibroblast phenotype (Darby et al. 1990). This reorganization dramatically improves the tensile strength in the new scar tissue, although the original strength of the uninjured skin is never recuperated (Singer and Clark 1999). Recently, skeletal muscle progenitors ($Pax7^{+}$) have been shown to contribute to the repair of the skin by adopting a myofibroblastic phenotype through a process mediated by β-catenin (Amini-Nik et al. 2011). These skeletal muscle stem cells may potentially utilized to increase tensile strength of scar tissue. Overall, this phase can last from weeks to years and results in the contraction of the wound and formation of scar tissue. The coordinated and timely response of these three phases is required for efficient skin healing, and any complications may result in the development of acute or chronic wounds that fail to heal.

9.2.2 *Signaling Pathways*

There are several key signaling pathways that are activated during cutaneous wound repair that are also active during embryonic skin development. These include, but are not limited to, Wnt/β-catenin, TGF-β, Notch, and Sonic hedgehog signaling pathways. Despite this similarity, there are a number of critical differences between the molecular mechanisms that control embryonic skin development and postnatal skin repair. Consequently, these differences may be at fault for the inability of injured skin to recapitulate the structure and function of uninjured skin. A number of studies have shown that these signaling pathways regulate the contribution and the fate of stem cells during the process of healing (Huelsken et al. 2001; Athar et al. 2006; St-Jacques et al. 1998; Karlsson et al. 1999; Niemann et al. 2003; Chiang et al. 1999); therefore these pathways may have significant role in understanding the mechanisms of stem cells in wound repair. We will briefly discuss these developmental pathways as they may have implications in the evolution of stem cell therapies for skin repair/regeneration.

The Wnt/β-catenin pathway plays a prominent role in cutaneous wound repair and development. Numerous Wnt = proteins are highly expressed in cutaneous wounds during the first 7 days of healing (Okuse et al. 2005), and Wnt is primarily responsible for the regeneration of hair follicles (Ito et al. 2007). On the other hand, β-catenin regulates fibroblasts during the proliferative phase of wound repair by altering fibroblast numbers, cellularity, and matrix production (Cheon et al. 2002, 2005, 2006; Sato 2006). The proliferative phase is also marked by an increase in the expression of β-catenin and its target genes fibronectin and MMP7 (Cheon et al. 2005). Upregulating β-catenin results in a fibrotic phenotype characterized by scarring, increased collagen deposition, and the formation of myofibroblasts (Cheon et al. 2006). At cellular level, an essential role for β-catenin in macrophages has been shown during wound repair. Moreover, as mentioned previously, β-catenin regulates dermal repair through regulation of $Pax7^+$ cells; knocking out β-catenin in Pax7-expressing cells in mice leads to fewer dermal fibroblast cells and a smaller scar size (Bielefeld et al. 2013; Amini-Nik et al. 2011).

TGF-β is one of numerous growth factors involved in cutaneous wound healing, and it is comprised of three isoforms: TGF-β1, TGF-β2, and TGF-β3. TGF-βs primarily exert their effects by activating (through phosphorylation) Smads 2 and 3 which subsequently allows them to translocate to the nucleus and alter gene expression (Schultz and Wysocki 2009; Owens et al. 2008; Biernacka et al. 2011). Similarly to the Wnt/β-catenin, the TGF-β pathways exert its healing effects by regulating fibroblast proliferation and behavior and matrix production (Bielefeld et al. 2013). TGF-β1 in particular has fibrosis and scar healing properties. In aged rats, treatment with topical TGF-β1 improves ECM deposition and fibroblast proliferation and migration (Puolakkainen et al. 1995). Moreover, research has demonstrated TGF-β1's role in the synthesis of collagen I and fibronectin (Varga et al. 1987; Hocevar et al. 1999), proliferation of fibroblasts (Schreier et al. 1993), alteration of fibroblasts to a myofibroblast phenotype (Desmouliere et al. 1993), and finally bolster wound contraction (Martinez-Ferrer et al. 2010). Unlike Wnt/β-catenin signaling, however, TGF-β has been shown to inhibit reepithelization. For example, treatment of porcine burn wounds with TGF-β1 antagonist increased the rate of complete reepithelization, and deletion of Smad3 in mice accelerates proliferation (Ashcroft et al. 1999). However, in double mutant TGF-β $^{-/-}$ Scid$^{-/-}$ mice, there is a significant delay in reepithelization (Crowe et al. 2000). These data suggest that TGF-β exerts some inhibitory effects on keratinocyte proliferation. Despite its positive effects in stimulating cutaneous wound healing, excessive TGF-β1 activity may lead to the formation of hypertrophic scars through a β-catenin-mediated mechanisms (Cheon et al. 2004). In contrast, higher TGF-β3 activity is associated with reduced scar formation as evidenced by scarless embryonic wounds in rats (Soo et al. 2003). A clinical trial which administered TGF-β3 prophylactically to prevent dermal scarring showed accelerated healing in phase I/II studies testing (Ferguson et al. 2009); however phase III trials were unsuccessful later.

The Notch signaling pathway regulates epidermal differentiation during skin homeostasis and development, and it is vital for vascular maintenance (Moriyama et al. 2008). In terms of wound healing, blocking the expression of Notch causes delayed healing, whereas the Notch ligand, Jagged, accelerates wound closure (Chigurupati et al. 2007). Furthermore, Notch interacts with Wnt/β-catenin and Sonic hedgehog, two other signaling pathways involved in cutaneous wound healing (Okuyama et al. 2008). The exact mechanisms by which Notch promotes healing is not as well defined as Wnt/β-catenin, but there is data indicating that Notch regulates macrophage behavior and matrix formation. The Notch signaling pathway is necessary during the inflammatory phase of wound healing. Outtz et al. (2010) observed reduced TNF-α expression and macrophage recruitment in Notch1$^{+/-}$ mice (Outtz et al. 2010). Furthermore, they found that Notch1 regulates vascular endothelial growth factor 1 (VEGF1) expression and inflammatory cytokine expression in macrophages in vitro (Outtz et al. 2010). With regard to matrix formation, an in vitro study has shown that Notch signaling mediates the adhesion of bone marrow-derived vascular precursor cells to the ECM (Caiado et al. 2008), an essential step during wound healing.

The Sonic hedgehog (Shh) signaling pathway plays an important role in skin development (Athar et al. 2006), but like Notch its role in wound healing is not fully understood. Shh is primarily involved in dermal reconstruction. Treating diabetic mice with Shh increases wound vascularity and cellularity and even increases VEGF expression and recruitment of endothelial progenitors (Asai et al. 2006). Interestingly, the Shh inhibitor cyclopamine hinders wound vascularity and cell proliferation leading to delayed wound closure in mice (Le et al. 2008). Further research is needed to elucidate the mechanisms that allow Shh to influence dermal wound repair.

9.3 Acute and Complex Wounds

Acute wounds will generally heal within a few weeks in healthy individuals, and the healing process strictly follows the aforementioned wound healing phases. In contrast, complex wounds do not conform to the cellular and molecular events that lead to healthy wound healing. Instead, there may be difficulties in the phases of repair that result in the wound failing to heal. As such, complex wounds are defined as wounds that have failed to proceed through an orderly and timely reparative process to produce anatomic and functional integrity over a period of 3 months (Mustoe et al. 2006). A common theory is that complex wounds remain in the inflammatory stage for too long resulting in a wound that is unable to progress to the proliferative and remodeling phases of healing. Additionally, there may be increased free radical production, ischemia, impaired growth factor and cytokine production, infection, and diminished cellular infiltration that disrupt the healing process (Agren et al. 2000; Mast and Schultz 1996). A robust understanding of the mechanisms involved in chronic wound healing is necessary to develop effective stem cell-based therapies that go beyond the treatments currently available. The following section will discuss wound healing in diabetic patients and the elderly, two groups of people that may suffer from complex wounds that result in deficient healing.

9.3.1 Deficient Healing

The prolonged inflammatory response in chronic wounds causes reepithelization to stall as granulation tissue is defective and does not foster healing (Martin and Nunan 2015). In diabetic ulcer wounds, fibroblasts become defective as they are unresponsive to growth factor signals, and their ability to migrate to the wound bed is reduced (Brem et al. 2007; Lerman et al. 2003; Seibold et al. 1985). Ulcers also display a significant decrease in the number of TGF-β receptors and the TGF-β signaling cascade (Pastar et al. 2010). These aberrations combined with elevated levels of degrading MMPs reduce collagen deposition and thus affecting ECM integrity and tissue remodeling in diabetic patients (Trengove et al. 1999). Moreover, the effectiveness of immune cells such as neutrophils and macrophages to phagocytose debris is hindered resulting in a buildup of necrotic debris at the wound edge. There is an enrichment of pro-inflammatory cytokines such as TNF-α and IL-1β in the wound bed which can inhibit the growth and phenotype of fibroblasts (Zykova et al. 2000). There is also a decrease in IL-6, a cytokine that is responsible for mobilizing fibroblasts and keratinocytes to the wound bed (Werner and Grose 2003). Accordingly, this exaggerated and unbalanced production of pro-inflammatory cytokines adversely affects wound closure. In elderly patients, wound healing follows a similar trajectory to diabetic wound healing. There are several factors that lead to deficient healing in the elderly; however there are two that are well described in the literature. First, there is impaired formation of new blood vessels (neovascularization) (Chang et al. 2007). Second, aged stem cells may be dysfunctional compared to their younger counterparts, and this dysfunction may possibly be due to the microenvironment aged stem cells that are exposed and/or cell intrinsic alterations (Blau et al. 2015). Research has shown that aged bone marrow-derived stem cells (BM-MSCs) have reduced wound healing, angiogenesis, and proliferation capabilities compared to younger BM-MSCs (Choudhery et al. 2012). Moreover, it has been shown that mesenchymal stem cells (MSC) isolated from the skin of elderly burn patients have diminished cell proliferation and deficient migration (Jeschke et al. 2015).

9.3.2 Excessive Healing

In contrast to deficient healing, excessive healing leads to the formation of a keloid or hypertrophic scar. As mentioned previously, there is a delicate balance between ECM deposition and degradation in normal wound healing. In both keloids and hypertrophic scars, there is increased collagen deposition that is oriented in thick bundles rather than basket weave-like fibrils in normal dermis (Martin and Nunan 2015). In human hypertrophic scars, the number of macrophage cells is positively correlated with scar size and the level of β-catenin, suggesting that macrophages play a role in the development of excessive healing (Amini-Nik et al. 2014). Furthermore, deletion of β-catenin in macrophage-specific cells resulted in impaired migration, cell adhesion, and inability to produce TGF-β1 leading to inadequate wound healing (Amini-Nik et al. 2014). Moreover, there is a positive correlation between the number of inflammatory cells and expression of β-catenin (Cheon et al. 2002). Considering that β-catenin plays a role during the proliferative phase of wound healing and that increased β-catenin activity associates hypertrophic scars, it is plausible that β-catenin is partly responsible for the excessive healing. TGF-β signaling may also play a role in excessive healing as blocking the activation of TGF-β1/2 receptors prevents scarring (Ferguson and O'Kane 2004). The difference between keloids and hypertrophic scars arises primarily from the arrangement of the collagen fibers and the composition of the wound (Martin and Nunan 2015). Keloids feature collagen fibers that spill beyond the wound margin, while hypertrophic scars do not (Martin and Nunan 2015). Keloids are also characterized by occluded blood vessels and fewer fibroblasts when compared with hypertrophic scars (Martin and Nunan 2015).

9.4 Current Treatments of Acute and Complex Wounds

As mentioned previously, diabetes mellitus is a condition in which insufficient wound healing can result in foot ulcers, hence the term "diabetic foot" (Singh et al. 2005). Foot ulcers are a common problem in diabetic patients with some experiencing significant epidermis damage that can expose the dermis and even deeper layers like the muscle and bone (Singh et al. 2005). The International Wound Bed Preparation Advisory Board provides a systemic approach to the management of chronic wounds such as diabetic ulcers: debridement, management of exudate, and management of infection (Schultz et al. 2003; Falanga 2000). Following a severe cutaneous wound, there may be an accumulation of necrotic tissue which can lead to infection. Debridement and drainage of wound fluid is done to remove infected tissue, clean the wound, ensure adequate blood flow, and provide a moist wound environment for optimal wound healing (Tsourdi et al. 2013). After the wound is contained, frequent dressing changes and use of topical antibiotics help maintain a clean environment and curtail infection. In some cases the necrosis and infection may be uncontrollable, and as such an amputation of the leg is necessary to prevent system infection (Margolis et al. 2005). Debridement strategies can be a strenuous task and can be painful for the patient. Some diabetic patients and sufferers of chronic wounds require negative-pressure wound therapy (NPWT). This is a treatment option that uses a vacuum to reduce swelling in the wound area which allows for blood and nutrients to reach the wound site resulting in improved blood flow and removed bacterial fluid (Guffanti 2014).

Chronic wounds are often lacking in growth factors necessary for wound healing. Growth factors control many key cellular processes in normal tissue repair such as cell migration, proliferation, angiogenesis, and production of ECM components (Branski et al. 2007). Therefore, increasing growth factor concentrations in chronic wounds may enhance wound healing. Applying growth factors directly to the wound bed is a common method of delivery. For instance, human recombinant platelet-derived growth factor BB (PDGF-BB) was the first cytokine growth factor approved by the FDA (Snyder 2005). PDGF-BB is topically applied to the wound where it promotes proliferation of cells involved in cutaneous wound repair. When PDGF-BB is used in conjunction with the wound healing practices described above, there is a reduction in healing time in patients with diabetic ulcers (Signorini and Clementoni 2007). An alternative to growth factors is applying or transplanting cultured keratinocytes and/or fibroblasts to the wound in order to stimulate repair. In fact, implanting cultured keratinocytes into the wound bed has been shown to promote reepithelization in diabetic pigs (Karagoz et al. 2009).

In some cases, grafting skin from cadavers onto the wound bed may be necessary to prevent infection and promote healing; the transplanted skin provides a scaffold for epithelial cells that boost cell proliferation (Snyder 2005). The skin transplanted from a member of the same species is called an allograft, whereas an autograft is a technique in which a patient's own skin is used for transplantation significantly reducing the risk of rejection and may be required if allografts are unsuccessful (Snyder 2005). Over the decades, bioengineering has produced dermal substitutes that can facilitate wound healing when traditional methods cannot be used or have failed. In severe burns, for example, a dermal substitute named Integra (Integra Limit Uncertainty, Plainsboro, New Jersey, USA) serves as temporary epidermis and is often used to assist wound closure (Branski et al. 2007; Jeschke et al. 2004). Integra is an acellular bilaminar structure made of silicone that provides a provisional dermal replacement which prevents fluid loss and acts as barricade against pathogens. Integra is composed of two layers: the bottom layer is made of a cross-linked matrix of bovine collagen and chondroitin-6-sulfate that is vital to preventing hypertrophic scarring, while the top layer is the silicon layer described above (Branski et al. 2007; Jeschke et al. 2004). The matrix in the bottom layer is vascularized and populated by host cells as the wound heals which forms a new dermis layer. Clinical trials have demonstrated the efficacy of Integra in promoting wound closure in addition to its ability to lessen hypertrophic scarring (Branski et al. 2007; Jeschke et al. 2004). Despite its advantages and success in the clinic, Integra is unable to recapitulate the structural properties of intact skin due to its lack of cellularity. Furthermore, it is expensive to produce, especially in the case of large total surface area burns (TBSA).

In excessive healing like hypertrophic scars and keloids, nonsurgical techniques are the most common methods of reducing scar size. The most accepted treatment of hypertrophic scars currently is the topical application of silicone gels or sheets onto the scar. The use of silicone has become standard practice among plastic surgeons, and it is effective in treating hypertrophic scars (Signorini and Clementoni 2007). Silicone decreases scar size mainly through wound hydration; silicone decreases water vapor transmission at the wound site resulting in a build of moisture in the wound (Signorini and Clementoni 2007). While it is a simple and effective treatment, silicone is difficult to apply, requires multiple applications per day, and may cause adverse skin reactions (Karagoz et al. 2009). An alternative method of reducing scar size is the use of mechanical force exerted by pressure garments. The theory behind this treatment is that the pressure limits collagen synthesis by creating a hypoxic environment that prevents the delivery of blood, oxygen, and nutrients (Puzey 2002). Pressure garments remain a prevalent treatment despite there being no conclusive data illustrating a reduction in scar size (Macintyre and Baird 2006). Lastly, an interesting and unusual treatment has emerged in the use of topical onion extract gel. The healing properties of onion extract are derived from its ability to inhibit fibroblast proliferation and hinder ECM production by increasing expression of MMP-1 (Cho et al. 2010). The efficacy of onion extract in decreasing scar size, however, is lacking in literature. Two studies observed no significant difference in scar appearance and size in patients treated with a 10% onion extract gel (Willital and Heine 1994; Chanprapaph et al. 2012). Stem cells hold great potential in producing treatments that are simple, safe, prevent complications, and go beyond typical treatments of acute and chronic wounds by fully regenerating skin. We will now describe the characteristics of different types of stem cells and how they have used in generating exciting new therapies for skin regeneration.

9.5 Stem Cells in Wound Healing

The goal of stem cell therapy is to restore skin back to its functional state after cutaneous wound injury. Unfortunately, dermal substitutes lack the biological activity necessary to regenerate the skin despite the outstanding physical support they provide to the wound bed. Incorporating stem cells and growth factors into these dermal substitutes would add the cellularity needed to accelerate skin regeneration. It is believed that the addition of these stem cells will enhance wound healing through trophic and paracrine activity that will regenerate the skin by neutralizing the damaging effects of the healing process and repairing skin by replacing lost or damaged cells (Falanga 2012).

Stem cells are groups of undifferentiated cells capable of transforming all types of cells in the human body. Using these cells as a renewable source for new tissue is the basis of regenerative medicine. The ability of these cells to self-renew and differentiate into different cell types is termed "stemness," and it is these unique features that make stem cell-based therapies an attractive alternative to traditional treatments (Cha and Falanga 2007). Not all stem cells demonstrate true "stemness" however as only embryonic stem cells are capable of differentiating into all cell lineages. This is a serious problem for regenerative medicine as the most commonly used stem cells, adult stem cells, only partially fulfill the requirements of stemness since their differentiation potential is limited to a few cell lineages. In the following section, we will discuss recent progress in the application of different types of stem cells for cutaneous wound healing.

9.5.1 Embryonic Stem Cells

Embryonic stem cells (ESCs) are derived from the inner cell mass of the blastocyst of a human embryo (Cha and Falanga 2007). ECSs are the standard to which all stem cells are held due to their ability to differentiate into all three germ layers (ectoderm, mesoderm, and endoderm) given the correct stimulus, an ability termed pluripotency (Cha and Falanga 2007). ECSs can be maintained in culture in an undifferentiated or be differentiated into specialized cell types. With regard to wound healing, ESCs have been differentiated into basal keratinocytes which were used to construct a stratified epidermis. Nevertheless, there are no stem cell therapies in wound healing that utilize ESCs despite their unlimited potential for self-renewal and plasticity. This is due to ethical and legal restrictions that limit the application of ESCs in regenerative medicine. Moreover, ESCs are restricted by their allogenic nature and safety concerns regarding the risk of developing tumors (Cha and Falanga 2007). As a result, regenerative medicine has primarily focused on adult stem cells since they are able to provide some of advantages of ESCs without the rigorous ethical and legal constraints.

9.5.2 Bone Marrow-Derived Stem Cells

The bone marrow (BM) is comprised of a heterogeneous cell population which includes hematopoietic stem cells (HSCs) and mesenchymal stem cells (MSCs), two groups of cells that are highly plastic progenitor cells (Salem and Thiemermann

2010). HSCs are responsible for the bone marrow and blood, while MSCs are non-hematopoietic stem cells that are self-renewing and differentiate into various cell types such as osteoblasts, adipocytes, chondrocytes, and even dermis (Wu et al. 2007a). MSCs are marked by their ability to adhere to a plastic surface, and cultured MSCs are strongly positive for CD105, CD90, and CD73 but negative for CD34 and CD45 (Boxall and Jones 2012). During wound healing, HSCs and MSCs mobilize to the wound site where they modulate the inflammatory response so that the wound does not progress to a chronic state. BM-MSCs attenuate the inflammatory response by decreasing the secretion of pro-inflammatory cytokines while amplifying the production of anti-inflammatory cytokines in the wound bed (Aggarwal and Pittenger 2005). In addition to diminishing the inflammatory response, BM-MSCs demonstrate antimicrobial activity (Maxson et al. 2012) and secrete growth factors such as FGF and VEGF that promote fibroblast proliferation, collagen deposition, and angiogenesis (Gnecchi et al. 2008). An alternative explanation for the wound healing properties of BM MSCs is that these stem cells migrate to the wound site and differentiate into cells necessary for the renewal and repair of damaged tissue (Wu et al. 2007a). BM-MSCs are well characterized and most widely studied stem cell population; however the isolation procedure is painful for patients, and there are few stem cells present in bone marrow aspirate.

Stem cell therapies in cutaneous wound healing have relied heavily on BM-MSCs, and numerous studies have demonstrated the usefulness of BM-MSCs in treating deficient and excessive healing in animals and humans (Table 9.1). In humans, autologous BM-MSCs delivered via a fibrin spray reduced wound size in patients with acute surgical excisional wounds (Falanga et al. 2007). In another study, direct injection of BM followed by topical application of cultured MSCs directly onto the wound bed led to complete wound closure, dermal rebuilding, and less scarring in three patients with chronic ulcer wounds (Badiavas and Falanga 2003). Moreover, examining biopsies from the healed tissue revealed significant vascularization and dermal thickness (Badiavas and Falanga 2003). The clinical benefits of BM-MSCs have also been illustrated in several trials involving diabetic ulcers where traditional therapies have failed. In one study, intramuscular injection of autologous BM-MSCs in patients with diabetic ulcers resulted in improved blood flow to the wound area (Lu et al. 2011). The injected was well tolerated, and patients experienced a significant reduction in pain and a marked improvement in healing rate of the ulcer at 24 weeks postinjection (Lu et al. 2011). Another study observed improved circulation and complete wound closure after transplantation of BM-MSCs in 18 out of 22 patients with diabetic ulcers (Kirana et al. 2012). Two other studies show that injection of BM-MSCs in patients with diabetic ulcers improves recovery time (Liu et al. 2014; Dash et al. 2009), decreases wound size (Dash et al. 2009), and enhances development of fibroblasts and inflammatory cells (Dash et al. 2009).

Animal studies involving BM-MSCs provide a more comprehensive and detailed view of the effect of these cells on cutaneous wound healing. In mouse acute excisional wounds, local injection of BM-MSCs accelerated wound closure and increased epithelization and angiogenesis (Wu et al. 2007b). Similar beneficial effects of BM-MSCs have been observed in other animals as transplantation of BM-MCSs improved tensile strength and healing in albino rat excisional wounds

Table 9.1 Published clinical studies utilizing stem cell therapies for the treatment of chronic wounds

Wound category	Type of stem cell	Delivery method	Patient outcome	Reference
Diabetic limb ischemia ($n = 41$)	BM-MSC and BM-MNC	Intramuscular injection	BM-MSC treated had improved healing rate compared to BM-MNC. Pain-free walking time was higher as well	Lu et al. (2011)
Chronic wound ($n = 3$)	BM-MSC	Topical application	Complete wound closure in all three patients. Improved inflammatory response and angiogenesis	Badiavas and Falanga (2003)
Chronic limb ischemia ($n = 6$) and acute surgical wound ($n = 4$)	BM-MSC	Topical application via fibrin spray	40% reduction in wound size in four CLI patients. Direct correlation with number of cells sprayed and decrease in wound size	Falanga et al. (2007)
Diabetic nonhealing ulcer ($n = 24$)	BM-MSC	Intramuscular injection and topical application	Ulcer size decreased by 73% in treated group. Improved reepithelization and pain-free walking distance	Dash et al. (2009)
Intractable dermatopathies ($n = 20$)	BM-MSC	BM-MSCs on collagen sponges applied to wound	Improved wound healing in 18 patients. Two patients died	Yoshikawa et al. (2008)
Radiation burn ($n = 1$)	BM-MSC	Local injection	Complete healing in one patient	Lataillade et al. (2007)
Critical limb ischemia ($n = 7$)	BM-MSC	Intramuscular injection	Six of seven patients experienced complete wound healing	Gupta et al. (2013)
Diabetic foot ($n = 1$)	BM-MSC	Direct application + injection	Wound size decreased and thickness of epidermis increased after 1 month in this single patient	Vojtassak et al. (2006)
Critical limb ischemia ($n = 4$)	BM-MSC	Intra-arterial administration	Significant pain relief and improved wound healing	Das et al. (2013)
Critical limb ischemia ($n = 9$)	ASC	Intramuscular injection	Improved wound healing in six of nine patients	Lee et al. (2012)
Ischemic ulcer ($n = 6$)	ASC	Intramuscular injection	Ulcer size was decreased	Bura et al. (2014)
Chronic ulcer lower limb ($n = 10$)	ASC	Direct injection	Six of ten patients experienced complete wound healing	Marino et al. (2013)

(Basiouny et al. 2013). As mentioned earlier, the dermis is remodeled in part through contribution of fibroblasts and deposition of collagen, and BM-MSCs may impact this process. In acute mouse wounds, BM-MSCs transcribe both collagen I and III and contributed to the production of 15–20% of the dermal fibroblast population (Fathke et al. 2004). Similar beneficial effects of BM-MSCs have been shown in diabetic wound healing in animals. Treatment with BM-MSCs has led to enhanced wound closure (Falanga et al. 2007), accelerated wound healing (Ha et al. 2010), and increased reepithelization (Fiorina et al. 2010) in various acute wound healing animal models.

The method of delivery of BM-MSCs may influence the healing effects observed. For instance, BM-MSCs seeded on a pullulan-collagen hydrogel scaffold accelerated healing and increased angiogenesis in a murine excisional wound model (Rustad et al. 2012). Interestingly, BM-MSCs seeded on this hydrogel expressed increased levels of VEGF and were able to differentiate into fibroblasts, endothelial cells, and pericytes, but not epithelial cells (Rustad et al. 2012). In contrast, injection of BM-MSCs after acute skin damage caused by irradiation speeds up wound healing and increases expression of growth factors such as TGF-β1 and SDF-1 in rats (Zheng et al. 2015). Lastly, one study observed increased healing rate, increased collagen fibers, increased tensile strength, and infiltration of CD68+ macrophages in mice after BM-MSCS were delivered topically via fibrin glue (Mehanna et al. 2015). Treatment of the BM-MSCs also has an effect on wound healing as BM-MSCs that undergo hypoxia pretreatment are more effective at healing in rats with diabetic hindlimb ischemia (Tong et al. 2015). Optimizing the method of delivery and the way BM-MSCs are prepared is crucial to developing effective stem cell therapies in cutaneous wound healing.

There are no clinical studies that examine the effect of BM-MSCs in treating hypertrophic scarring and keloids; however there are a couple animal studies that demonstrate the potential of BM-MSCs in reducing fibrosis. Infusion of human MSCs significantly reduced fibrosis in a dermal fibrosis mouse model through suppression of fibroblast proliferation (Wu et al. 2015). Another study showed that local injection of human MSCs prevented hypertrophic scar formation in rabbits through regulation of the anti-inflammatory protein, TSG-6, rather than modulating fibroblast activity like in the previous study. The efficacy of BM-MSCs in treating acute and complex wounds and the associated mechanisms need to be further explored.

9.5.3 Adipose-Derived Stem Cells

As mentioned earlier, harvesting BM-MSCs from patients is invasive and painful and yields relatively few cells. Adipose-derived stem cells (ASCs) have been identified as a new source of adult stem cells that appear to be promising in the field of regenerative medicine. Harvesting ASCs from adipose tissue is less invasive, less painful, and avoids the ethical and immunological concerns plaguing other stem cells (Zuk et al. 2002). Adipose tissue is also remarkable source of ASCs and can easily be isolated from a section of whole fat (biopsies) or lipoaspirate following surgery in patients (Zuk et al. 2002). For example, adipose tissue yields a 500-fold greater number of stem cells when compared to an equivalent amount of the bone marrow (Strioga et al. 2012). ASCs can differentiate into adipocytes, chondrocytes, and myogenic lineages; however one drawback is that ASCs do not offer as much plasticity as BM-MSCs (Gimble et al. 2012). Similarly to BM-MSCs, ASCs secrete a number of different cytokines and growth factors that promote wound healing (Kilroy et al. 2007). There are a couple drawbacks to ASCs however. First, the metabolic activity of ASCs (i.e., proliferation and differentiation capacity) depends on the anatomical location of the fat, age, and gender of the patient (Bailey et al. 2010). Second, it has been reported that ASCs undergo spontaneous malignant transformation after 4 months of culturing (Rubio et al. 2005); however researchers argue that this results from contamination rather than some inherent characteristic of ASCs (Garcia et al. 2010; Torsvik et al. 2010). Nevertheless, the above characteristics make ASCs an exciting alternative in the field of stem cell therapy, and a number of studies in various conditions have shown the therapeutic potential of ASCs in wound healing.

In 2012, Lee et al. (2012) showed that multiple intramuscular injections of ASCs resulted in clinical improvement in 66.7% of patients with critical limb ischemia. These patients exhibited decreased pain and improved claudication walking distance suggesting that multiple intramuscular injections of ASCs may be a viable therapeutic method of increasing blood flow in patients with critical limb ischemia (Lee et al. 2012). An interesting technique that may be used in the future to prevent the development of chronic wounds may be attaching ASCs (or other stem cells) to surgical sutures. Biodegradable surgical sutures filled with ASCs create a pro-regenerative environment in vitro with a steady release of cytokines and endothelial growth factors (Reckhenrich et al. 2014). Moreover, conditioned media from the supernatant of these cell-filled sutures significantly decreased wound area compared to controls in a wound healing assay (Reckhenrich et al. 2014).

Animal studies have demonstrated the wide variety of positive effects ASCs can have on cutaneous wound healing. Following acute wounds, injection of ASCs increases epithelization, decreases healing time, and reduces inflammation in fisher rats and rabbits (Pelizzo et al. 2015; Uysal et al. 2014). A recent study by Mendez et al. (2015) has shown that subcutaneous injection of ASCs decreases wound size and increases granulation and collagen synthesis in mice with full-thickness wounds. Interestingly, these researchers were able to develop a fibrin formulation that transforms ASCs into vascular tubes that express various endothelial markers (Mendez et al. 2015). It is thought that the growth of new blood vessels via the release of endothelial growth factors such as VEGF and HGF may be one mechanism by which ASCs exert their healing effects. In fact, preclinical studies have reported improved tissue hydration and angiogenesis after treatment with ASCs in patients with radiation-induced skin damage (Rigotti et al. 2007; Akita et al. 2010). Two animal studies also show increased vascularization and improved healing after acute wounds in mice and rats, respectively (Garg et al. 2014; Meruane et al. 2012).

One study has shown that ASCs may be more effective than BM-MSCs in treating acute wounds (find reference) and found that local injection of ASCs in rabbit excisional wounds was more effective than BM-MSCs in improving tensile strength and attenuating scar formation. With regard to deficient healing, allogenic transplantation of ASCs accelerated wound healing and vascularization in a diabetic rat wound model (Kato et al. 2015). Recently, injection of ASCs had positive effects in young and old mice with pressure ulcers (Strong et al. 2015): observed accelerated wound closure, increased adipogenesis, and improved epidermal structure in these mice.

There are no human studies examining the effect of ASCs in excessive healing; however there are couple of animal studies which are promising. In fibrosis rabbit ear model, it was shown that intralesional injection of ASCs decreased expression of α-SMA and collagen type I leading to diminished collagen deposition and hence reduced hypertrophic scarring (Zhang et al. 2015). Another study revealed that ASCs delivered through an ECM patch derived from porcine small intestine submucosa (SIS) significantly reduce fibrotic scar size in mice (Lam et al. 2013). In the future, more animal studies are needed to justify the use of ASCs in treating hypertrophic scars and keloids in humans. The unique trophic functions of ASCs combined with the studies described above make them an exciting alternative in stem cell-based therapies and support their applicability in cutaneous wound healing.

9.5.4 Umbilical and Placental Derived Stem Cells

It is believed that umbilical cord blood is the largest potential source of hematopoietic and non-hematopoietic stem cells with naïve immune status (Knudtzon 1974). These cord blood stem cells have been shown to differentiate into hepatic, pancreatic, and neural precursor cells (McGuckin et al. 2006; Denner et al. 2007). In fact, hematopoietic stem cells from cord blood have been used to treat many pediatric and adult diseases (Branski et al. 2009). Moreover, MSCs have been isolated from umbilical cord blood and successfully differentiated into epithelial cells in vitro (Sanmano et al. 2005; Kamolz et al. 2006). In addition to the blood, different layers of the umbilical cord such as Wharton's jelly (the gelatinous supporting matrix) and the outer lining are a source of stem cells. Stem cells isolated from the cord lining can be divided into two groups: epithelial cells and mesenchymal cells. These cells express typical stem cell markers such as Oct-4 and Nanog (In 't Anker et al. 2003) and are capable of differentiating into epithelial cells, hepatocytes, neural cells, and endothelium. These characteristics suggest that cord lining stem cells can potentially be used in the clinic to regenerate epithelial cells in cutaneous wound healing. While the umbilical cord is large reservoir of stem cells, isolating these cells is difficult as the umbilical cord provides 30% less cells when compared with MSCs isolated from the bone marrow (Wilson et al. 2011). Umbilical cord stem cells are a promising new source of new tissue in skin regeneration, and coadministration with other stem cells like BM-MSCs or ASCs may offer synergistic benefits in wound healing. Many clinical trials are ongoing to test the efficacy of umbilical cord-derived stem cells in treating diabetic wounds and burns. These studies have not reported any immunological rejection or tumor formation indicating that these cells pose no risk to patients.

Transplantation of Wharton's jelly MSCs increases expression of genes in reepithelization and promoted wound healing in mouse excisional wounds (Shi et al. 2015). Other studies have reported improved wound healing after administration of Wharton's jelly MSCs in addition to increased keratinocyte differentiation (Luo et al. 2010), reduced scar formation, and improved hair growth (Sabapathy et al. 2014). Delivering umbilical cord MSCs via a collagen-fibrin double-layered reduced healing time in a mouse full-thickness wound model further illustrating that cell delivery is an important aspect of developing stem cell therapies (Nan et al. 2015). The mechanism of action of umbilical cord-derived stem cells may be modulation of growth factors involved in the TGF-β signaling pathway. Interestingly, intraperitoneal injection of placenta-derived MSCs accelerates wound healing by increasing the expression of pro-angiogenic factors such as VEGF (Arno et al. 2014). In deficient healing, cord-derived MSCs delivered via topical application accelerated wound healing in db/db mice (Ha et al. 2010). Interestingly, TGF-β expression was significantly increased after the first week of treatment. Lastly, Wharton's jelly MSCs increase proliferation rate of keloid fibroblasts in vitro through regulation of TGF-β2 further supporting the notion that umbilical cord-derived MSCs modulate TGF-β activity (Arno et al. 2014).

9.5.5 *Epidermal Stem Cells*

The skin has become a promising source of adult stem cells that could potentially be seeded onto dermal substitutes to create a scaffold with biological activity. In contrast to ESCs and iPSCs, epidermal stem cells are non-oncogenic, have no ethical concerns, and are easily accessible (Fuchs and Nowak 2008). As a result, these cells may solve many problems in the field of skin engineering such as immune rejection and lack of skin appendages. Epidermal stem cells are located in the basal layer of the epidermis and are unipotent and able to regenerate the epidermis in adults (Morasso and Tomic-Canic 2005). In contrast, stem cells located within the bulge region of the hair follicle are multipotent cells capable of repopulating the epidermis, sebaceous glands, and hair follicles (Blanpain et al. 2004). Additionally, these cells are K15-positive and can differentiate into smooth muscle cells, neurons, and glial cells (Amoh et al. 2010). During injury, hair follicle stem cells rapidly mobilize and migrate to repair the epidermis in vivo by contributing to reepithelization (Ito et al. 2005). This response is tightly regulated and stops once the wound has healed. Lineage tracing studies have confirmed that hair follicle stem cells are normally unipotent as they maintain their niche compartment, the hair follicle, during homeostasis (Tumbar et al. 2004).

In the treatment of deep burn wounds, transplantation of cultured epithelial cells containing epidermal stem cells and keratinocytes via a fibrin matrix is well established (Fang et al. 2014). There are also reports of cultured epidermal cells being effective in treating chronic skin ulcers and deep dermal wounds (Oshima et al. 2002). Likewise, animal studies have illustrated the potential of epidermal stem cells in treating acute and complex wounds. Transplantation of epidermal stem cells via a collagen-chitin biomimetic membrane has been shown to regenerate whole skin in mice with full-thickness skin defects (Shen et al. 2014). Comparable healing effects were also found in diabetic mice treated with stem cells isolated from the hair follicle dermal sheath (Ma et al. 2015). Furthermore, in vitro examination relieved that these stem cells secreted paracrine factors such as IL-6 and enhanced wound healing. Recently, a novel epidermal stem cell population marked by Lgr6 was found in the isthmus region of the hair follicle (Snippert et al. 2010). Similar to other hair follicle stem cells, transplantation of Lgr6-positive cells gives rise to all epidermal lineages and also migrate to and repair the interfollicular epidermis during injury (Snippert et al. 2010). Unlike other hair follicle stem cells, however, Lgr6-positive cells contribute to the maintenance of the isthmus region and the sebaceous glands. Lastly, in mice with full-thickness wounds, local injection of Lgr6-expressing epithelial cells onto the wound bed promotes hair growth, reepithelization, and angiogenesis demonstrating the potential of these cells to facilitate cutaneous wound healing (Lough et al. 2014).

9.5.6 *Induced Pluripotent Stem Cells*

The ethical and legal concerns surrounding ESCs were quickly forgotten in 2007 when a new class of stem cells was discovered which combines the advantages of embryonic and adult stem cells. Induced pluripotent stem cells (iPSCs) were generated by exposing fibroblasts cultures to a cocktail of genes such as Oct3/Oct4, Sox2, and Klf4 by retroviral delivery (Takahashi et al. 2007) and Oct4, Sox2, Nanog, and Lin278 by lentiviral vectors (Yu et al. 2007). These iPSCs were immature, pluripotent, and nonimmunogenic cells generated from adult-differentiated tissue that exhibited similar plasticity to human ESCs. Their pluripotency was confirmed by their capability of differentiating into various cell types in vitro, expression of pluripotent marker genes, and their ability to produce all three germ layers (Takahashi and Yamanaka 2006). There are, however, safety concerns regarding iPSCs as they are cancerous in an undifferentiated state, and 20% of chimeric mice develop tumors due to expression of the well-known oncogene, c-Myc (Cartwright et al. 2005). Despite these concerns, iPSCs hold great promise in regenerative medicine. Human-derived iPSCs have been differentiated into epithelial cells resulting in the development of all hair follicle lineages (Yang et al. 2014). To take this concept further, Itoh et al. (2011) generated a 3-D skin equivalent in vitro that was comprised mostly of iPSC-derived keratinocytes and fibroblasts. Lastly, iPSC-derived keratinocytes have been used to create a stratified epithelium that was subsequently used to treat recessive dystopic epidermolysis bullosa, a type of deficient healing (Sebastiano et al. 2014). Overall, iPSCs will play an important role in stem cell therapies, but further research is needed to improve safety and refine the methods of reprogramming these cells.

9.6 Methods of Delivering Stem Cells

The therapeutic benefit of stem cells is limited by the method of delivery, and this is a unique challenge that regenerative medicine faces. There are three main techniques that are utilized in stem cell therapy: injection-based delivery, topical application via spray, and scaffold-based delivery.

9.6.1 Injection-Based Delivery

Injection-based delivery is a commonly used technique to deliver stem cells, and it is well described in preclinical and early clinical studies (Falanga et al. 2007). This method involves directly injecting stem cells within a suspension or hydrogel matrix into the site of injury. While local administration of stem cells via injection is the simplest method available, it may not be the optimal method to deliver cells to a cutaneous wound. Cell viability is a major issue as the immense pressure from the injection process results in massive cell death (Garg et al. 2014). Moreover, the wound environment increases cell death and impairs cell attachment resulting in stem cells becoming disorganized and delocalized (Zhang et al. 2008). This is problematic because these stem cells may migrate and integrate with tissue away from the wound site, thus creating unwanted harmful effects in other parts of the body such as cancer.

9.6.2 Spray-Based Delivery

Another local administration technique is spray-based delivery of stem cells. This technique involves isolating stem cells and mixing them with a fibrinogen that forms fibrin upon administration of thrombin (Zimmerlin et al. 2013). The fibrin spray prevents the degradation of the stem cells and also facilitates the adherence of the stem cells to the wound (Falanga et al. 2007). Fibrin spray can also distribute cells across a large area unlike injection-based therapies; however this can also be a potential weakness as there might be poor control of cell density and spacing (Duscher et al. 2015). Furthermore, because the stem cells are topically applied to a non-protective and harsh wound environment, cell viability may be problematic. Regardless, many studies have proven spray-based delivery as a safe and effective technique at promoting wound healing (Falanga et al. 2007; Wu et al. 2011).

9.6.3 Scaffold-Based Stem Cell Therapy

In the past decade, alternative techniques have been developed to overcome the weaknesses of spray- and injection-based delivery methods. Scaffold-based therapy has been the foundation of new stem cell delivery techniques that hope to improve therapeutic benefit. The goal of scaffold-based therapy is to seed stem cells on bioscaffolds in the hope that stem cell viability, differentiation, and engraftment into the wound are enhanced. Bioscaffolds are typically composed of collagen, hyaluronic acid, pullulan, or chitosan that mimic the native tissue environment (Jayakumar et al. 2010; Landsman et al. 2009). An important characteristic of these scaffolds is sufficient porosity to support adequate transport of nutrients and waste (Celiz et al. 2014). Moreover, these scaffolds provide protection for the stem cells and important spatial cues that mimic a natural environment, thus increasing the likelihood that these cells become functional parts of the wound. Indeed, seeding stem cells onto a pullulan-collagen matrix preserves the "stemness" of the cells and accelerate wound healing (Rustad et al. 2012).

While bioscaffolds have tremendous potential, they are still limited by poor nutrient transport and vascularization after engraftment (Meinel et al. 2004). Novel approaches need to be developed that will improve cell seeding efficiency, cell viability, cell uniformity, and cell integration. For instance, ASCs have been seeded more effectively on a hydrogel scaffold via capillary force than other methods such as injection and centrifuge and were effective at improving wound healing (Garg et al. 2014). As mentioned earlier, growth factors and cytokines play an important role in the wound healing process. Integrating these molecules, or even plasmids and vectors that modulate stem cell activity, within a scaffold could potentially help facilitate healing. Another novel approach to improving scaffold-based cell delivery is modifying the structure and composition of these scaffolds. An example would be altering the intermolecular bonds within the matrix to improve porosity, and interestingly one study has demonstrated that this can modulate cell proliferation and differentiation (Jeon and Alsberg 2013). These examples illustrate the importance of optimizing the delivery method so that stem cell-based therapies in wound healing can be readied for clinical trials. Unfortunately, bioscaffolds are limited by manufacturing costs and safety concerns (Duscher et al. 2015), but future advancements in biology and engineering may eventually produce a viable stem cell-based treatment option for acute and complex wounds.

9.7 Challenges and Future Directions in Stem Cell Therapy

Most stem cell-based therapies discussed in this chapter have not been proven clinically. While stem cell therapy in cutaneous wound healing has shown tremendous promise, there are still significant challenges that need to be addressed before these therapies are clinically relevant. Firstly, there is no evidence that skin and appendages regenerated by stem cell therapies are functional – flawless regeneration without scarring has yet to be achieved. There needs to be studies demonstrating confirming whether regenerated skin recapitulates the function of intact skin. Second, harvesting enough adult stem cells with high purity is obstructing the progress of developing new therapies. Lastly, the differentiation potential of stem cells may give rise to malignant cells as the niche microenvironment is crucial in directing tumor growth. Optimizing cell growth to replace damaged tissue while at the same time preventing disproportionate cell proliferation and differentiation remains a challenge for researchers. To accomplish this a robust understanding of the cellular and molecular mechanisms underlying stem cell action is necessary. Therefore, innovative approaches are needed to refine the manipulation of stem cells, improve the methods of delivering these cells, and finally validate these therapies through clinical trials so that patients can reap the benefits of regenerative medicine.

9.8 Conclusion

In this chapter we addressed the potential of stem cells in the treatment of acute and complex cutaneous wounds. There are various aspects to consider when developing stem cell therapies such as the signaling pathways involved in wound healing, the source of stem cells, and the method of delivery. While there may be significant barriers limiting the clinical translation of stem cell-based therapies, the clinical and animal studies discussed illustrate the therapeutic potential of stem cell therapy in regenerative medicine. More clinical studies are needed so that the plethora of research in basic science can be translated into clinical wound care.

References

Abd-Allah SH, El-Shal AS, Shalaby SM, Abd-Elbary E, Mazen NF, Abdel Kader RR (2015) The role of placenta-derived mesenchymal stem cells in healing of induced full-thickness skin wound in a mouse model. IUBMB Life 67(9):701–709. doi:10.1002/iub.1427. Epub 2015/09/01; PubMed PMID: 26315141,https://www.ncbi.nlm.nih.gov/pubmed/26315141

Aggarwal S, Pittenger MF (2005) Human mesenchymal stem cells modulate allogeneic immune cell responses. Blood 105(4):1815–1822. doi:10.1182/blood-2004-04-1559. Epub 2004/10/21; PubMed PMID: 15494428

Agren MS, Eaglstein WH, Ferguson MW, Harding KG, Moore K, Saarialho-Kere UK et al (2000) Causes and effects of the chronic inflammation in venous leg ulcers. Acta Derm Venereol Suppl 210:3–17. Epub 2000/07/08. PubMed PMID: 10884942

Akita S, Akino K, Hirano A, Ohtsuru A, Yamashita S (2010) Mesenchymal stem cell therapy for cutaneous radiation syndrome. Health Phys 98(6):858–862. doi:10.1097/HP.0b013e3181d3d52c. Epub 2010/05/07; PubMed PMID: 20445394

Amini-Nik S, Glancy D, Boimer C, Whetstone H, Keller C, Alman BA (2011) Pax7 expressing cells contribute to dermal wound repair, regulating scar size through a beta-catenin mediated process. Stem Cells (Dayton, Ohio) 29(9):1371–1379. doi:10.1002/stem.688. Epub 2011/07/09; PubMed PMID: 21739529

Amini-Nik S, Cambridge E, Yu W, Guo A, Whetstone H, Nadesan P et al (2014) Beta-catenin-regulated myeloid cell adhesion and migration determine wound healing. J Clin Invest 124(6):2599–2610. doi:10.1172/jci62059. Epub 2014/05/20; PubMed PMID: 24837430; PubMed Central PMCID: PMCPmc4089463

Amoh Y, Katsuoka K, Hoffman RM (2010) The advantages of hair follicle pluripotent stem cells over embryonic stem cells and induced pluripotent stem cells for regenerative medicine. J Dermatol Sci 60(3):131–137. doi:10.1016/j.jdermsci.2010.09.007. Epub 2010/11/03; PubMed PMID: 21036545

Arno AI, Amini-Nik S, Blit PH, Al-Shehab M, Belo C, Herer E et al (2014) Effect of human Wharton's jelly mesenchymal stem cell paracrine signaling on keloid fibroblasts. Stem Cells Transl Med 3(3):299–307. doi:10.5966/sctm.2013-0120. Epub 2014/01/18; PubMed PMID: 24436441; PubMed Central PMCID: PMCPmc3952928

Asai J, Takenaka H, Kusano KF, Ii M, Luedemann C, Curry C et al (2006) Topical sonic hedgehog gene therapy accelerates wound healing in diabetes by enhancing endothelial progenitor cell-mediated microvascular remodeling. Circulation 113(20):2413–2424. doi:10.1161/circulationaha.105.603167. Epub 2006/05/17; PubMed PMID: 16702471

Ashcroft GS, Yang X, Glick AB, Weinstein M, Letterio JL, Mizel DE et al (1999) Mice lacking Smad3 show accelerated wound healing and an impaired local inflammatory response. Nat Cell Biol 1(5):260–266. doi:10.1038/12971. Epub 1999/11/13; PubMed PMID: 10559937

Athar M, Tang X, Lee JL, Kopelovich L, Kim AL (2006) Hedgehog signalling in skin development and cancer. Exp Dermatol 15(9):667–677. doi:10.1111/j.1600-0625.2006.00473.x. Epub 2006/08/03; PubMed PMID: 16881963

Badiavas EV, Falanga V (2003) Treatment of chronic wounds with bone marrow-derived cells. Arch Dermatol 139(4):510–516. doi:10.1001/archderm.139.4.510. Epub 2003/04/23; PubMed PMID: 12707099

Bailey AM, Kapur S, Katz AJ (2010) Characterization of adipose-derived stem cells: an update. Curr Stem Cell Res Ther 5(2):95–102. Epub 2009/11/28. PubMed PMID: 19941461

Basiouny HS, Salama NM, Maadawi ZM, Farag EA (2013) Effect of bone marrow derived mesenchymal stem cells on healing of induced full-thickness skin wounds in albino rat. Int J Stem Cells 6(1):12–25. Epub 2013/12/04. PubMed PMID: 24298370; PubMed Central PMCID: PMCPmc3840998

Bielefeld KA, Amini-Nik S, Alman BA (2013) Cutaneous wound healing: recruiting developmental pathways for regeneration. Cell Mol Life Sci: CMLS 70(12):2059–2081. doi:10.1007/s00018-012-1152-9. Epub 2012/10/12; PubMed PMID: 23052205; PubMed Central PMCID: PMCPmc3663196

Biernacka A, Dobaczewski M, Frangogiannis NG (2011) TGF-beta signaling in fibrosis. Growth Factors (Chur, Switzerland) 29(5):196–202. doi:10.3109/08977194.2011.595714. Epub 2011/07/12; PubMed PMID: 21740331; PubMed Central PMCID: PMCPmc4408550

Blanpain C, Lowry WE, Geoghegan A, Polak L, Fuchs E (2004) Self-renewal, multipotency, and the existence of two cell populations within an epithelial stem cell niche. Cell 118(5):635–648. doi:10.1016/j.cell.2004.08.012. Epub 2004/09/02; PubMed PMID: 15339667

Blau HM, Cosgrove BD, Ho AT (2015) The central role of muscle stem cells in regenerative failure with aging. Nat Med 21(8):854–862. doi:10.1038/nm.3918. Epub 2015/08/08; PubMed PMID: 26248268

Boxall SA, Jones E (2012) Markers for characterization of bone marrow multipotential stromal cells. Stem Cells Int 2012:975871. doi:10.1155/2012/975871. Epub 2012/06/06; PubMed PMID: 22666272; PubMed Central PMCID: PMCPmc3361338

Branski LK, Herndon DN, Pereira C, Mlcak RP, Celis MM, Lee JO et al (2007) Longitudinal assessment of Integra in primary burn management: a randomized pediatric clinical trial. Crit Care Med 35(11):2615–2623. doi:10.1097/01.ccm.0000285991.36698.e2. Epub 2007/09/11; PubMed PMID: 17828040

Branski LK, Gauglitz GG, Herndon DN, Jeschke MG (2009) A review of gene and stem cell therapy in cutaneous wound healing. Burn J Int Soc Burn Inj 35(2):171–180. doi:10.1016/j.burns.2008.03.009. Epub 2008/07/08; PubMed PMID: 18603379; PubMed Central PMCID: PMCPmc3899575

Brem H, Stojadinovic O, Diegelmann RF, Entero H, Lee B, Pastar I et al (2007) Molecular markers in patients with chronic wounds to guide surgical debridement. Mol Med (Cambridge, Mass) 13(1–2):30–39. doi:10.2119/2006-00054.Brem. Epub 2007/05/23; PubMed PMID: 17515955; PubMed Central PMCID: PMCPmc1869625

Bura A, Planat-Benard V, Bourin P, Silvestre JS, Gross F, Grolleau JL et al (2014) Phase I trial: the use of autologous cultured adipose-derived stroma/stem cells to treat patients with non-revascularizable critical limb ischemia. Cytotherapy 16(2):245–257. doi:10.1016/j.jcyt.2013.11.011. Epub 2014/01/21; PubMed PMID: 24438903

Caiado F, Real C, Carvalho T, Dias S (2008) Notch pathway modulation on bone marrow-derived vascular precursor cells regulates their angiogenic and wound healing potential. PLoS ONE 3(11):e3752. doi:10.1371/journal.pone.0003752. Epub 2008/11/19; PubMed PMID: 19015735; PubMed Central PMCID: PMCPmc2582964

Cartwright P, McLean C, Sheppard A, Rivett D, Jones K, Dalton S (2005) LIF/STAT3 controls ES cell self-renewal and pluripotency by a Myc-dependent mechanism. Development 132(5):885–896. doi:10.1242/dev.01670. Epub 2005/01/28; PubMed PMID: 15673569

Celiz AD, Smith JG, Langer R, Anderson DG, Winkler DA, Barrett DA et al (2014) Materials for stem cell factories of the future. Nat Mater 13(6):570–579. doi:10.1038/nmat3972. Epub 2014/05/23; PubMed PMID: 24845996

Cha J, Falanga V (2007) Stem cells in cutaneous wound healing. Clin Dermatol 25(1):73–78. doi:10.1016/j.clindermatol.2006.10.002. Epub 2007/02/06; PubMed PMID: 17276204

Chang EI, Loh SA, Ceradini DJ, Chang EI, Lin SE, Bastidas N et al (2007) Age decreases endothelial progenitor cell recruitment through decreases in hypoxia-inducible factor 1alpha stabilization during ischemia. Circulation 116(24):2818–2829. doi:10.1161/circulationaha.107.715847. Epub 2007/11/28; PubMed PMID: 18040029

Chanprapaph K, Tanrattanakorn S, Wattanakrai P, Wongkitisophon P, Vachiramon V (2012) Effectiveness of onion extract gel on surgical scars in Asians. Dermatol Res Pract 2012:212945. doi:10.1155/2012/212945. Epub 2012/08/28; PubMed PMID: 22924037; PubMed Central PMCID: PMCPmc3423794

Cheon SS, Cheah AY, Turley S, Nadesan P, Poon R, Clevers H et al (2002) Beta-catenin stabilization dysregulates mesenchymal cell proliferation, motility, and invasiveness and causes aggressive fibromatosis and hyperplastic cutaneous wounds. Proc Natl Acad Sci U S A 99(10):6973–6978. doi:10.1073/pnas.102657399. Epub 2002/05/02; PubMed PMID: 11983872; PubMed Central PMCID: PMCPmc124513

Cheon SS, Nadesan P, Poon R, Alman BA (2004) Growth factors regulate beta-catenin-mediated TCF-dependent transcriptional activation in fibroblasts during the proliferative phase of wound healing. Exp Cell Res 293(2):267–274. Epub 2004/01/20. PubMed PMID: 14729464

Cheon S, Poon R, Yu C, Khoury M, Shenker R, Fish J et al (2005) Prolonged beta-catenin stabilization and tcf-dependent transcriptional activation in hyperplastic cutaneous wounds. Lab Inv J Tech Methods Pathol 85(3):416–425. doi:10.1038/labinvest.3700237. Epub 2005/01/18; PubMed PMID: 15654359

Cheon SS, Wei Q, Gurung A, Youn A, Bright T, Poon R et al (2006) Beta-catenin regulates wound size and mediates the effect of TGF-beta in cutaneous healing. FASEB J: Off Publ Fed Am Soc Exp Biol 20(6):692–701. doi:10.1096/fj.05-4759com. Epub 2006/04/04; PubMed PMID: 16581977

Chiang C, Swan RZ, Grachtchouk M, Bolinger M, Litingtung Y, Robertson EK et al (1999) Essential role for Sonic hedgehog during hair follicle morphogenesis. Dev Biol 205(1):1–9. doi:10.1006/dbio.1998.9103. Epub 1999/01/12; PubMed PMID: 9882493

Chigurupati S, Arumugam TV, Son TG, Lathia JD, Jameel S, Mughal MR et al (2007) Involvement of notch signaling in wound healing. PLoS ONE 2(11):e1167. doi:10.1371/journal.pone.0001167. Epub 2007/11/15; PubMed PMID: 18000539; PubMed Central PMCID: PMCPmc2048753

Cho JW, Cho SY, Lee SR, Lee KS (2010) Onion extract and quercetin induce matrix metalloproteinase-1 in vitro and in vivo. Int J Mol Med 25(3):347–352. Epub 2010/02/04. PubMed PMID: 20127038

Choudhery MS, Khan M, Mahmood R, Mehmood A, Khan SN, Riazuddin S (2012) Bone marrow derived mesenchymal stem cells from aged mice have reduced wound healing, angiogenesis, proliferation and anti-apoptosis capabilities. Cell Biol Int 36(8):747–753. doi:10.1042/cbi20110183. Epub 2012/02/23; PubMed PMID: 22352320

Clark RA (1990) Fibronectin matrix deposition and fibronectin receptor expression in healing and normal skin. J Invest Dermatol 94(6 Suppl):128s–134s. Epub 1990/06/01. PubMed PMID: 2161886

Crowe MJ, Doetschman T, Greenhalgh DG (2000) Delayed wound healing in immunodeficient TGF-beta 1 knockout mice. J Invest Dermatol 115(1):3–11. doi:10.1046/j.1523-1747.2000.00010.x. Epub 2000/07/08; PubMed PMID: 10886500

Darby I, Skalli O, Gabbiani G (1990) Alpha-smooth muscle actin is transiently expressed by myofibroblasts during experimental wound healing. Lab Inv J Tech Methods Pathol 63(1):21–29. Epub 1990/07/01; PubMed PMID: 2197503

Das AK, Bin Abdullah BJ, Dhillon SS, Vijanari A, Anoop CH, Gupta PK (2013) Intra-arterial allogeneic mesenchymal stem cells for critical limb ischemia are safe and efficacious: report of a phase I study. World J Surg 37(4):915–922. doi:10.1007/s00268-012-1892-6. Epub 2013/01/12; PubMed PMID: 23307180

Dash NR, Dash SN, Routray P, Mohapatra S, Mohapatra PC (2009) Targeting nonhealing ulcers of lower extremity in human through autologous bone marrow-derived mesenchymal stem cells. Rejuvenation Res 12(5):359–366. doi:10.1089/rej.2009.0872. Epub 2009/11/26; PubMed PMID: 19929258

Denner L, Bodenburg Y, Zhao JG, Howe M, Cappo J, Tilton RG et al (2007) Directed engineering of umbilical cord blood stem cells to produce C-peptide and insulin. Cell Prolif 40(3):367–380. doi:10.1111/j.1365-2184.2007.00439.x. Epub 2007/05/29; PubMed PMID: 17531081

Desmouliere A, Geinoz A, Gabbiani F, Gabbiani G (1993) Transforming growth factor-beta 1 induces alpha-smooth muscle actin expression in granulation tissue myofibroblasts and in quiescent and growing cultured fibroblasts. J Cell Biol 122(1):103–111. Epub 1993/07/01. PubMed PMID: 8314838; PubMed Central PMCID: PMCPmc2119614

Dreifke MB, Jayasuriya AA, Jayasuriya AC (2015) Current wound healing procedures and potential care. Mater Sci Eng C Mater Biol Appl 48:651–662. doi:10.1016/j.msec.2014.12.068. Epub 2015/01/13; PubMed PMID: 25579968; PubMed Central PMCID: PMCPmc4443476

Duscher D, Barrera J, Wong VW, Maan ZN, Whittam AJ, Januszyk M et al (2015) Stem cells in wound healing: the future of regenerative medicine? A mini-review. Gerontology. doi:10.1159/000381877. Epub 2015/06/06; PubMed PMID: 26045256

Falanga V (2000) Classifications for wound bed preparation and stimulation of chronic wounds. Wound Repair Regen: Off Publ Wound Healing Soc Eur Tissue Repair Soc 8(5):347–352. Epub 2000/12/15. PubMed PMID: 11115147

Falanga V (2012) Stem cells in tissue repair and regeneration. J Invest Dermatol 132(6):1538–1541. doi:10.1038/jid.2012.77. Epub 2012/05/16; PubMed PMID: 22584501; PubMed Central PMCID: PMCPmc4084617

Falanga V, Iwamoto S, Chartier M, Yufit T, Butmarc J, Kouttab N et al (2007) Autologous bone marrow-derived cultured mesenchymal stem cells delivered in a fibrin spray accelerate healing in murine and human cutaneous wounds. Tissue Eng 13(6):1299–1312. doi:10.1089/ten.2006.0278. Epub 2007/05/24; PubMed PMID: 17518741

Fang T, Lineaweaver WC, Sailes FC, Kisner C, Zhang F (2014) Clinical application of cultured epithelial autografts on acellular dermal matrices in the treatment of extended burn injuries. Ann Plast Surg 73(5):509–515. doi:10.1097/SAP.0b013e3182840883. Epub 2013/12/11; PubMed PMID: 24322642

Fathke C, Wilson L, Hutter J, Kapoor V, Smith A, Hocking A et al (2004) Contribution of bone marrow-derived cells to skin: collagen deposition and wound repair. Stem Cells (Dayton, Ohio) 22(5):812–822. doi:10.1634/stemcells.22-5-812. Epub 2004/09/03; PubMed PMID: 15342945; PubMed Central PMCID: PMCPmc1388268

Ferguson MW, O'Kane S (2004) Scar-free healing: from embryonic mechanisms to adult therapeutic intervention. Philos Trans R Soc Lond Ser B Biol Sci 359(1445):839–850. doi:10.1098/rstb.2004.1475. Epub 2004/08/06; PubMed PMID: 15293811; PubMed Central PMCID: PMCPmc1693363

Ferguson MW, Duncan J, Bond J, Bush J, Durani P, So K et al (2009) Prophylactic administration of avotermin for improvement of skin scarring: three double-blind, placebo-controlled, phase I/II studies. Lancet (London, England) 373(9671):1264–1274. doi:10.1016/s0140-6736(09)60322-6. Epub 2009/04/14; PubMed PMID: 19362676

Fiorina P, Pietramaggiori G, Scherer SS, Jurewicz M, Mathews JC, Vergani A et al (2010) The mobilization and effect of endogenous bone marrow progenitor cells in diabetic wound healing. Cell Transplant 19(11):1369–1381. doi:10.3727/096368910x514288. Epub 2010/10/28; PubMed PMID: 20977829

Fuchs E, Nowak JA (2008) Building epithelial tissues from skin stem cells. Cold Spring Harb Symp Quant Biol 73:333–350. doi:10.1101/sqb.2008.73.032. Epub 2008/11/22; PubMed PMID: 19022769; PubMed Central PMCID: PMCPmc2693088

Garcia S, Bernad A, Martin MC, Cigudosa JC, Garcia-Castro J, de la Fuente R (2010) Pitfalls in spontaneous in vitro transformation of human mesenchymal stem cells. Exp Cell Res 316(9):1648–1650. doi:10.1016/j.yexcr.2010.02.016. Epub 2010/02/23; PubMed PMID: 20171963

Garg RK, Rennert RC, Duscher D, Sorkin M, Kosaraju R, Auerbach LJ et al (2014) Capillary force seeding of hydrogels for adipose-derived stem cell delivery in wounds. Stem Cells Transl Med 3(9):1079–1089. doi:10.5966/sctm.2014-0007. Epub 2014/07/20; PubMed PMID: 25038246; PubMed Central PMCID: PMCPmc4149299

Gimble JM, Bunnell BA, Guilak F (2012) Human adipose-derived cells: an update on the transition to clinical translation. Regen Med 7(2):225–235. doi:10.2217/rme.11.119. Epub 2012/03/09; PubMed PMID: 22397611; PubMed Central PMCID: PMCPmc3321837

Gnecchi M, Zhang Z, Ni A, Dzau VJ (2008) Paracrine mechanisms in adult stem cell signaling and therapy. Circ Res 103(11):1204–1219. doi:10.1161/circresaha.108.176826. Epub 2008/11/26; PubMed PMID: 19028920; PubMed Central PMCID: PMCPmc2667788

Guffanti A (2014) Negative pressure wound therapy in the treatment of diabetic foot ulcers: a systematic review of the literature. J Wound Ostomy Continence Nurs: Off Publ Wound Ostomy Continence Nurses Soc WOCN 41(3):233–237. doi:10.1097/won.0000000000000021. Epub 2014/05/09; PubMed PMID: 24805174

Guo S, Dipietro LA (2010) Factors affecting wound healing. J Dent Res 89(3):219–229. doi:10.1177/0022034509359125. Epub 2010/02/09; PubMed PMID: 20139336; PubMed Central PMCID: PMCPmc2903966

Gupta PK, Chullikana A, Parakh R, Desai S, Das A, Gottipamula S et al (2013) A double blind randomized placebo controlled phase I/II study assessing the safety and efficacy of allogeneic bone marrow derived mesenchymal stem cell in critical limb ischemia. J Transl Med 11:143. doi:10.1186/1479-5876-11-143. Epub 2013/06/14; PubMed PMID: 23758736; PubMed Central PMCID: PMCPmc3688296

Gurtner GC, Werner S, Barrandon Y, Longaker MT (2008) Wound repair and regeneration. Nature 453(7193):314–321. doi:10.1038/nature07039. Epub 2008/05/16; PubMed PMID: 18480812

Ha X, Yin Q, Dong F, Jia Q, Lv T (2010) Study on bone marrow mesenchymal stem cells transfected with adenovirus hepatocyte growth factor gene promoting wounds repair in diabetic rats. Zhongguo Xiu Fu Chong Jian Wai Ke Za Zhi = Zhongguo Xiufu Chongjian Waike Zazhi = Chin J Reparative Reconstr Surg 24(12):1520–1524. Epub 2011/01/26. PubMed PMID: 21261106

Heng MC (2011) Wound healing in adult skin: aiming for perfect regeneration. Int J Dermatol 50(9):1058–1066. doi:10.1111/j.1365-4632.2011.04940.x. Epub 2011/12/01; PubMed PMID: 22126865

Hocevar BA, Brown TL, Howe PH (1999) TGF-beta induces fibronectin synthesis through a c-Jun N-terminal kinase-dependent, Smad4-independent pathway. EMBO J 18(5):1345–1356. doi:10.1093/emboj/18.5.1345. Epub 1999/03/04; PubMed PMID: 10064600; PubMed Central PMCID: PMCPmc1171224

Hostetler SG, Xiang H, Gupta S, Sen C, Gordillo JM (2006) Discharge patterns of injury-related hospitalizations with an acute wound in the United States. Wounds 18(12):340–351

Huelsken J, Vogel R, Erdmann B, Cotsarelis G, Birchmeier W (2001) Beta-catenin controls hair follicle morphogenesis and stem cell differentiation in the skin. Cell 105(4):533–545. Epub 2001/05/24. PubMed PMID: 11371349

In 't Anker PS, Scherjon SA, Kleijburg-van der Keur C, Noort WA, Claas FH, Willemze R et al (2003) Amniotic fluid as a novel source of mesenchymal stem cells for therapeutic transplantation. Blood 102(4):1548–1549. doi:10.1182/blood-2003-04-1291. Epub 2003/08/06; PubMed PMID: 12900350

Ito M, Liu Y, Yang Z, Nguyen J, Liang F, Morris RJ et al (2005) Stem cells in the hair follicle bulge contribute to wound repair but not to homeostasis of the epidermis. Nat Med 11(12):1351–1354. doi:10.1038/nm1328. Epub 2005/11/17; PubMed PMID: 16288281

Ito M, Yang Z, Andl T, Cui C, Kim N, Millar SE et al (2007) Wnt-dependent de novo hair follicle regeneration in adult mouse skin after wounding. Nature 447(7142):316–320. doi:10.1038/nature05766. Epub 2007/05/18; PubMed PMID: 17507982

Itoh M, Kiuru M, Cairo MS, Christiano AM (2011) Generation of keratinocytes from normal and recessive dystrophic epidermolysis bullosa-induced pluripotent stem cells. Proc Natl Acad Sci U S A 108(21):8797–8802. doi:10.1073/pnas.1100332108. Epub 2011/05/11; PubMed PMID: 21555586; PubMed Central PMCID: PMCPmc3102348

Jayakumar R, Prabaharan M, Nair SV, Tamura H (2010) Novel chitin and chitosan nanofibers in biomedical applications. Biotechnol Adv 28(1):142–150. doi:10.1016/j.biotechadv.2009.11.001. Epub 2009/11/17; PubMed PMID: 19913083

Jeon O, Alsberg E (2013) Regulation of stem cell fate in a three-dimensional micropatterned dual-crosslinked hydrogel system. Adv Funct Mater 23(38):4765–4775. doi:10.1002/adfm.201300529. Epub 2014/03/01; PubMed PMID: 24578678; PubMed Central PMCID: PMCPmc3933204

Jeschke MG, Rose C, Angele P, Fuchtmeier B, Nerlich MN, Bolder U (2004) Development of new reconstructive techniques: use of Integra in combination with fibrin glue and negative-pressure therapy for reconstruction of acute and chronic wounds. Plast Reconstr Surg 113(2):525–530. doi:10.1097/01.prs.0000100813.39746.5a. Epub 2004/02/06; PubMed PMID: 14758212

Jeschke MG, Patsouris D, Stanojcic M, Abdullahi A, Rehou S, Pinto R et al (2015) Pathophysiologic response to burns in the elderly. Biomedicine 2(10):1536–1548. doi:10.1016/j.ebiom.2015.07.040. PubMed PMID: 26629550; PubMed Central PMCID: PMC4634201

Kamolz LP, Kolbus A, Wick N, Mazal PR, Eisenbock B, Burjak S et al (2006) Cultured human epithelium: human umbilical cord blood stem cells differentiate into keratinocytes under in vitro conditions. Burns J Int Soc Burn Inj 32(1):16–19. doi:10.1016/j.burns.2005.08.020. Epub 2005/12/22; PubMed PMID: 16368194

Karagoz H, Yuksel F, Ulkur E, Evinc R (2009) Comparison of efficacy of silicone gel, silicone gel sheeting, and topical onion extract including heparin and allantoin for the treatment of postburn hypertrophic scars. Burns J Int Soc Burn Inj 35(8):1097–1103. doi:10.1016/j.burns.2009.06.206. Epub 2009/09/22; PubMed PMID: 19766399

Karlsson L, Bondjers C, Betsholtz C (1999) Roles for PDGF-A and sonic hedgehog in development of mesenchymal components of the hair follicle. Development 126(12):2611–2621. Epub 1999/05/20. PubMed PMID: 10331973

Kato Y, Iwata T, Morikawa S, Yamato M, Okano T, Uchigata Y (2015) Allogeneic transplantation of an adipose-derived stem cell sheet combined with artificial skin accelerates wound healing in a rat wound model of type 2 diabetes and obesity. Diabetes 64(8):2723–2734. doi:10.2337/db14-1133. Epub 2015/03/22; PubMed PMID: 25795216

Kilroy GE, Foster SJ, Wu X, Ruiz J, Sherwood S, Heifetz A et al (2007) Cytokine profile of human adipose-derived stem cells: expression of angiogenic, hematopoietic, and pro-inflammatory factors. J Cell Physiol 212(3):702–709. doi:10.1002/jcp.21068. Epub 2007/05/05; PubMed PMID: 17477371

Kirana S, Stratmann B, Prante C, Prohaska W, Koerperich H, Lammers D et al (2012) Autologous stem cell therapy in the treatment of limb ischaemia induced chronic tissue ulcers of diabetic foot patients. Int J Clin Pract 66(4):384–393. doi:10.1111/j.1742-1241.2011.02886.x. Epub 2012/01/31; PubMed PMID: 22284892

Knudtzon S (1974) In vitro growth of granulocytic colonies from circulating cells in human cord blood. Blood 43(3):357–361. Epub 1974/03/01. PubMed PMID: 4811820

Lam MT, Nauta A, Meyer NP, Wu JC, Longaker MT (2013) Effective delivery of stem cells using an extracellular matrix patch results in increased cell survival and proliferation and reduced scarring in skin wound healing. Tissue Eng A 19(5–6):738–747. doi:10.1089/ten.TEA.2012.0480. Epub 2012/10/18; PubMed PMID: 23072446; PubMed Central PMCID: PMCPmc3566655

Landsman A, Taft D, Riemer K (2009) The role of collagen bioscaffolds, foamed collagen, and living skin equivalents in wound healing. Clin Podiatr Med Surg 26(4):525–533. doi:10.1016/j.cpm.2009.08.012. Epub 2009/09/26; PubMed PMID: 19778686

Lataillade JJ, Doucet C, Bey E, Carsin H, Huet C, Clairand I et al (2007) New approach to radiation burn treatment by dosimetry-guided surgery combined with autologous mesenchymal stem cell therapy. Regen Med 2(5):785–794. doi:10.2217/17460751.2.5.785. Epub 2007/10/03; PubMed PMID: 17907931

Le H, Kleinerman R, Lerman OZ, Brown D, Galiano R, Gurtner GC et al (2008) Hedgehog signaling is essential for normal wound healing. Wound Repair Regen: Off Publ Wound Healing Soc Eur Tissue Repair Soc 16(6):768–773. doi:10.1111/j.1524-475X.2008.00430.x. Epub 2009/01/09; PubMed PMID: 19128247

Lee HC, An SG, Lee HW, Park JS, Cha KS, Hong TJ et al (2012) Safety and effect of adipose tissue-derived stem cell implantation in patients with critical limb ischemia: a pilot study. Circ J: Off J Jpn Circ Soc 76(7):1750–1760. Epub 2012/04/14. PubMed PMID: 22498564

Lerman OZ, Galiano RD, Armour M, Levine JP, Gurtner GC (2003) Cellular dysfunction in the diabetic fibroblast: impairment in migration, vascular endothelial growth factor production, and response to hypoxia. Am J Pathol 162(1):303–312. doi:10.1016/s0002-9440(10)63821-7. Epub 2003/01/01; PubMed PMID: 12507913; PubMed Central PMCID: PMCPmc1851127

Liu Y, Liu Y, Wang P, Tian H, Ai J, Liu Y et al (2014) Autologous bone marrow stem cell transplantation for the treatment of postoperative hand infection with a skin defect in diabetes mellitus: a case report. Oncol Lett 7(6):1857–1862. doi:10.3892/ol.2014.1998. Epub 2014/06/17; PubMed PMID: 24932248; PubMed Central PMCID: PMCPmc4049741

Lough DM, Yang M, Blum A, Reichensperger JD, Cosenza NM, Wetter N et al (2014) Transplantation of the LGR6+ epithelial stem cell into full-thickness cutaneous wounds results in enhanced healing, nascent hair follicle development, and augmentation of angiogenic analytes. Plast Reconstr Surg 133(3):579–590. doi:10.1097/prs.0000000000000075. Epub 2014/02/28; PubMed PMID: 24572851

Lu D, Chen B, Liang Z, Deng W, Jiang Y, Li S et al (2011) Comparison of bone marrow mesenchymal stem cells with bone marrow-derived mononuclear cells for treatment of diabetic critical limb ischemia and foot ulcer: a double-blind, randomized, controlled trial. Diabetes Res Clin Pract 92(1):26–36. doi:10.1016/j.diabres.2010.12.010. Epub 2011/01/11; PubMed PMID: 21216483

Luo G, Cheng W, He W, Wang X, Tan J, Fitzgerald M et al (2010) Promotion of cutaneous wound healing by local application of mesenchymal stem cells derived from human umbilical cord blood. Wound Repair Regen: Off Publ Wound Healing Soc Eur Tissue Repair Soc 18(5):506–513. doi:10.1111/j.1524-475X.2010.00616.x. Epub 2010/09/16; PubMed PMID: 20840520

Ma D, Kua JE, Lim WK, Lee ST, Chua AW (2015) In vitro characterization of human hair follicle dermal sheath mesenchymal stromal cells and their potential in enhancing diabetic wound healing. Cytotherapy 17(8):1036–1051. doi:10.1016/j.jcyt.2015.04.001. Epub 2015/05/20; PubMed PMID: 25981558

Macintyre L, Baird M (2006) Pressure garments for use in the treatment of hypertrophic scars – a review of the problems associated with their use. Burns J Int Soc Burn Inj 32(1):10–15. doi:10.1016/j.burns.2004.06.018. Epub 2006/01/18; PubMed PMID: 16413399

Mahdavian Delavary B, van der Veer WM, van Egmond M, Niessen FB, Beelen RH (2011) Macrophages in skin injury and repair. Immunobiology 216(7):753–762. doi:10.1016/j.imbio.2011.01.001. Epub 2011/02/02; PubMed PMID: 21281986

Margolis DJ, Allen-Taylor L, Hoffstad O, Berlin JA (2005) Diabetic neuropathic foot ulcers and amputation. Wound Repair Regen: Off Publ Wound Healing Soc Eur Tissue Repair Soc 13(3):230–236. doi:10.1111/j.1067-1927.2005.130303.x. Epub 2005/06/15; PubMed PMID: 15953040

Marino G, Moraci M, Armenia E, Orabona C, Sergio R, De Sena G et al (2013) Therapy with autologous adipose-derived regenerative cells for the care of chronic ulcer of lower limbs in patients with peripheral arterial disease. J Surg Res 185(1):36–44. doi:10.1016/j.jss.2013.05.024. Epub 2013/06/19; PubMed PMID: 23773718

Martin P, Nunan R (2015) Cellular and molecular mechanisms of repair in acute and chronic wound healing. Br J Dermatol 173(2):370–378. doi:10.1111/bjd.13954. Epub 2015/07/16; PubMed PMID: 26175283; PubMed Central PMCID: PMCPmc4671308

Martinez-Ferrer M, Afshar-Sherif AR, Uwamariya C, de Crombrugghe B, Davidson JM, Bhowmick NA (2010) Dermal transforming growth factor-beta responsiveness mediates wound contraction and epithelial closure. Am J Pathol 176(1):98–107. doi:10.2353/ajpath.2010.090283. Epub 2009/12/05; PubMed PMID: 19959810; PubMed Central PMCID: PMCPmc2797873

Mascre G, Dekoninck S, Drogat B, Youssef KK, Brohee S, Sotiropoulou PA et al (2012) Distinct contribution of stem and progenitor cells to epidermal maintenance. Nature 489(7415):257–262. doi:10.1038/nature11393. Epub 2012/09/04; PubMed PMID: 22940863

Mast BA, Schultz GS (1996) Interactions of cytokines, growth factors, and proteases in acute and chronic wounds. Wound Repair Regen: Off Publ Wound Healing Soc Eur Tissue Repair Soc 4(4):411–420. doi:10.1046/j.1524-475X.1996.40404.x. Epub 1996/10/01; PubMed PMID: 17309691

Maxson S, Lopez EA, Yoo D, Danilkovitch-Miagkova A, Leroux MA (2012) Concise review: role of mesenchymal stem cells in wound repair. Stem Cells Transl Med 1(2):142–149. doi:10.5966/sctm.2011-0018. Epub 2012/12/01; PubMed PMID: 23197761; PubMed Central PMCID: PMCPmc3659685

McGuckin C, Forraz N, Baradez MO, Basford C, Dickinson AM, Navran S et al (2006) Embryonic-like stem cells from umbilical cord blood and potential for neural modeling. Acta Neurobiol Exp 66(4):321–329. Epub 2007/02/03. PubMed PMID: 17269167

Mehanna RA, Nabil I, Attia N, Bary AA, Razek KA, Ahmed TA et al (2015) The effect of bone marrow-derived mesenchymal stem cells and their conditioned media topically delivered in fibrin glue on chronic wound healing in rats. Biomed Res Int 2015:846062. doi:10.1155/2015/846062. Epub 2015/08/04; PubMed PMID: 26236740; PubMed Central PMCID: PMCPmc4508387

Meinel L, Karageorgiou V, Fajardo R, Snyder B, Shinde-Patil V, Zichner L et al (2004) Bone tissue engineering using human mesenchymal stem cells: effects of scaffold material and medium flow. Ann Biomed Eng 32(1):112–122. Epub 2004/02/18. PubMed PMID: 14964727

Mendez JJ, Ghaedi M, Sivarapatna A, Dimitrievska S, Shao Z, Osuji CO et al (2015) Mesenchymal stromal cells form vascular tubes when placed in fibrin sealant and accelerate wound healing in vivo. Biomaterials 40:61–71. doi:10.1016/j.biomaterials.2014.11.011. Epub 2014/12/01; PubMed PMID: 25433608; PubMed Central PMCID: PMCPmc4268422

Meruane MA, Rojas M, Marcelain K (2012) The use of adipose tissue-derived stem cells within a dermal substitute improves skin regeneration by increasing neoangiogenesis and collagen synthesis. Plast Reconstr Surg 130(1):53–63. doi:10.1097/PRS.0b013e3182547e04. Epub 2012/03/16; PubMed PMID: 22418720

Morasso MI, Tomic-Canic M (2005) Epidermal stem cells: the cradle of epidermal determination, differentiation and wound healing. Biol Cell Auspices Eur Cell Biol Organ 97(3):173–183. doi:10.1042/bc20040098. Epub 2005/02/18; PubMed PMID: 15715523; PubMed Central PMCID: PMCPmc1283090

Moriyama M, Durham AD, Moriyama H, Hasegawa K, Nishikawa S, Radtke F et al (2008) Multiple roles of Notch signaling in the regulation of epidermal development. Dev Cell 14(4):594–604. doi:10.1016/j.devcel.2008.01.017. Epub 2008/04/16; PubMed PMID: 18410734

Mustoe TA, O'Shaughnessy K, Kloeters O (2006) Chronic wound pathogenesis and current treatment strategies: a unifying hypothesis. Plast Reconstr Surg 117(7 Suppl):35s–41s. doi:10.1097/01.prs.0000225431.63010.1b. Epub 2006/06/27; PubMed PMID: 16799373

Nan W, Liu R, Chen H, Xu Z, Chen J, Wang M et al (2015) Umbilical cord mesenchymal stem cells combined with a collagen fibrin double-layered membrane accelerates wound healing. Wounds 27(5):134–140. Epub 2015/05/13. PubMed PMID: 25965183

National Diabetes Statistical Report [Internet] (2014) Available from: http://www.cdc.gov/diabetes/pubs/statsreport14/national-diabetes-report-web.pdf

Niemann C, Unden AB, Lyle S, Zouboulis Ch C, Toftgard R, Watt FM (2003) Indian hedgehog and beta-catenin signaling: role in the sebaceous lineage of normal and neoplastic mammalian epidermis. Proc Natl Acad Sci U S A 100(Suppl 1):11873–11880. doi:10.1073/pnas.1834202100. Epub 2003/08/15; PubMed PMID: 12917489; PubMed Central PMCID: PMCPmc304101

Okuse T, Chiba T, Katsuumi I, Imai K (2005) Differential expression and localization of WNTs in an animal model of skin wound healing. Wound Repair Regen: Off Publ Wound Healing Soc Eur Tissue Repair Soc 13(5):491–497. doi:10.1111/j.1067-1927.2005.00069.x. Epub 2005/09/24; PubMed PMID: 16176457

Okuyama R, Tagami H, Aiba S (2008) Notch signaling: its role in epidermal homeostasis and in the pathogenesis of skin diseases. J Dermatol Sci 49(3):187–194. doi:10.1016/j.jdermsci.2007.05.017. Epub 2007/07/13; PubMed PMID: 17624739

Oshima H, Inoue H, Matsuzaki K, Tanabe M, Kumagai N (2002) Permanent restoration of human skin treated with cultured epithelium grafting – wound healing by stem cell based tissue engineering. Hum Cell 15(3):118–128. Epub 2003/04/22. PubMed PMID: 12703542

Outtz HH, Wu JK, Wang X, Kitajewski J (2010) Notch1 deficiency results in decreased inflammation during wound healing and regulates vascular endothelial growth factor receptor-1 and inflammatory cytokine expression in macrophages. J Immunol (Baltimore, MD: 1950) 185(7):4363–4373. doi:10.4049/jimmunol.1000720. Epub 2010/08/27; PubMed PMID: 20739676; PubMed Central PMCID: PMCPmc3887523

Owens P, Han G, Li AG, Wang XJ (2008) The role of Smads in skin development. J Invest Dermatol 128(4):783–790. doi:10.1038/sj.jid.5700969. Epub 2008/03/14; PubMed PMID: 18337711

Pastar I, Stojadinovic O, Krzyzanowska A, Barrientos S, Stuelten C, Zimmerman K et al (2010) Attenuation of the transforming growth factor beta-signaling pathway in chronic venous ulcers. Mol Med (Cambridge, Mass) 16(3–4):92–101. doi:10.2119/molmed.2009.00149. Epub 2010/01/14; PubMed PMID: 20069132; PubMed Central PMCID: PMCPmc2804290

Pelizzo G, Avanzini MA, Icaro Cornaglia A, Osti M, Romano P, Avolio L et al (2015) Mesenchymal stromal cells for cutaneous wound healing in a rabbit model: pre-clinical study applicable in the pediatric surgical setting. J Transl Med 13:219. doi:10.1186/s12967-015-0580-3. Epub 2015/07/15; PubMed PMID: 26152232; PubMed Central PMCID: PMCPmc4495634

Puolakkainen PA, Reed MJ, Gombotz WR, Twardzik DR, Abrass IB, Sage HE (1995) Acceleration of wound healing in aged rats by topical application of transforming growth factor-beta(1). Wound Repair Regen: Off Publ Wound Healing Soc Eur Tissue Repair Soc 3(3):330–339. doi:10.1046/j.1524-475X.1995.t01-1-30314.x. Epub 1995/07/01; PubMed PMID: 17173560

Puzey G (2002) The use of pressure garments on hypertrophic scars. J Tissue Viability 12(1):11–15. Epub 2002/03/13. PubMed PMID: 11887386

Reckhenrich AK, Kirsch BM, Wahl EA, Schenck TL, Rezaeian F, Harder Y et al (2014) Surgical sutures filled with adipose-derived stem cells promote wound healing. PLoS ONE 9(3):e91169. doi:10.1371/journal.pone.0091169. Epub 2014/03/15; PubMed PMID: 24625821; PubMed Central PMCID: PMCPmc3953386

Rigotti G, Marchi A, Galie M, Baroni G, Benati D, Krampera M et al (2007) Clinical treatment of radiotherapy tissue damage by lipoaspirate transplant: a healing process mediated by adipose-derived adult stem cells. Plast Reconstr Surg 119(5):1409–1422. doi:10.1097/01.prs.0000256047.47909.71. discussion 23––4. Epub 2007/04/07; PubMed PMID: 17415234

Rubio D, Garcia-Castro J, Martin MC, de la Fuente R, Cigudosa JC, Lloyd AC et al (2005) Spontaneous human adult stem cell transformation. Cancer Res 65(8):3035–3039. doi:10.1158/0008-5472.can-04-4194. Epub 2005/04/19; PubMed PMID: 15833829

Rustad KC, Wong VW, Sorkin M, Glotzbach JP, Major MR, Rajadas J et al (2012) Enhancement of mesenchymal stem cell angiogenic capacity and stemness by a biomimetic hydrogel scaffold. Biomaterials 33(1):80–90. doi:10.1016/j.biomaterials.2011.09.041. Epub 2011/10/04; PubMed PMID: 21963148; PubMed Central PMCID: PMCPmc3997302

Sabapathy V, Sundaram B, Sreelakshmi VM, Mankuzhy P, Kumar S (2014) Human Wharton's jelly mesenchymal stem cells plasticity augments scar-free skin wound healing with hair growth. PLoS ONE 9(4):e93726. doi:10.1371/journal.pone.0093726. Epub 2014/04/17; PubMed PMID: 24736473; PubMed Central PMCID: PMCPmc3988008

Salem HK, Thiemermann C (2010) Mesenchymal stromal cells: current understanding and clinical status. Stem Cells (Dayton, Ohio) 28(3):585–596. doi:10.1002/stem.269. Epub 2009/12/08; PubMed PMID: 19967788; PubMed Central PMCID: PMCPmc2962904

Sanmano B, Mizoguchi M, Suga Y, Ikeda S, Ogawa H (2005) Engraftment of umbilical cord epithelial cells in athymic mice: in an attempt to improve reconstructed skin equivalents used as epithelial composite. J Dermatol Sci 37(1):29–39. doi:10.1016/j.jdermsci.2004.10.008. Epub 2004/12/28; PubMed PMID: 15619432

Sato M (2006) Upregulation of the Wnt/beta-catenin pathway induced by transforming growth factor-beta in hypertrophic scars and keloids. Acta Derm Venereol 86(4):300–307. doi:10.2340/00015555-0101. Epub 2006/07/29; PubMed PMID: 16874413

Schreier T, Degen E, Baschong W (1993) Fibroblast migration and proliferation during in vitro wound healing. A quantitative comparison between various growth factors and a low molecular weight blood dialysate used in the clinic to normalize impaired wound healing. Res Exp Med Z Gesamten Exp Med Einschliesslich Exp Chir 193(4):195–205. Epub 1993/01/01. PubMed PMID: 8235072

Schultz GS, Wysocki A (2009) Interactions between extracellular matrix and growth factors in wound healing. Wound Repair Regen: Off Publ Wound Healing Soc Eur Tissue Repair Soc 17(2):153–162. doi:10.1111/j.1524-475X.2009.00466.x. Epub 2009/03/27; PubMed PMID: 19320882

Schultz GS, Sibbald RG, Falanga V, Ayello EA, Dowsett C, Harding K et al (2003) Wound bed preparation: a systematic approach to wound management. Wound Repair Regen: Off Publ Wound Healing Soc Eur Tissue Repair Soc 11(Suppl 1):S1–28. Epub 2003/03/26; PubMed PMID: 12654015

Schultz GS, Davidson JM, Kirsner RS, Bornstein P, Herman IM (2011) Dynamic reciprocity in the wound microenvironment. Wound Repair Regen: Off Publ Wound Healing Soc Eur Tissue Repair Soc 19(2):134–148. doi:10.1111/j.1524-475X.2011.00673.x. Epub 2011/03/03; PubMed PMID: 21362080; PubMed Central PMCID: PMCPmc3051353

Sebastiano V, Zhen HH, Haddad B, Bashkirova E, Melo SP, Wang P et al (2014) Human COL7A1-corrected induced pluripotent stem cells for the treatment of recessive dystrophic epidermolysis bullosa. Sci Transl Med 6(264):264ra163. doi:10.1126/scitranslmed.3009540. Epub 2014/11/28; PubMed PMID: 25429056; PubMed Central PMCID: PMCPmc4428910

Seibold JR, Uitto J, Dorwart BB, Prockop DJ (1985) Collagen synthesis and collagenase activity in dermal fibroblasts from patients with diabetes and digital sclerosis. J Lab Clin Med 105(6):664–667. Epub 1985/06/01; PubMed PMID: 2987379

Sen CK, Gordillo GM, Roy S, Kirsner R, Lambert L, Hunt TK et al (2009) Human skin wounds: a major and snowballing threat to public health and the economy. Wound Repair Regen 17(6):763–771. doi:10.1111/j.1524-475X.2009.00543.x. PubMed PMID: 19903300; PubMed Central PMCID: PMC2810192

Shen Y, Dai L, Li X, Liang R, Guan G, Zhang Z et al (2014) Epidermal stem cells cultured on collagen-modified chitin membrane induce in situ tissue regeneration of full-thickness skin defects in mice. PLoS ONE 9(2):e87557. doi:10.1371/journal.pone.0087557. Epub 2014/02/12; PubMed PMID: 24516553; PubMed Central PMCID: PMCPmc3917838

Shi S, Jia S, Liu J, Chen G (2015) Accelerated regeneration of skin injury by co-transplantation of mesenchymal stem cells from Wharton's jelly of the human umbilical cord mixed with microparticles. Cell Biochem Biophys 71(2):951–956. doi:10.1007/s12013-014-0292-y. Epub 2014/11/06; PubMed PMID: 25370297

Signorini M, Clementoni MT (2007) Clinical evaluation of a new self-drying silicone gel in the treatment of scars: a preliminary report. Aesthet Plast Surg 31(2):183–187. doi:10.1007/s00266-005-0122-0. Epub 2006/12/16; PubMed PMID: 17171514

Singer AJ, Clark RA (1999) Cutaneous wound healing. N Engl J Med 341(10):738–746. doi:10.1056/nejm199909023411006. Epub 1999/09/02; PubMed PMID: 10471461

Singh N, Armstrong DG, Lipsky BA (2005) Preventing foot ulcers in patients with diabetes. JAMA 293(2):217–228. doi:10.1001/jama.293.2.217. Epub 2005/01/13; PubMed PMID: 15644549

Snippert HJ, Haegebarth A, Kasper M, Jaks V, van Es JH, Barker N et al (2010) Lgr6 marks stem cells in the hair follicle that generate all cell lineages of the skin. Science (New York, NY) 327(5971):1385–1389. doi:10.1126/science.1184733. Epub 2010/03/13; PubMed PMID: 20223988

Snyder RJ (2005) Treatment of nonhealing ulcers with allografts. Clin Dermatol 23(4):388–395. doi:10.1016/j.clindermatol.2004.07.020. Epub 2005/07/19; PubMed PMID: 16023934

Soo C, Beanes SR, Hu FY, Zhang X, Dang C, Chang G et al (2003) Ontogenetic transition in fetal wound transforming growth factor-beta regulation correlates with collagen organization. Am J Pathol 163(6):2459–2476. Epub 2003/11/25. PubMed PMID: 14633618; PubMed Central PMCID: PMCPmc1892380

St-Jacques B, Dassule HR, Karavanova I, Botchkarev VA, Li J, Danielian PS et al (1998) Sonic hedgehog signaling is essential for hair development. Curr Biol: CB 8(19):1058–1068. Epub 1998/10/13. PubMed PMID: 9768360

Strioga M, Viswanathan S, Darinskas A, Slaby O, Michalek J (2012) Same or not the same? Comparison of adipose tissue-derived versus bone marrow-derived mesenchymal stem and stromal cells. Stem Cells Dev 21(14):2724–2752. doi:10.1089/scd.2011.0722. Epub 2012/04/04; PubMed PMID: 22468918

Strong AL, Bowles AC, MacCrimmon CP, Frazier TP, Lee SJ, Wu X et al (2015) Adipose stromal cells repair pressure ulcers in both young and elderly mice: potential role of adipogenesis in skin repair. Stem Cells Transl Med 4(6):632–642. doi:10.5966/sctm.2014-0235. Epub 2015/04/23; PubMed PMID: 25900728; PubMed Central PMCID: PMCPmc4449094

Takahashi K, Yamanaka S (2006) Induction of pluripotent stem cells from mouse embryonic and adult fibroblast cultures by defined factors. Cell 126(4):663–676. doi:10.1016/j.cell.2006.07.024. Epub 2006/08/15; PubMed PMID: 16904174

Takahashi K, Tanabe K, Ohnuki M, Narita M, Ichisaka T, Tomoda K et al (2007) Induction of pluripotent stem cells from adult human fibroblasts by defined factors. Cell 131(5):861–872. doi:10.1016/j.cell.2007.11.019. Epub 2007/11/24; PubMed PMID: 18035408

Tong C, Hao H, Xia L, Liu J, Ti D, Dong L et al (2015) Hypoxia pretreatment of bone marrow-derived mesenchymal stem cells seeded in a collagen-chitosan sponge scaffold promotes skin wound healing in diabetic rats with hindlimb ischemia. Wound Repair Regen: Off Publ Wound Healing Soc Eur Tissue Repair Soc. doi:10.1111/wrr.12369. Epub 2015/10/16; PubMed PMID: 26463737

Torsvik A, Rosland GV, Svendsen A, Molven A, Immervoll H, McCormack E et al (2010) Spontaneous malignant transformation of human mesenchymal stem cells reflects cross-contamination: putting the research field on track – letter. Cancer Res 70(15):6393–6396. doi:10.1158/0008-5472.can-10-1305. Epub 2010/07/16; PubMed PMID: 20631079

Trengove NJ, Stacey MC, MacAuley S, Bennett N, Gibson J, Burslem F et al (1999) Analysis of the acute and chronic wound environments: the role of proteases and their inhibitors. Wound Repair Regen: Off Publ Wound Healing Soc Eur Tissue Repair Soc 7(6):442–452. Epub 2000/01/13; PubMed PMID: 10633003

Tsourdi E, Barthel A, Rietzsch H, Reichel A, Bornstein SR (2013) Current aspects in the pathophysiology and treatment of chronic wounds in diabetes mellitus. Biomed Res Int 2013:385641. doi:10.1155/2013/385641. Epub 2013/05/09; PubMed PMID: 23653894; PubMed Central PMCID: PMCPmc3638655

Tumbar T, Guasch G, Greco V, Blanpain C, Lowry WE, Rendl M et al (2004) Defining the epithelial stem cell niche in skin. Science (New York, NY) 303(5656):359–363. doi:10.1126/

science.1092436. Epub 2003/12/13; PubMed PMID: 14671312; PubMed Central PMCID: PMCPmc2405920

Uysal CA, Tobita M, Hyakusoku H, Mizuno H (2014) The effect of bone-marrow-derived stem cells and adipose-derived stem cells on wound contraction and epithelization. Adv Wound Care 3(6):405–413. doi:10.1089/wound.2014.0539. Epub 2014/06/19; PubMed PMID: 24940554; PubMed Central PMCID: PMCPmc4048969

Vagnozzi AN, Reiter JF, Wong SY (2015) Hair follicle and interfollicular epidermal stem cells make varying contributions to wound regeneration. Cell Cycle (Georgetown, Tex) 14(21):3408–3417. doi:10.1080/15384101.2015.1090062. Epub 2015/09/24; PubMed PMID: 26398918

Varga J, Rosenbloom J, Jimenez SA (1987) Transforming growth factor beta (TGF beta) causes a persistent increase in steady-state amounts of type I and type III collagen and fibronectin mRNAs in normal human dermal fibroblasts. Biochem J 247(3):597–604. Epub 1987/11/01. PubMed PMID: 3501287; PubMed Central PMCID: PMCPmc1148454

Vojtassak J, Danisovic L, Kubes M, Bakos D, Jarabek L, Ulicna M et al (2006) Autologous biograft and mesenchymal stem cells in treatment of the diabetic foot. Neurol Endocrinol Lett 27(Suppl 2):134–137. Epub 2006/12/13. PubMed PMID: 17159798

Watt FM, Fujiwara H (2011) Cell-extracellular matrix interactions in normal and diseased skin. Cold Spring Harb Perspect Biol 3(4):a005124. doi:10.1101/cshperspect.a005124. Epub 2011/03/29; PubMed PMID: 21441589; PubMed Central PMCID: PMCPmc3062212

Werner S, Grose R (2003) Regulation of wound healing by growth factors and cytokines. Physiol Rev 83(3):835–870. doi:10.1152/physrev.00031.2002. Epub 2003/07/05; PubMed PMID: 12843410

Willital GH, Heine H (1994) Efficacy of Contractubex gel in the treatment of fresh scars after thoracic surgery in children and adolescents. Int J Clin Pharmacol Res 14(5–6):193–202. Epub 1994/01/01. PubMed PMID: 7672876

Wilson A, Butler PE, Seifalian AM (2011) Adipose-derived stem cells for clinical applications: a review. Cell Prolif 44(1):86–98. doi:10.1111/j.1365-2184.2010.00736.x. Epub 2011/01/05; PubMed PMID: 21199013

Wu Y, Wang J, Scott PG, Tredget EE (2007a) Bone marrow-derived stem cells in wound healing: a review. Wound Repair Regen: Off Publ Wound Healing Soc Eur Tissue Repair Soc 15(Suppl 1):S18–S26. doi:10.1111/j.1524-475X.2007.00221.x. Epub 2008/09/10; PubMed PMID: 17727462

Wu Y, Chen L, Scott PG, Tredget EE (2007b) Mesenchymal stem cells enhance wound healing through differentiation and angiogenesis. Stem Cells 25(10):2648–2659. Epub 2007/07/07. doi: 2007-0226 [pii] 1634/stemcells.2007-0226. PubMed PMID: 17615264

Wu X, Wang G, Tang C, Zhang D, Li Z, Du D et al (2011) Mesenchymal stem cell seeding promotes reendothelialization of the endovascular stent. J Biomed Mater Res A 98(3):442–449. doi:10.1002/jbm.a.33133. Epub 2011/06/11; PubMed PMID: 21661093

Wu Y, Peng Y, Gao D, Feng C, Yuan X, Li H et al (2015) Mesenchymal stem cells suppress fibroblast proliferation and reduce skin fibrosis through a TGF-beta3-dependent activation. Int J Low Extrem Wounds 14(1):50–62. doi:10.1177/1534734614568373. Epub 2015/04/11; PubMed PMID: 25858630

Yang R, Zheng Y, Burrows M, Liu S, Wei Z, Nace A et al (2014) Generation of folliculogenic human epithelial stem cells from induced pluripotent stem cells. Nat Commun 5:3071. doi:10.1038/ncomms4071. Epub 2014/01/29; PubMed PMID: 24468981; PubMed Central PMCID: PMCPmc4049184

Yoshikawa T, Mitsuno H, Nonaka I, Sen Y, Kawanishi K, Inada Y et al (2008) Wound therapy by marrow mesenchymal cell transplantation. Plast Reconstr Surg 121(3):860–877. doi:10.1097/01.prs.0000299922.96006.24. Epub 2008/03/05; PubMed PMID: 18317135

Yu J, Vodyanik MA, Smuga-Otto K, Antosiewicz-Bourget J, Frane JL, Tian S et al (2007) Induced pluripotent stem cell lines derived from human somatic cells. Science (New York, NY) 318(5858):1917–1920. doi:10.1126/science.1151526. Epub 2007/11/22; PubMed PMID: 18029452

Zhang G, Hu Q, Braunlin EA, Suggs LJ, Zhang J (2008) Enhancing efficacy of stem cell transplantation to the heart with a PEGylated fibrin biomatrix. Tissue Eng A 14(6):1025–1036. doi:10.1089/ten.tea.2007.0289. Epub 2008/05/15; PubMed PMID: 18476809

Zhang Q, Liu LN, Yong Q, Deng JC, Cao WG (2015) Intralesional injection of adipose-derived stem cells reduces hypertrophic scarring in a rabbit ear model. Stem Cell Res Ther 6:145. doi:10.1186/s13287-015-0133-y. Epub 2015/08/19; PubMed PMID: 26282394; PubMed Central PMCID: PMCPmc4539671

Zheng K, Wu W, Yang S, Huang L, Chen J, Gong C et al (2015) Bone marrow mesenchymal stem cell implantation for the treatment of radioactivity induced acute skin damage in rats. Mol Med Rep 12(5):7065–7071. doi:10.3892/mmr.2015.4270. Epub 2015/09/02; PubMed PMID: 26323987

Zimmerlin L, Rubin JP, Pfeifer ME, Moore LR, Donnenberg VS, Donnenberg AD (2013) Human adipose stromal vascular cell delivery in a fibrin spray. Cytotherapy 15(1):102–108. doi:10.1016/j.jcyt.2012.10.009. Epub 2012/12/25; PubMed PMID: 23260090; PubMed Central PMCID: PMCPmc4159959

Zuk PA, Zhu M, Ashjian P, De Ugarte DA, Huang JI, Mizuno H et al (2002) Human adipose tissue is a source of multipotent stem cells. Mol Biol Cell 13(12):4279–4295. doi:10.1091/mbc.E02-02-0105. Epub 2002/12/12; PubMed PMID: 12475952; PubMed Central PMCID: PMCPmc138633

Zykova SN, Jenssen TG, Berdal M, Olsen R, Myklebust R, Seljelid R (2000) Altered cytokine and nitric oxide secretion in vitro by macrophages from diabetic type II-like db/db mice. Diabetes 49(9):1451–1458. Epub 2000/09/02. PubMed PMID: 10969828

Chapter 10
Wound Treatment by Stem Cells

Leyla Türker Şener, Hakan Darici, Işil Albeniz, and Erdal Karaöz

10.1 Introduction

Stem cells are capable to form an organism starting from a fertilized egg and have the ability of unlimited proliferation, self-renewal, and differentiation into the cell types of target tissues they were transferred. One of the most important aspects of stem cells is differentiation which plays the key role in development of multicellular organisms. Differentiation is defined as the overall set of changes of cell phenotype with the effect of cytokines, growth factors, extracellular matrix (ECM) proteins, and intercellular signaling pathways. Stem cell transplantation is more accurately defined as cell therapy is the use of stem cells or their products, derived from their patient's own tissues or from another donor for the treatment of tissue damage, various diseases, or loss of function. Cell therapies provide hope for many cases we cannot adequately treat with conventional medical methods so far. Along with their differentiation ability, stem cells have immunoregulatory, anti-inflammatory, anti-apoptotic, anti-scarring, and neovascularization induction functions which are used for joint problems, some neurological and autoimmune diseases, as well as for the treatment of muscular-neurodegenerative disorders. More recently some nanoparticles secreted by stem cells such as exosomes or secretomes were shown to induce different pathways such as transdifferentiation when transferred into tissue microenvironment.

L.T. Şener • I. Albeniz
Department of Biophysics, Istanbul University, Istanbul Faculty of Medicine, 34093 Capa-Istanbul, Turkey
e-mail: leylaturker@hotmail.com; ialbeniz@istanbul.edu.tr

H. Darici • E. Karaöz (✉)
Department of Histology and Embryology, Istinye University Hospital, Faculty of Medicine, 34093 Topkapı-Istanbul, Turkey
e-mail: hakan.darici@yahoo.com; ekaraoz@hotmail.com

P.V. Pham (ed.), *Pancreas, Kidney and Skin Regeneration*, Stem Cells in Clinical Applications, DOI 10.1007/978-3-319-55687-1_10

Use of stem cells in wound healing is among the numerous medical methods where the cell therapies have benefits. Wounds can be acquired during surgical incisions or accidents or may occur due to various conditions like aging, burns, or diseases such as diabetes. Healing of the wound is a highly complex process, requires cooperation of various cell types from different tissues such as epithelia, connective tissue, and blood cells. Coagulation, inflammation, and anti-inflammatory processes follow each other with the continuous degradation and formation of ECM. Growth factors, cytokines, and ECM components play important roles during wound healing, and interruption of these steps may cause nonhealing chronic wounds. This chapter will focus on existing stem cell technologies and their application in the treatment of wound healing along with recently developed methods such as use of stem cells' own products, exosomes. To understand mechanisms of stem cells, we will briefly go over the structure of most commonly wounded tissues, namely, skin, which consist of epithelia and connective tissue. We will explain the roles of tissue components during wound healing, describe common wound types with causes, and finally explain the results of stem cell therapies on wounds with most recent examples on human trials.

10.2 Histology of Common Wound Areas

Wounds generally occur externally through the skin, although some wounds such as ulcers may develop internally. Other limited numbers of wound types may happen in cartilage tissues, tendons, or ligaments (Paz and West 2014). Almost all internal or external wounds occur in the epithelia and underlying connective tissue (Marieb 1995). Deep wounds may reach other tissues beneath the connective tissue layer such as muscle, deeper blood vessels, and bone, which are also surrounded by connective tissue layers. Serious bone fractures create wounds that may reach the skin, made by the bone itself (Ertekin et al. 2005). Therefore, knowledge of the histologic structure of epithelia, underlying connective tissue, and, more specifically, skin, plays a key role in cellular therapies of both acute and chronic wounds (Metcalfe and Ferguson 2007).

10.2.1 Histology of Skin

Skin and its derivatives (hair, nails, glands) are the largest organ of the human body. It covers the external surface of and constitutes up to 15–20% of the total body mass. Other than its barrier function, skin has immunologic, sensory, and endocrine functions, provides excretion via various glands, and helps maintain homeostasis (Weinzweig 1999). Skin consists of two main layers, the epidermis layer, formed of epithelia, and the dermis layer, which is formed by connective tissue (Vaccari et al. 2005).

10.2.1.1 Epidermis

The epidermis is the upper, epithelial part of the skin, composed of keratinized stratified squamous epithelium. Epidermal cells called *keratinocytes* are arranged as five layers, distinct by their shapes and characteristics although they are from the same origin. Proliferation starts from the *stratum basale*, the deepest of the five layers of epidermis, which lies on the basal lamina. Keratinocytes at this layer continuously proliferate, and daughter cells move upward and change shape and cellular features to maintain their functions (Singer et al. 2013).

Wound regeneration requires the stratum basale layer for the formation of other parts of the epidermis (Blumberg et al. 2012). Proliferation remarkably increases with the wound formation, and basal cells begin migration across the wound surface within 8–18 h of wound formation. The speed of migration may be as high as 0.5 mm/day.

The upper layers provide a barrier function to water, light, bacteria, and other particles. Keratin filaments begin accumulating as cells and they move upward and finally lose their nuclei and organelles and become filled with keratin at the topmost layers.

The epidermis also contains other cell types such as melanocytes, Langerhans, and Merkel's cells. Langerhans cells are associated with the immune system and function as the antigen-presenting cells of the skin. Merkel's cells are sensory cells responsible for cutaneous sensation; therefore, they are mostly accumulated at the fingertips. Free nerve endings are also intersparsed between keratinocytes of epidermis to receive various sensory modalities such as heat, cold, and fine touch. Melanocytes are dendritic cells located among the keratinocytes of the stratum basale but extend their cytoplasmic processes to the upper parts of the epidermis. The main function of melanocytes is producing the pigment, melanin, which protects the organism against the damaging effects of ultraviolet radiation (Pawlina 2016; Norton et al. 2003). Newly formed skin at the wound area cannot perform its all duties such as sensory, immunologic, or (UV) barrier functions, without these cell types, and shows discoloration, which causes unwanted cosmetic problems. Therefore, appropriate addition of these cell types among keratinocytes must be considered in cellular therapy methods.

10.2.1.2 Dermis

The dermis is the connective tissue part of the skin, which nourishes the avascular epidermis via its numerous small blood vessels and increases the thickness of the skin for mechanical support. Fibroblasts are the main cell type of connective tissue, which conduct the main functions such as extracellular matrix (ECM) production and tissue regeneration. The ECM primarily consists of type I collagen fibers along with less present collagen types such as type III (Pawlina 2016). Type I collagen is always the most important component; however, the thinner but stronger type III collagen becomes more prominent during wound healing and plays key roles. Synthesis and posttranslational modifications of collagen fibrils depend on various systemic and local factors such as oxygen support; adequate nutrition, especially with vitamin C; and the local micro-environment (Charles Brunicardi et al. 2005).

Another fiber type is the elastic fibers, which are responsible of the elasticity of the skin. Glycosaminoglycans (GAGs) are made of protein and carbohydrate components and perform many functions of the ECM such as entangling water molecules, thereby creating a highly liquid-like environment. Other important functions of the ECM are the formation of a 3D environment for cell migration, facilitating diffusion of nutrients, and helping tissue regeneration and development via attached growth factors and cytokines (Pawlina 2016; Norton et al. 2003). The most important GAGs are dermatan sulfate and chondroitin sulfate, which are synthesized by fibroblasts at increasing amounts in the 3 months following an injury. GAGs attach proteins to form proteoglycans. Collagens and proteoglycans actively interact, and proteoglycans become entrapped within collagen fibers because the amount of collagen increases with time, especially at the last step of wound healing (Xue and Jackson 2015).

The epidermis overlays and sticks to the dermis via the basement membrane. Additionally, wavelike protrusions called dermal papillae, which contain numerous capillaries to nourish the avascular epidermis layer, increase the attachment. The dermis is also formed by two layers, the papillary and reticular layers. The superficial papillary layer is composed of loose connective tissue with abundant cell types, collagen type I and III, and elastic fibers. The reticular layer contains less cells, changes in thickness throughout the body, but always remains thicker than the upper papillary layer. Type I collagen and elastic fibers are also thicker and present as irregular bundles in this layer. However, these bundles form regular lines called *Langer's lines* in regions of tension. Langer's lines are especially important for skin surgeries; incisions made parallel to Langer's lines leave the least scar marks while healing. Adipose tissue layers, smooth muscle, and sometimes striated muscle can be found beneath the reticular layer. Encapsulated nerve endings such as Pacinian corpuscles are also present within the dermis, which detect pressure, Meissner's corpuscles, which are sensitive to light touch, and Ruffini's corpuscles, which detect stretching of the skin.

The hypodermis, also known as the *subcutaneous fascia* in anatomy, is the adipose tissue-containing part of the skin, which lies underneath of dermis layer. The main function of the hypodermis is mechanical protection and heat isolation. Adipose tissue within the hypodermis also functions as energy storage units. Therefore, the hypodermis thickens with nourishment or in cold climates. The hypodermis also participates in hormonal regulation via its adipose tissue-secreted factors.

The skin also contains epidermal appendages such as nails, hair follicles, sweat glands, and sebaceous glands, which produce an oily substance called sebum that is thought to have bacteriostatic, barrier, and pheromone functions. Hair follicles cover almost the entire body. Even though their roots may be in very deep regions of the dermis, they are continuous parts of the epidermis. Epidermal stem cells residing within the root sheath of hairs are responsible for hair growth (Pawlina 2016; Norton et al. 2003). These stem cells may form new epidermis in some cases if the epidermal layer is destroyed by trauma or removed during surgery, while being protected due to their deeper location. However, extensive wounds such as third-degree burns may destroy the dermis and epidermis completely and require grafting for healing. Other than hair transplant surgeries, hair follicle stem cells are also important for their clinical implication in skin regeneration (Singer et al. 2013; Leirósa et al. 2014).

Skin thickness varies from 1 to 5 mm. The skin on the palms of the hands and soles of the feet are hairless and have thicker epidermis and are therefore called thick skin, whereas other parts of the skin have thinner epidermal layers and contain hair follicles (Pawlina 2016). Superficial wounds of the epithelium heal without scar formation, but wounds of the dermis may require surgical intervention or other medical treatments to avoid scar formation (Weinzweig 1999).

10.2.2 Histology of Other Epithelia

Although most injuries harm the body from outside to inside, some acute or chronic wounds can occur within the body going outward. Usually the first affected tissue is again the epithelia, even though the injury may have occurred within a very large organ like the stomach, which could result from swallowing sharp materials or as tiny as inside blood capillaries (Ertekin et al. 2005). Epithelia cover all open surfaces, canals, glands, and ducts of the body. Similar to skin, various epithelia of the body contain unipotent stem cells, which reside in the basal compartment and constantly regenerate epithelia. The height and lamination of epithelia differ depending on the organ, which can be as thin as a single thin layer of squamous cells like alveolar epithelia of lungs, or multiple layers of cells of various shapes, such as can be seen within the oral cavity, bladder, vagina, or anus. The shape of the cells at the top layer gives epithelia its names, e.g., columnar, cuboidal, or squamous epithelia, along with being simple with only one layer or stratified with many layers. Most of the digestive tract is formed by simple columnar epithelia, whereas the skin and oral cavity are formed by stratified squamous epithelia. The regeneration rate of these various epithelia also differs according to organ and function. Epithelia of the stomach and intestines continuously regenerate via their stem cells, which reside in the deeper parts of gastric pits or intestinal villi; the regeneration rate of glandular epithelia in other parts of the body is generally much slower.

Like the skin, internal epithelia require a connective tissue support that connects it to the deeper layers of the organ and also provides nourishment. The underlying connective tissue of epithelia called *lamina propria,* is similar to the dermis of the skin, except being thinner. The amount of vascular support effects the regeneration rate of epithelia, as well as the lamina propria. Poorly oxygenated tissues such as cartilage regenerate much more slowly than epithelial wounds because they receive less nutrition (Pawlina 2016; Norton et al. 2003). Vascularization becomes higher than normal during wound regeneration, which decreases to the normal levels at the last step of wound healing (Blumberg et al. 2012). The lamina propria lies on top of various tissues according to the region and function, such as smooth or stratified muscle, glands, cartilage, or bone. Each tissue has its own regeneration system. However, mesenchymal stem cells (MSC) and pericytes play a major role along with tissue-specific unipotent stem cells like the osteoprogenitor cells of bone (Pawlina 2016).

10.3 Wound Healing Process

Wound is defined as the disruption of tissue integrity, which consists of a series of cellular and biochemical events, starts with trauma, and ends with new tissue formation. These events overlap each other and follow certain steps. These steps of wound healing are defined as:

1. Hemostasis and inflammation
2. Proliferation
3. Maturation (remodeling)

Although these stages are sequential, there is no distinct border between them, and various stages can be observed simultaneously in different regions of the wound area (Ertekin et al. 2005; Norton et al. 2003).

10.3.1 Hemostasis and Inflammation

Clean surgical incisions and many other wounds begin the healing process with naturally occurring blood clotting. Platelets of damaged blood vessels make direct contact with the ECM at the injured site. Contact of platelets with subendothelial collagen causes platelets to aggregate and degranulate and starts the coagulation cascade. The clot contains an important fiber called fibrin, along with blood cells. This structure closes the wound surface, protects against infection, and helps subsequent inflammatory and healing processes by providing a 3D ultrastructure for movement of migrating cells. Increased capillary permeability, local release of chemoattractants such as prostaglandins, complement factors, interleukin 1 (IL-1), transforming growth factor-beta (TGFβ), and tumor necrosis factor-alpha (TNFα) induce neutrophil growth. Bacterial products also serve as chemoattractants if the wound site is infected (Duque and Descoteaux 2014).

The next phase, inflammation, starts around 24 h after injury. During the inflammatory step, dead tissues are cleared along with fibrin, followed by the deposition of new extracellular matrix (ECM) molecules. Neutrophils and monocytes infiltrate into the injured area at the beginning of inflammation, which peaks around the first to second day after injury. These leukocytes are the main source of TNFα, one of the key factors for angiogenesis and collagen synthesis; however, leukocytes do not involve in collagen synthesis or contribute toward mechanical strength. Later at this step, migrated monocytes differentiate into macrophages, which play a key role in successful healing. Macrophages replace neutrophils, reaching maximum numbers around 48–96 h after injury, increase the removal rate of dead tissues, and remain until the end of the healing process. Other functions of macrophages are cell proliferation regulation, matrix production, and angiogenesis through secreted mediators such as vascular endothelial growth factor (VEGF), insulin-like growth factor (IGF), epithelial growth factor (EGF), and TGFβ (Norton et al. 2003; Nandy and Mukhopadhyay 2011).

T lymphocytes are present in lower numbers than macrophages in the wound area and peak around one week after injury. They mediate transition from the inflammatory phase to the proliferation phase. The amount and strength of collagen in a wound decrease with the increase of T lymphocytes. The selective decrease of $CD8^+$ cells increases wound healing; the decrease of $CD4^+$ cells has no effect (Barbul et al. 1989).

10.3.2 Proliferation

Proliferation is the second step of wound healing, which starts between the 4th and 12th days. Fibroblasts and endothelial cells join the healing process, attracted mainly via PDGF and VEGF, respectively. After they reach the wound area, fibroblasts must proliferate and become activated in order to execute their primary function of matrix production. Cytokines and growth factors secreted by macrophages induce fibroblast activation. Fibroblasts isolated from wounds have been shown to synthesize more collagen than controls, proliferate less, and perform active matrix contraction. Endothelial cells of the damaged capillaries also start to proliferate, migrate, and contribute to angiogenesis under the influence of various cytokines and growth factors such as TNFα, TGFβ, and VEGF. Wounds appear to be in reddish color at this stage and are called granulation tissue due to the numerous newly formed capillaries. These vessels accelerate healing until the end of the process, and then redundant vessels disappear, leaving tissue with a normal vascularization rate.

10.3.3 Contraction of Wound

All wounds contract with time in order to bring wound sites together. Granulation tissue contains mainly fibroblasts, myofibroblasts, and various other connective tissue cells along with numerous newly formed small blood vessels. TGFβ stimulates the differentiation of fibroblasts into myofibroblasts, which have a key role in wound gap closure with their contractile abilities provided by alpha-smooth muscle actin molecules (α-SMA) within their cytoplasm. Fibroblasts and myofibroblasts recognize the stress lines of the wound and localize on this lines. Myofibroblasts apply steady force along the gap, shorten current ECM fibers, and slowly pull the sides of the wound together. Meanwhile, fibroblasts and myofibroblasts produce new collagen and other ECM molecules.

After one week, inflammatory cells, fibroblasts, and myofibroblasts undergo apoptosis, which leaves scar tissue containing mostly connective tissue fibers and few cells. New-forming epidermis covers the surface of the scar during normal healing of small wounds. However some pathologic conditions, such as hypertrophic scar or keloid formation, connective tissue-forming cells cannot completely undergo apoptosis and remain at the wound area. Hypertrophic scars have excessive connective tissue formation with little or no epithelial cover. These types of scars form white, protruding scar marks; however, the borders remain within the wound area. Keloids, on the other hand, are scar tissues that go beyond the wound borders of initial scar with no reduction in connective tissue formation. Another difference in keloids is the absence of myofibroblasts. Persistent myofibroblast activity also causes continuous pulling of tissue, which may cause fibromatosis like in Dupuytren's disease (Pawlina 2016). Hypertrophic scars are not related with skin pigmentation; however, keloid formation is more common in the African-American population. Scar formation risk is also higher in individuals with darker skin than lighter-skinned people (Ertekin et al. 2005; Weinzweig 1999).

10.3.4 Maturation and Remodeling

Maturation and remodeling of wound tissue start at the fibroblastic stage and are related with the rearranging of the previously synthesized collagen fibers. Matrix metalloproteinases are responsible for the breakdown of collagen, and net amount of collagen is balanced between collagenolysis and new collagen formation. During the remodeling phase, degradation of the ECM decreases, and the balance shifts in favor of production. However, deposition of large amounts of collagen also causes relatively acellular scar tissue formation. EMC deposition in the wound area follows certain steps: fibronectin and type III collagen constitute the early matrix, followed by glycosaminoglycans, proteoglycans, and, finally, type I collagen (Ghatak et al. 2015). The amount of collagen plateaus a few weeks after wound formation, but the tensile strength continues to rise for a few more mouths. However, the maximum rate of tensile strength of a healed wound remains at 80% of the pre-scar tissue. Remodeling of scar tissue continues its maturation for 6 to 12 more months (Weinzweig 1999).

10.3.5 Epithelization

While rebuilding the integrity of the connective tissue, the outlining epithelial barrier must also be completed. Incisional wounds re-epithelize within 48 h if the wound sides are stitched together; however, major epidermal and dermal loses take much longer to heal. Regeneration of the epidermal layer is characterized by the migration of neighboring epithelial cells. The epidermis at the rim of the wound area thickens following wound formation, some keratinocytes at the stratum basale lose cellular adhesion with the neighbor cells and basal membrane, enlarge, and migrate along the temporary wound matrix, while others rapidly proliferate without leaving their position. This proliferation continues until the defect has been closed. After closure, shapes of proliferating cells become columnar again, and epidermal layers reform. The last step of epithelization is keratin deposition (Weinzweig 1999; Pawlina 2016; Norton et al. 2003) Fig. 10.1.

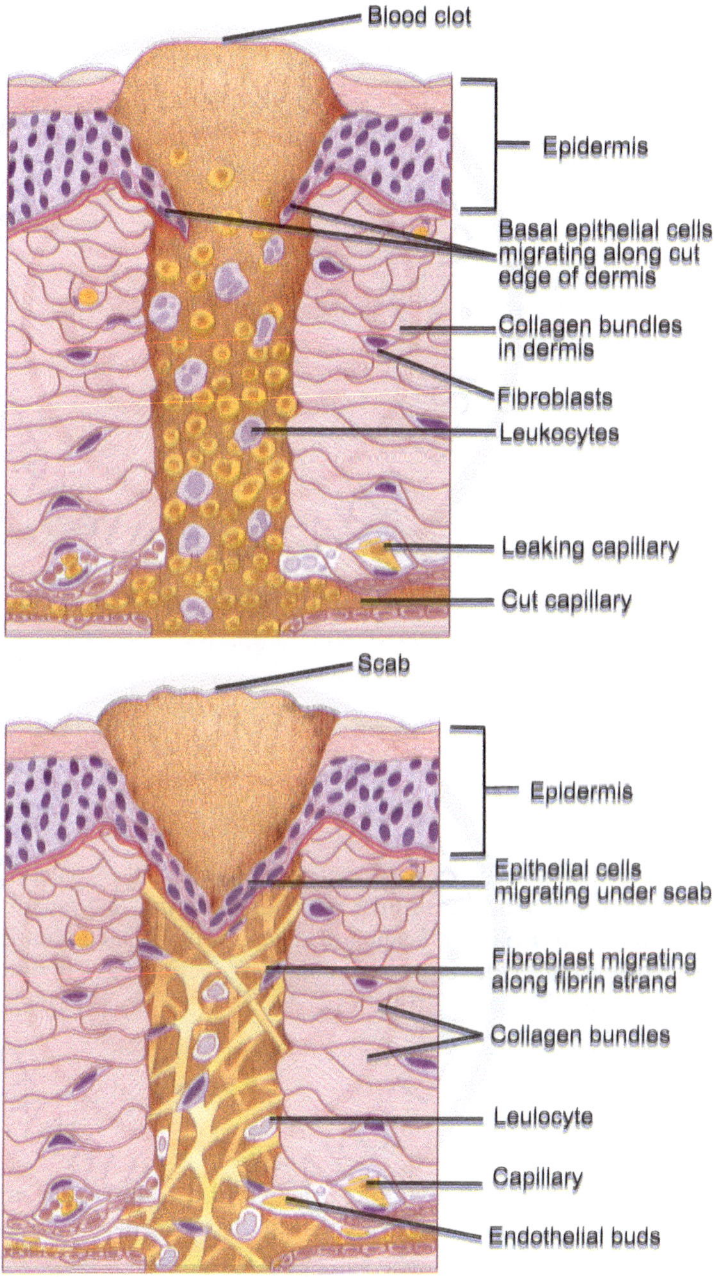

Fig. 10.1 Illustration of wound healing stages; (*A*) hemostasis and inflammation, (*B*) proliferation, and (*C*) maturation (remodeling). (**a**) During the first step of wound healing, blood clot fills the wound gap and entraps leaked erythrocytes. Leucocytes migrate to initiate inflammation, while epithelial cells of the adjacent stratum basale layer start to proliferate. (**b**) Macrophages phagocyte fibrin, dead cells, and bacteria, while proliferated fibroblasts produce new collagen bundles. (**c**) Epithelia reform with all layers, and underlying dermis matures into the functioning connective tissue

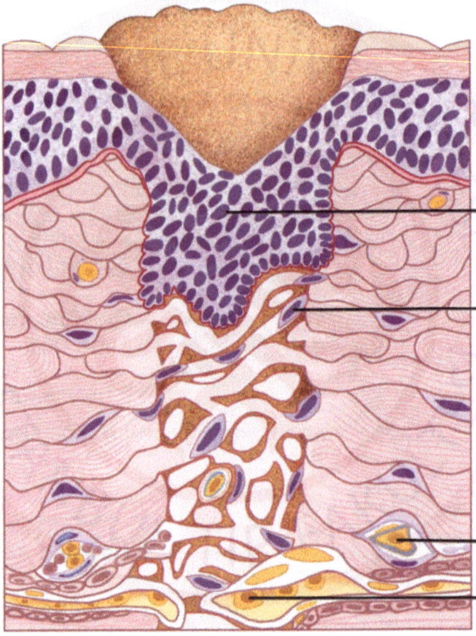

Fig. 10.1 (continued)

10.4 Wound Types

Wounds can be classified under two categories as acute and chronic. Acute wounds heal on time more regularly at cellular level, and anatomic and functional integrity is better maintained. However, healing of chronic wounds takes consderably more time than acute wounds; therefore, anatomic and functional amelioration cannot be properly repaired (Arya et al. 2014).

10.4.1 Acute Wounds

An acute wound can be formed during a surgery procedure which are done under sterile conditions, neat, and more likely to heal. However, most wounds are not that cleanly formed, cover more surface, and fail to heal properly, proportional with the deepness and surface area of the wound. Burns, occurred by heat or other causes like chemicals or electricity, usually fall into this category and carry a serious, even life-threatening risk if not medically tended due to the fluid loss or infection. We will briefly discuss these wound types with their treatment methods before explaining the outcomes of stem cell treatment on acute wounds.

10.4.1.1 Cuts and Surgical Incisions

Cuts that may occur under surgery, called incisions, are usually made with a sterile scalpel, cautery pen, or laser which forms neat wounds. Cuts with sharp objects like knife or glass may also cause to formation of neat wounds. Damage done by scalpels and lasers is minimal; however, despite their convenience, incisions made with cautery pens which can be categorized under burns cause more damage to the tissues (Ertekin et al. 2005; Charles Brunicardi et al. 2005).

Medical treatment of cuts universally made via stitching which brings the two sides of the wound together minimalizes the gap, which should be filled with inflammatory tissue if not stitched therefore skips the pulling process of myofibroblasts and causes faster and effective healing. Different stitching methods and techniques exist which are selected by the surgeon according to wound type, location, and shape (Weinzweig 1999).

10.4.1.2 Burns

Various causes such as heat, cold, electricity, acids, bases, and various other chemicals, such as phosphorus compounds, may cause burns. Therefore, the reaction of the body varies along with treatment methods (Weinzweig 1999).

Burns are dynamic and invasive wounds that can cause maximum tissue loss due to necrosis; therefore, they are not considered as ordinary wounds. Lesser but serious damage such as ischemia may occur outside regions of necrosis. Although the exact mechanism of ischemia transition to necrosis remains unknown, it is probably caused by disruption of the cell cycle due to oxidative stress or infection (Cavanagh et al. 2012).

Cells can tolerate relatively high temperatures; however, cellular changes overwhelm repair mechanisms over 45 °C due to protein denaturation, even though cells do not have a static response and react differently to temperature increases. The depth and surface area of burns also affect wound healing. Burns covering more than 15–20% of the body surface affect other non-burnt areas as well as underlying organs, as such they are considered systemic injuries or disease. The standard classification of burns is well known as first- to third-degree burns. First-degree burns affect the epidermis, and second-degree burns reach the dermis layer. These types of burns are painful due to the affected nerve endings and generally leave a permanent scar. Third-degree burns reach beneath the skin. Patients do not feel pain because of the complete destruction of nerve endings but lose all sensory stimulation from these areas.

Burns wound tissues in two steps. The first step is the immediate cell damage and coagulation necrosis caused by heat. The second step is delayed progressive cell damage caused by ischemia, which occurs 24–48 h after the initial damage. These two steps can be observed in the burnt area. The necrotic area is covered by an ischemic (stasis) zone. Stasis zone also becomes necrotic during delayed damage step. Another third area, called hyperemia zone, can also be observed around the stasis zone which is characterized by inflammation and vasodilatation. The hyperemia zone heals faster, within 7–10 days. If delayed damage of the ischemia zone can be prevented via pharmaceutical agents or another method such as stem cell treatment. Necrosis and deepening of progressive burns can be prevented and can be healed faster. Burns smaller than 3–5 mm heal within 3–4 weeks, but larger burns require grafting. Chemical burns can form through external contact (Ertekin et al. 2005) or accidental ingestion of harmful substances such as alkali cleaning agents and cause epithelial burns in digestive tract, especially at esophagus epithelia. Chemical burns may be classified under four groups as acid, alkali, phosphorus burns, and chemical injections, which require slightly different treatment methods (Norton et al. 2003).

10.4.1.3 Grafting Methods

Grafting techniques, used in burns or other wound types, include transplantation of a part of a patient's own skin, taken from another appropriate area of the body or artificial grafts. However, grafting is not 100% successful and may cause the formation of two scarred areas instead of having just one original wounded site. On the other hand, artificial grafts, or artificial skin, are quite useful when donor skin areas are small or need to be decreased. Artificial grafts are matrices that constitute GAGs and collagen fibers, which function as dermis and therefore create a 3D network for wound healing while protecting the wound area from heat and fluid loss and infections via additional materials such as silver. Epithelization of artificially grafted areas can be enhanced by grafting thinner epidermal grafts of a patient's own tissue, which is considerably smaller than full skin grafts. Artificial grafts can be obtained as allografts and xenografts (Ertekin et al. 2005; Norton et al. 2003).

10.4.2 Chronic Wounds

Wounds that do not heal within 3 months are called chronic wounds. They can be classified under three categories as diabetic wounds, decubitus ulcers, and wounds caused by venous hypertension.

Evidence for the treatment of chronic ulcers in the elderly is still limited. Dressings containing calcium alginate may shorten healing period of pressure ulcers in the elderly; however, other dressings in this age group are not studied sufficiently (Madec et al. 2009a).

10.4.3 Diabetic Wounds

Diabetes mellitus (DM) is a complex, chronic metabolic disorder which affects almost all age groups. Diabetes mellitus affects several organs, including the muscle, skin, heart, brain, and kidneys, and requires continuous medical care beyond glycemic control due to prolonged and uncontrolled diabetes may lead various complications which are divided generally into microvascular complications and macrovascular complications due to damages to small blood vessels and arteries, respectively.

Patients with diabetes are prone to the development of chronic wounds, especially diabetic foot ulcers (DFUs), due to the deficiencies in either peripheral tissue homing and engraftment of bone marrow or endothelial progenitor cells. Additionally, diabetes impairs wound healing at various steps. During the proliferation step, macrophage number and activation decrease which results in the reduced lymphatic vessel formation. Diabetes mellitus also affects signaling mechanism, responsible for coordinating/regulating angiogenesis and vasculogenesis. DFU is considered as a major source of morbidity and a primer cause of hospitalization in diabetic patients. DFUs are caused by neuropathic, ischemic, or combined neuroischemic abnormalities. The most common pathway to develop foot problems in diabetic patients is autonomic and peripheral sensorimotor neuropathy that causes foot deformities, high foot pressure, and gait instability, which increases the risks of developing ulcers. Primary management goal for DFU is to obtain wound closure as fast as possible. Proper education of the patient and blood sugar control are the initial techniques for DFU therapies. Most common therapy method is debridement, which is the removal of necrotic tissues as well as foreign and infected materials from the wound. Different kinds of debridement exist such as surgical, enzymatic, autolytic, mechanical, and biological (use of maggots of the green bottle fly). Debridement decreases in the possibility of limb amputation; however, all debridement methods have their own drawbacks. Dressings are highly important in the treatment of DFU, which vary according to the classification and degree of DFU. Other advanced therapies such as hyperbaric oxygen treatment, negative pressure, or artificial skin grafts are also in use for DFU (Arya et al. 2014; Yazdanpanah et al. 2015). We will discuss in detail the use of stem cells in DFU treatment with examples from both literature and our studies.

10.5 Stem Cells

Stem cells can be classified according to their source or differentiation potential. Stem cells are acquired from early-stage embryos called embryonic stem cells (ESCs), whereas adult stem cells (ASCs) can be obtained from adult tissues, as well as later-stage embryos or fetuses (Ozturk and Karagoz 2015). ESCs, which are acquired from later-stage embryos, generally from the inner cell mass of blastocysts, are characterized by their ability to form colonies when cultured, superior proliferation ability, and, most importantly, in vitro and in vivo differentiation into all types of cells of three germ layers. They can form chimeric organisms when transferred into early-stage embryos (not applicable for humans) and can form embryonic bodies when cultured in appropriate conditions. In contrast, ASCs have a limited differentiation capacity but still can self-renew (Maehr et al. 2009; Melton and Cowan 2009; Draper et al. 2007).

Another classification of stem cells is based on their differentiation ability, which is called potency. Most potent cells are called totipotent stem cells, including zygote and first blastomeres, which can differentiate into all types of cells present within the body and also placenta. Pluripotent stem cells (PSCs), like ESCs, can differentiate into all cell types except placental cells. PSCs can be acquired more conveniently by reprogramming somatic cells back to the embryonic stage via viruses (Takahashi and Yamanaka 2006) or other methods (Polo et al. 2010). Different types of pluripotent cells also exist such as embryonic carcinoma cells and embryonic germ cells. Cells generated via reprogramming are called induced pluripotent stem cells (iPSCs) and do not carry the ethical problems of ESCs because they do not require human embryos (Balbach et al. 2009).

Pluripotent human stem cells are characterized through their surface markers, stage-specific embryonic antigen-3 (SSEA-3), SSEA-4, and keratan sulfate proteoglycans TRA-1-60 and TRA-1-81, which are also useful for sorting these cells. Pluripotent stem cells also express key pluripotency genes; Oct4 and Nanog (Draper et al. 2007; Harrison et al. 2011). A limited but increasing number of clinical trials can be found for PSC; however, a less potent but more easily accessible stem cell type, multipotent stem cells, has been used for a quite more number of clinical trials (Yolanda et al. 2014).

Multipotent stem cells are more specialized stem cells that can still differentiate into various cell types such as neural, adipogenic, osteogenic, and chondrogenic cells, as well as blood cells. Hematopoietic stem cells (HSC) reside within bone marrow and continuously produce all types of blood cells. MSCs are the most common multipotent stem cell type within the body, except bone marrow (Gupta et al. 2016 and Xue et al. 2013). Other than differentiation into various cell types, MSCs play key roles in the regeneration of adult tissues and have functions such as immunoregulation, angiogenesis, and epithelialization augmentation. MSCs have been shown to accelerate wound closure and correct chronic wound inflammation, as such they carry great promise in wound healing (Khosrotehrani 2013; Duscher et al. 2016; Lee et al. 2016). Though stem cells are known for their ability to home damaged areas within the body and differentiate into required cell types, only a small percentage of these survive in the host organism. However, latest research emphasizes the paracrine effects of MSCs, which will be detailed later in this chapter (Jayaraman et al. 2013).

10.6 Stem Cell Therapies on Acute Wounds

Shumakov et al. were the first to use mesenchymal bone marrow-derived stem cells (BM-MSC) in burn wound healing and compared them to embryonic fibroblasts on rats (Ghieh et al. 2015). Where BM-MSCs were applied, wounds showed decreased cell infiltration of the wound and an accelerated formation of new vessels and granulation tissue in comparison with embryonic fibroblasts and controls. Another study done by Rasulov et al. on rats also showed the superiority of stem cells in burn wound healing (Ozturk and Karagoz 2015). In another study, the application of MSC on burns reduced cell infiltration, improved neoangiogenesis, and reduced the formation of granulation tissue (van Zuijlen et al. 2015).

10.7 Stem Cell Therapies on Chronic Wounds

Various experimental or clinical studies in the literature have demonstrated the benefits of stem cells, especially MSCs. Here we will give examples of various studies on various wound types with or without underlying diseases.

10.7.1 Stem Cell Therapies for Ischemic Wounds

Ischemia, related to the various diseases such as Buerger's, may cause the occlusion of arteries at the extremities and conclusively, ischemia and necrosis of related tissues. Untreated ischemia may cause chronic wounds in these patients, which are difficult to heal due to low nourishment and cellular support. Stem cell therapies have shown clinical improvement in these disease in both double-blinded randomized studies (Shumakov et al. 2003), phase I (Lu et al. 2011a) or phase II clinical trials (Bura et al. 2014).

In a study study included 15 male patients with critical limb ischemia (CLI) Rutherford's class II-4, III-5, or III-6 and patients with ischemic resting pain in one limb with/without nonhealing ulcers and necrotic foot. Adipose tissue-derived mesenchymal stem cells (ATMSCs) were isolated from adipose tissue of patients with thromboangiitis obliterans (TAO), otherwise known as Buerger's disease (B-ATMSC), patients with diabetes (D-ATMSC), and healthy donors (control ATMSC). In a colony-forming unit assay, the stromal vascular fraction of patients with TAO and diabetes yielded lesser colonies than those of healthy donors. D-ATMSCs showed lower proliferation ability than B-ATMSCs and control ATMSCs but showed similar angiogenic factor expression to control ATMSCs and B-ATMSCs. Multiple intramuscular ATMSC injections caused no complications during the follow-up period (mean follow-up time: 6 months). Clinical improvement occurred in 66.7% of patients. Five patients required a minor amputation during follow-up, and all amputation sites healed completely. At 6 months, significant improvement was noted on pain rating scales and in claudication walking distance. Digital subtraction angiography before and 6 months after ATMSC implantation showed formation of numerous vascular collateral networks across affected arteries (Lee et al. 2012).

10.7.2 Stem Cell Therapies for Diabetic Wounds

Diabetes is a common medical problem in modern communities. Diabetic foot ulcers (DFUs) are the significant complications of diabetes, affect the life quality, and can be life-threatening. DFUs are a chronic significant health problem which presents with nonhealing wounds on foot (Fritschi 2001; Sener and Albeniz 2015). Various studies indicated that the prevalence of diabetic foot ulcer is around 2% (Besse et al. 2011). About 15% of diabetic patients have a tendency of developing diabetic foot ulcer throughout their life. Therefore, clinical studies carry great importance in fighting against diabetes. MSC has been used to produce insulin-releasing cells, fighting against autoimmunity, to provide islet compatibility and their living, and in treatment of diabetic ulcers and arm and leg ischemia (Karnieli et al. 2007; Madec et al. 2009b; Fiorina et al. 2009; Ding et al. 2009; Berman et al. 2010; Lu et al. 2011b). Studies have also shown that MSCs may also repair the glomerular cells which are damaged due to acute renal failure (Morigi et al. 2004; Black and Woodbury 2001; Jiang et al. 2010). These findings show that clinically, the MSCs may be used in treatment of diabetic foot ulcers.

Li XY et al. transplanted human cord blood mesenchymal stem cells (hCB-MSC) to patients diagnosed as having type 2 diabetes and diabetic foot ulcer and evaluated the Tregs/Th17/Th1 cell distribution. hCB-MSC was injected directly into quadriceps muscle in diabetic foot ulcer patients, and the rates of Treg/Th17, Treg/Th1, and Th17/Th1 cells were calculated in flow cytometry. In addition, their correlation with various cytokines (FoxP3, IL-17, INF-γ, C-RP, TNFα, and VEGF) was evaluated. There was a significant increase in the rates of CD4+CD25hiFoxP3+ Treg/Th17 and CD4+CD25hiFoxP3+ Treg/Th1 cells 4 months after transplant; the rates of Th17/Th1 cells did not change (Li et al. 2013). This data is supportive for Treg T cells in the initiation and proceeding of T2D (Lau et al. 2009a).

Plasticity of MSC is the basic feature of stem cells for use in treatment of disorders. hCB-MSC has differentiated into epidermal cells within the microenvironment of skin injuries. Also several existing bioactive factors, including cytokines and chemokines, can transform stem cells into skin cells in the wound region. Thus, both methods may enhance the recovery rate of injury and the quality. Conversely, MSC are reported to inhibit wound inflammatory reaction. MSCs have been reported to regulate IL-10, TGF-B1, IL-6, and TSP-1 constructs via paracrine pathway in the chemically burned cornea model, which does not result in healing enhancement.

Type 2 diabetes mellitus (T2DM) is one of the most complex and common types of diabetes. The most frequently linked factors to insulin resistance that progress to T2DM include obesity, aging, cell dysfunction, tissue lipid accumulation, oxidative stress, endoplasmic reticulum stress (ER stress) in cells, tissue inflammation, and physical inactivity (Akash et al. 2013). In a diabetes study, Li et al. transplanted hCB-MSCs into patients with T2DM who had diabetic foot ulcers and evaluated the distribution of Tregs/Th17/Th1 cells. hCB-MSCs were injected directly into the patients' quadriceps femoris muscle (2×10^6 cells per point). Proportions of Treg/Th17, Treg/Th1, and Th17/Th1cells were measured using flow cytometry, and their correlations with various cytokines (FoxP3, IL-17, INF-g, C-RP, TNF-a, and VEGF) were analyzed. While the proportions of CD4 + CD25hiFoxP3+ Treg/Th17 and CD4 + CD25hiFoxP3+ Treg/Th1 cells displayed a significant increase at 4 weeks posttransplantation, the percentage of Th17/Th1 cells did not change. These data suggest a role for Treg cells in the initiation and progression of T2DM (Li et al. 2014).

Xue et al. transplanted human BM-MSCs carrying yellow fluorescent protein into a mouse burn model to enhance healing with 30 mice in each group. BM-MSCs carry minimal to no immunogenetic markers, which is their advantage. Obvious wound healing was observed in 14 days, whereas controls took approximately 25 days to reach the same stage. Allogenic MSC increased wound healing as well as blood vessel intensity and angiogenesis (Metcalfe and Ferguson 2007).

Oskouei et al. showed that cardiac stem cells (C-kit, SCA1, anbg2 positive) had ameliorating effects and reduced scar formation, which was enhanced when they were used with BM-MSC (Stem Cells Transl Med 2012).

Singer et al. used BM-MSCs in a totally necrotic mice burn model with 48 animals in their experimental group. Burns of the 54 animals of control group completely necrotized, and 29 animals in the experimental group survived. Tissue amount and angiogenesis increased 20% in the area of necrosis. The authors also tested the effects of MSCs on tail biopsy wounds of diabetic mice and observed a near-complete regain of physical abilities (Lau et al. 2009b).

Zhao et al. isolated MSCs from human cord blood, which helped treatment of foot ulcers of rats (Lau et al. 2009b; Reed et al. 2000). hCB-MSCs were injected through left femoral artery into streptozotocin-induced diabetic rats. The foot ulcers of the rats significantly reduced when evaluated on the 7th and 14th days of the experiment. The number of inflammatory cells and development of new blood vessels on the third day, granulation on the seventh day, and increase in epithelization on the 14th day were significantly higher than in the controls. This study showed that MSCs could target and localize at wounded sites and increase ECM production and via cytokeratin 19-induced epithelization (Zhao et al. 2014).

10.7.3 Stem Cell Therapies for Nondiabetic Chronic Wounds

Chronic wounds may also occur due to the continued pressure at patients with movement disabilities, who are hospitalized for very long periods, or had inadequate medical care. This type of wounds, known as bed sores, are called decubitus ulcers. Decubitus ulcers are often seen at elder patients whom tissues regenerate very slowly and paralyzed or comatose patients. Several studies (Leung 2007; Game et al. 2012) were published, and one patented product also exists for the treatment of decubitus ulcers (Van Koppen and Hartmann 2015) which show clinical improvement of patients' chronic wounds.

10.7.4 *Grafts Used in Combination with Stem Cells*

The most recent in vivo studies reported the power of dermal autologous micrografts in ameliorating the healing of venous, diabetic, pressure, and post-traumatic ulcers. Various wounds are completely healed after a few weeks of treatment meanwhile increased the quality of life for the patients. In vitro final analyses showed that these micrografts express MSC markers such as CD34, CD73, CD90, and CD105 and are able to form a viable and proliferative biocomplex with collagen sponge. Finally, the sites of ulcers displayed varying expressions of several cytokines such as epidermal growth factors, insulin-like growth factors, platelet-derived growth factors and their receptors, and tumor necrosis factor-β, each one is different than normal tissues (De Francesco et al. 2016).

10.7.5 *Exosome Applications in Wound Healing*

Exosomes are nano-sized vesicles, usually between 30 and 100 nm in diameter. They are secreted by many cell types. Exosomes contain various cell components such as proteins, mRNA, and microRNAs, which are packed with a membrane which also contains specific receptors and markers. Exosomes are transported at the extracellular matrix where they incorporate to target cells and therefore deliver their contents into target cells (Shabbir et al. 2015; Zhang et al. 2015a).

Zhang et al. showed that human hCB-MSC-derived exosomes accelerated re-epithelization and enhanced proliferation and migration of keratinocytes via Wnt4 pathway on experimental acute burn (Zhang et al. 2015b).

We are currently using hCB-MSC-derived exosomes in our GMP standard laboratory for the treatment of various diseases such as muscular dystrophies, neurodegenerative diseases, and chronic wounds. Figure 10.2a shows that the diabetic foot

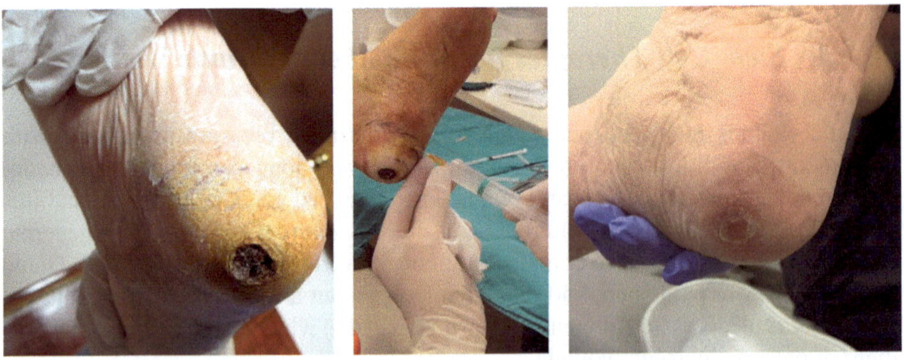

Fig. 10.2 (**a**) Diabetic foot ulcer of an 86-year-old female patient. (**b**) Application of exosomes around the wound area. (**c**) Five months later, completely healed wound

ulcer of an 86-year-old female patient. Exosomes are applied into connective tissue peripheral to the wound area (Fig. 10.2b). Exactly 5 months later, Fig. 10.2c shows complete healing of wound which otherwise could progress to amputation of foot. Our unpublished data also showed anti-inflammatory and antibacterial effects of MSC-derived exosomes in chronic wounds. We believe that the exosome treatment will be a future therapy method in many diseases.

10.8 Future Projections

Wound healing is a highly complex process with a combination of steps, which vary according to the cause and the amount of damage. Other factors such as age, malnutrition, diabetes, infection, kidney failure, hepatitis, chemotherapy, radiotherapy, and smoking negatively affect the healing process. Clinical trials with stem cells, conducted after many successful animal experiments have produced positive results, and therefore they are increasing in numbers. MSCs appear to be the best stem cell source for both availability and effectiveness. On the other hand, MSC-derived exosomes emerge as an alternative source in wound treatment. The development of new techniques for both acute and chronic wounds has shown that cell therapies are becoming the treatment method of the future. Various methods such as 3D-printed materials and robotic surgery methods that use stem cells may be the wound treatment methods of the next decade.

Acknowledgements Authors thank Prof. Dr. Murat AKSOY from Liv Hospital, Istanbul, for his sharing of patient photographs.

References

Akash MS, Rehman K, Chen S (2013) Role of inflammatory mechanisms in pathogenesis of type 2 diabetes mellitus. J Cell Biochem 114(3):525–531

Arya AK, Tripathi R, Kumar S, Tripathi K (2014) Recent advances on the association of apoptosis in chronic non healing diabetic wound. World J Diabetes 5(6):756–762

Balbach ST, Cavaleri FM, Gentile L, Araúzo-Bravo MJ, Schöler HR, Crosetto N, Boiani M (2009) Observing and manipulating pluripotency in normal and cloned mouse embryos. In: Baharvand H (ed) Trends in stem cell biology and technology. Humana Press, New York, p 101

Barbul A, Breslin JR, Woodyard JP, Wasserkrug HL, Efron G (1989) The effect of in vivo T helper and T suppressor lymphocyte depletion on wound healing. Ann Surg 209:479–483

Berman DM, Willman MA, Han D et al (2010) Mesenchymal stem cells enhance allogeneic islet engraftment in nonhuman primates. Diabetes 59:2558–2568

Besse JL, Leemrijse T, Deleu PA (2011) Diabetic foot: the orthopedic surgery angle. Orthop Traumatol Surg Res 97:314–329

Black IB, Woodbury D (2001) Adult rat and human bone marrow stromal stem cells differentiate into neurons. Blood Cells Mol Dis 27:632–636

Blumberg SN, Berger A, Hwang L, Pastar I, Warren SM, Chen W (2012) The role of stem cells in the treatment of diabetic foot ulcers. Diabetes Res Clin Pract 96(1):1–9

Bura A, Planat-Benard V, Bourin P, Silvestre JS, Gross F, Grolleau JL, Saint-Lebese B, Peyrafitte JA, Fleury S, Gadelorge M, Taurand M, Dupuis-Coronas S, Leobon B, Casteilla L (2014) Phase I trial: the use of autologous cultured adipose-derived stroma/stem cells to treat patients with non-revascularizable critical limb ischemia. Cytotherapy 16(2):245–257

Cavanagh P, Attinger C, Abbas Z, Bal A, Rojas N, Xu ZR (2012) Cost of treating diabetic foot ulcers in five different countries. Diabetes Metab Res Rev 28(Suppl 1):107–111

Charles Brunicardi F, Andersen DK, Billiar TR, Dunn DL, Hunter JG, Matthews JB, Raphael E (2005). Pollock, Schwartz's principles of surgery, Chapter 9. Wound healing

De Francesco F, Graziano A, Trovato L, Ceccarelli G, Romano M, Marcarelli M, Cusella De Angelis GM, Cillo U, Riccio M, Ferraro GA (2016) A regenerative approach with dermal micrografts in the treatment of chronic ulcers. Stem Cell Rev 13(1):139–148

Ding Y, Xu D, Feng G, Bushell A, Muschel RJ, Wood KJ (2009) Mesenchymal stem cells prevent the rejection of fully allogenic islet grafts by the immunosuppressive activity of matrix metalloproteinase-2 and -9. Diabetes 58:1797–1806

Draper JS, Seguin CA, Andrews PW (2007) Phenotypic analyses of human embryonic stem cells. In: Sullivan S, Cowan CA, Eggan K (eds) Human embryonic stem cells: the practical handbook. Wiley, Cambridge, MA, pp 93–106

Duque GA, Descoteaux A (2014) Macrophage cytokines: involvement in immunity and infectious diseases. Front Immunol 5:491

Duscher D, Barrera J, Wong VW, Maan ZN, Whittam AJ, Januszyk M, Gurtner GC (2016) Stem cells in wound healing: the future of regenerative medicine? Mini-Rev Gerontol 62(2):216–225

Ertekin C, Taviloğlu K, Güloğlu R, Kurtoğlu M (2005) İstanbul Medikal Yayıncılık, İstanbul. Travma pp 488–501

Fiorina P, Jurewicz M, Augello A et al (2009) Immunomodulatory function of bone marrow-derived mesenchymal stem cells in experimental autoimmune type 1 diabetes. J Immunol 183:993–1004

Fritschi C (2001) Preventive care of the diabetic foot. Nurs Clin North Am 36:303–320

Game FL, Hinchliffe RJ, Apelqvist J, Armstrong DG, Bakker K, Hartemann A, Löndahl M, Price PE, Jeffcoate WJ (2012) A systematic review of interventions to enhance the healing of chronic ulcers of the foot in diabetes. Diabetes Metab Res Rev 28(Suppl 1):119–141

Ghatak S, Maytin EV, Mack JA, Hascall VC, Atanelishvili I, Rodriguez RM, Markwald RR, Misra S (2015) Roles of proteoglycans and glycosaminoglycans in wound healing and fibrosis. Int J Cell Biol http://dx.doi.org/10.1155/2015/834893

Ghieh F, Jurjus R, Ibrahim A, Geagea AG, Daouk H, El Baba B, Chams S, Matar M, Zein W, Jurjus A (2015) The use of stem cells in burn wound healing: a review. Biomed Res Int 2015:1–9

Gupta PK, Krishna M, Chullikana A, Desai S, Murugesan R, Dutta S, Sarkar U, Raju R, Dhar A, Parakh R, Jeyaseelan L, Viswanathan P, Vellotare PK, Seetharam RN, Thej C, Rengasamy M, Balasubramanian S, Majumdar AS (2016) Administration of adult human bone marrow-derived, cultured, pooled, allogeneic mesenchymal stromal cells in critical limb ischemia due to Buerger's disease: phase II study report suggests clinical efficacy. Stem Cells Transl Med. pii: sctm.2016–0237

Harrison NJ, Barbaric I, Andrews PW (2011) Human embryonic stem cell characterization: similarities and differences between cell lines and sources. In: Bongso A, Lee EH (eds) Stem cells: from bench to bedside, 2nd edn. World Scientific, Singapore, pp 1–22

Jayaraman P, Nathan P, Vasanthan P, Musa S, Govindasamy V (2013) Stem cells conditioned medium: a new approach to skin wound healing management. Cell Biol Int 37(10):1122–1128

Jiang J, Lv Z, Gu Y, Li J, Xu L, Xu W et al (2010) Adult rat mesenchymal stem cells differentiate into neuronal-like phenotype and express a variety of neuro-regulatory molecules in vitro. Neurosci Res 66:46–52

Karnieli O, Izhar-Prato Y, Bulvik S, Efrat S (2007) Generation of insulin-producing cells from human bone marrow mesenchymal stem cells by genetic manipulation. Stem Cells 25:2837–2844

Khosrotehrani K (2013) Mesenchymal stem cell therapy in skin: why and what for? Exp Dermatol 22(5):307–310

Lau TW, Lam FF, Lau KM, Chan YW, Lee KM, Sahota DS et al (2009) Pharmacological investigation on the wound healing effects of Radix Rehmanniae in an animal model of diabetic foot ulcer. J Ethnopharmacol 123:155–162

Lee HC, An SG, Lee HW, Park JS, Cha KS, Hong TJ, Park JH, Lee SY, Kim SP, Kim YD, Chung SW, Bae YC, Shin YB, Kim JI, Jung JS (2012) Safety and effect of adipose tissue-derived stem cell implantation in patients with critical limb ischemia: a pilot study. Circ J 76(7):1750–1760. Epub 2012 Apr 12

Lee DE, Ayoub N, Agrawal DK (2016) Mesenchymal stem cells and cutaneous wound healing: novel methods to increase cell delivery and therapeutic efficacy. Stem Cell Res Ther 7:37

Leirósa GJ, Kusinskya AG, Dragob H, Bossib S, Sturlab F, Castellanosa FL, Stellac IY, Balañáa ME (2014) Dermal papilla cells improve the wound healing process and generate hair Bud-like structures in grafted skin substitutes using hair follicle stem cells. Stem Cells Trans Med 3(10):1209–1219

Leung PC (2007) Diabetic foot ulcers – a comprehensive review. Surgeon 5(4):219–231

Li XY, Zheng ZH, Li XY, Guo J, Zhang Y, Li H, Wang YW, Ren J, Wu ZB (2013) Treatment of foot disease in patients with type 2 diabetes mellitus using human umbilical cord blood mesenchymal stem cells: response and correction of immunological anomalies. Curr Pharm Des 19(27):4893–4899

Li XY, Zheng ZH, Guo J, Zhang Y, Li H, Wang YW, Ren J, Wu ZB (2014) Treatment of foot disease in patients with type 2 diabetes mellitus using human umbilical cord blood mesenchymal stem cells: response and correction of immunological anomalies. Curr Pharm Des 19(27):4893–4899

Lu D, Chen B, Liang Z, Deng W, Jiang Y, Li S, Xu J, Wu Q, Zhang Z, Xie B, Chen S (2011a) Comparison of bone marrow mesenchymal stem cells with bone marrow-derived mononuclear cells for treatment of diabetic critical limb ischemia and foot ulcer: a double-blind, randomized, controlled trial. Diabetes Res Clin Pract 92(1):26–36

Madec AM, Mallone R, Afonso G et al (2009) Mesenchymal stem cells protect NOD mice from diabetes by inducing regulatory T cells. Diabetologia 52:1391–1399

Maehr R, Chen S, Snitow M, Ludwig T, Yagasaki L, Goland R, Leibel RL, Melton DA (2009) Generation of pluripotent stem cells from patients with type 1 diabetes. Proc Natl Acad Sci U S A 106:15768–15773

Marieb EN (1995) Human anatomy and physiology. Hodder & Stoughton, London, p 103

Melton DA, Cowan C (2009) "Stemness": definitions, criteria and standards. In: Lanza R, Gaerhart J, Hogan B, Melton D, Thomson J, Wilmut I (eds) Essentials of stem cell biology, 2nd edn. Academic Press, San Diego, pp 23–29

Metcalfe AD, Ferguson MWJ (2007) Tissue engineering of replacement skin: the crossroads of biomaterials, wound healing, embryonic development, stem cells and regeneration. J R Soc Interface 4(14):413–437

Morigi M, Imberti B, Zoja C, Corna D, Tomasoni S, Abbate M et al (2004) Mesenchymal stem cells are renotropic, helping to repair the kidney and improve function in acute renal failure. J Am Soc Nephrol 15:1794–1804

Nandy D, Mukhopadhyay D (2011) Growth factor mediated signaling in pancreatic pathogenesis. Cancers Basel 3(1):841–871

Norton JA, Bollinger RR, Chang AE, Lowry SF, Mulvihill SJ, Pass HI, Thompson RW (2003) Essential practice of surgery: basic science and clinical evidence. In: Lorenz HP, Longaker MT (eds) Chapter 7: Wounds: biology, pathology and management. Springer, New York, pp 89–94

Oskouei BN et al (2012) Increased potency of cardiac stem cells compared with bone marrow mesenchymal stem cells in cardiac repair. Stem Cells Transl Med 1(2):116–124

Ozturk S, Karagoz H (2015) Experimental stem cell therapies on burn wound: do source, dose, timing and method matter? Burns 41:1133

Pawlina W Histology: a text and atlas: with correlated cell and molecular biology. 7th Ed. . Philadelphia Wolters Kluwer. China. Int. Ed. 2016. p. 488–516.

Paz JC, West MP. (2014) Acute care handbook for phsical therapists, 4th edn. Saunders, Philadelphia, pp 297–299.

Polo JM, Liu S, Figueroa MA, Kulalert W, Eminli S, Tan KY, Apostolou E, Stadtfeld M, Li Y, Shioda T, Natesan S, Wagers AJ, Melnick A, Evans T, Hochedlinger K (2010) Cell type of origin influences the molecular and functional properties of mouse induced pluripotent stem cells. Nat Biotechnol 28:848–855

Reed MJ, Meszaros K, Entes LJ, Claypool MD, Pinkett JG, Gadbois TM et al (2000) A new rat model of type 2 diabetes: the fat-fed, streptozotocin-treated rat. Metabolism 49:1390–1394

Sener LT, Albeniz I (2015) Challenge of mesenchymal stem cells against diabetic foot ulcer. Curr Stem Cell Res Ther 10:530–534

Shabbir A, Cox A, Rodriguez-Menocal L, Salgado M, Van Badiavas E (2015) Mesenchymal stem cell exosomes induce proliferation and migration of normal and chronic wound fibroblasts, and enhance angiogenesis in vitro. Stem Cells Dev 24(14):1635–1647

Shumakov VI, Onishchenko NA, Rasulov MF, Krasheninnikov ME, Zaidenov VA (2003) Mesenchymal bone marrow stem cells more effectively stimulate regeneration of deep burn wounds than embryonic fibroblasts. Bull Exp Biol Med 136(2):192–195

Singer DD, Singer AJ, Gordon C, Brink P (2013) The effects of rat mesenchymal stem cells on injury progression in a rat model. Acad Emerg Med 20(4):398–402

Takahashi K, Yamanaka S (2006) Induction of pluripotent stem cells from mouse embryonic and adult fibroblast cultures by defined factors. Cell 126(4):663–676

Vaccari DA, Strom PF, Alleman JE (2005) Environmental biology for engineers and scientists. Wiley-Interscience, Hoboken

Van Koppen CJ, Hartmann RW (2015) Advances in the treatment of chronic wounds: a patent review. Expert Opin Ther Pat 25(8):931–937

van Zuijlen PPM, Gardien KLM, Jaspers MEH, Bos EJ, Baas DC, van Trier AJM, Middelkoop E (2015) Tissue engineering in burn scar reconstruction. Burns Trauma 3:18

Jeffrey Weinzweig (1999) Plastic surgery secrets. Hanley & Belfus Inc. 156053219X, 9781560532194

Xue M, Jackson CJ (2015) Extracellular matrix reorganization during wound healing and its impact on abnormal scarring. Adv Wound Care (New Rochelle) 4(3):119–136

Xue L, Xu YB, Xie JL, Tang JM, Shu B, Chen L, Qi SH, Liu XS (2013) Effects of human bone marrow mesenchymal stem cells on burn injury healing in a mouse model. Int J Clin Exp Pathol 6(7):1327–1336

Yazdanpanah L, Nasiri M, Adarvishi S (2015) Literature review on the management of diabetic foot ulcer. World J Diabetes 6(1):37–53

Yolanda MM, Maria AV, Amaia FG, Marcos PB, Silvia PL et al (2014) Adult stem cell therapy in chronic wound healing. J Stem Cell Res Ther 4:162

Zhang J, Guan J, Niu X, Hu G, Guo S, Li Q, Xie Z, Zhang C, Wang Y (2015a) Exosomes released from human induced pluripotent stem cells-derived MSCs facilitate cutaneous wound healing by promoting collagen synthesis and angiogenesis. J Transl Med 13:49

Zhang B, Wang M, Gong A, Zhang X, Wu X, Zhu Y, Shi H, Wu L, Zhu W, Qian H, Xu W (2015b) HucMSC-exosome mediated-Wnt4 signaling is required for cutaneous wound healing. Stem Cells 33:2158–2168

Zhao QS, Xia N, Zhao N, Li M, Bi CL, Zhu Q, Qiao GF, Cheng ZF (2014) Localization of human mesenchymal stem cells from umbilical cord blood and their role in repair of diabetic foot ulcer in rat. Int J Biol Sci 10(1):80–89

Chapter 11
Adipose-Derived Stem Cells for Wound Healing: An Update

Ilaria Zollino and Paolo Zamboni

Abbreviations

ADMSC	Adipose tissue-derived mesenchymal stem cell
ADRCs	Purified adipose-derived stem and regenerative cells
ADSCs	Adipose-derived stem cells
ASCs	Adipose-derived stromal cells
ASCs	Adipose-derived stem cells
bFGF	Basic fibroblast growth factor
e-PRP	Enhanced PRP
ECM	Extracellular matrix
FGF	Fibroblast growth factor
GSK-3β/Fyn/Nrf2	Glycogen synthase kinase-3β/Fyn kinase/nuclear factor erythroid 2-related factor 2
HA	Hyaluronic acid
HGF	Hepatocyte growth factor
IGF	Insulin-like growth factor
KGF	Keratinocyte growth factor
MMPs	Matrix metalloproteinases
MMP9	Matrix metalloproteinase
PDGF	Platelet-derived growth factor
PGA	Polyglycolic acid
PRP	Platelet-rich plasma

I. Zollino, MD (✉)
Department of Morphology, Surgery and Experimental Medicine, Section of Translational of Medicine and Surgery, University of Ferrara, Via Aldo Moro 8, Ferrara 44124, Italy
e-mail: zlllri@unife.it

P. Zamboni, MD
Department of Morphology, Surgery and Experimental Medicine, Section of Translational of Medicine and Surgery, University of Ferrara, Via Aldo Moro 8, Ferrara 44124, Italy

Sant'Anna University Hospital, Vascular Diseases Center and Institute of Translational Medicine and Surgery, University of Ferrara, Via Aldo Moro, 8, Ferrara 44124, Italy
e-mail: zambo@unife.it

P.V. Pham (ed.), *Pancreas, Kidney and Skin Regeneration*, Stem Cells in Clinical Applications, DOI 10.1007/978-3-319-55687-1_11

ROS	Reactive oxygen species
SDF	Stromal cell-derived factor
TGF-A1 and TGF-2	Transforming growth factor-A1 and Transforming growth factor-2
TGF-β	Transforming growth factor-beta
TIME protocol	Tissue, infection, moisture, environment protocol
TNF	Tumor necrosis factor
TNF-α	Tumor necrosis factor-alpha
VEGF	Vascular endothelial growth factor

11.1 Introduction

Wound healing constitutes an intricate process, which requires coordinated interaction between diverse immunological and biological systems and includes different cell types and molecules. When the healing process fails to result in structural integrity or to follow an orderly and timely sequence, a chronic wound develops (Yolanda et al. 2014).

Chronic wounds are a major problem in medicine today as their incidence is continuously increasing due to an aging population and a growth in the incidence of underlying diseases (Fromm-Dornieden and Koenen 2013), causing a reduction in patient quality of life and rising healthcare costs. The incidence of severe burns in the United States is estimated at 70,000 per year by studies (Fromm-Dornieden and Koenen 2013; Phillips 2001); venous leg ulcers occur at a level of between 600,000 and 1,500,000 (Phillips 2001), and the prevalence of chronic foot wounds in diabetics is 15–20 % (Reiber et al. 1995). The cost of dressings alone to care for the above mentioned cases has been estimated at $5 billion per year from healthcare budgets (Eisenbud et al. 2004).

These wounds fail to heal due to several factors through which can be identified: prolonged or excessive inflammation, persistent infections, the formation of drug-resistant microbial biofilms, and the inability of dermal and/or epidermal cells to respond to reparative stimuli (Woo et al. 2007; Stojadinovic et al. 2008; Attinger et al. 2006; Demidova-Rice et al. 2007). In such situations, conventional therapies which incorporate surgery, dressings, and topical negative pressure among other treatments are increasingly reaching their limits, motivating the use of skin grafts, advanced therapies, and the search for substitute treatment options, including stem cell-based therapies (Fromm-Dornieden and Koenen 2013).

Over the last few years, stem cell application has been put forward as a promising novel therapy for regenerative medicine. This is due both to the stem cells' infinite capacity for self-renewal and the ability to differentiate into multiple cell types under appropriate stimuli (Lauritano et al. 2014). In addition, stem cells are pluripotent and secrete a variety of growth factors. Following initial attention paid to embryonic pluripotent cells, however, different types of adult stem cells have since been studied as a valid and continuous source of stem cells that are readily obtained. The most common source is bone marrow, probably because it can be easily and swiftly accessed and also because various devices are applicable for marrow harvesting (from bone marrow transplantation). The marrow contains hematopoietic cells, mesenchymal cells, and other cell types which may contribute to the promotion of tissue regeneration. Yet the limited amount of source material harvested and the low yield of cell isolation protocols are a major limitation to intraoperative stem cell therapy approaches. To surmount these challenges and obviate the requirement for highly invasive bone marrow harvesting, a procedure causing pain at the aspiration site, alternative sources should be sought from which to isolate autologous stem cells (Coelho et al. 2012). In this field, adipose-derived stem cells (ADSCs) are the stem cell of choice, being abundant, harvestable via a minimally invasive procedure, providing an elevated yield when isolated, and proving suitable for clinical application without prior manipulation (Raposio et al. 2016). The mechanism for healing chronic wounds using ADSCs is built around direct differentiation toward lineage-committed cells or on the production of angiogenic growth factors such as platelet-derived growth factor (PDGF), basic fibroblast growth factor (bFGF), and vascular endothelial growth factor (VEGF): ADSCs migrate to the wound site through paracrine effects and catalyze wound healing, as well as fusion and differentiation (Mizuno and Nambu 2011).

The focus of this chapter is to present an overview of the promising application of autologous cell therapy by using ADSCs approach. Herein, the isolation technique, in which cell products are harvested and delivered to the patient on the same day, is described, and preclinical and clinical applications of ADSCs on chronic wounds caused by several pathologies are reviewed.

Although stem cells therapy is a comparatively new tool, there have been several cases where their capacity to heal wounds has been established as well as several studies which have shown that these types of cells can be used safely.

11.1.1 Physiological Wound Healing

Cutaneous wound healing represents a complex biological process involving regenerating dermal and epidermal tissues consisting of four different but temporarily and spatially overlapping steps: hemostasis, inflammation, proliferation (with formation of granulation tissue), and remodeling (Uysal et al. 2014). Hemostasis occurs immediately after injury and is characterized by vasoconstriction and blood-clotting cascade, preventing excessive bleeding, and providing the provisional matrix for cell migration to ensure temporary protection of the wound area. PDGF and transforming growth factors A1 and A2 (TGF-A1 and TGF-A2) are released during this process, attracting inflammatory cells such as leukocytes and macrophages. Cytokines initiate the healing process by attracting fibroblasts, endothelial cells, and immune cells. The subsequent inflammation phase lasts up to 7 days, involving apoptosis of inflammatory cells such as neutrophils and macrophages.

Proteases and reactive oxygen species (ROS) are released by neutrophils, preventing contamination of bacteria and cleansing the wound of cellular debris. Blood monocytes reach the wound site and differentiate into macrophages. These macrophages remove bacteria and nonviable tissue through phagocytosis and also release various growth factors and cytokines. Endothelial cells, keratinocytes, and fibroblasts are recruited to repair the damaged vessels (Frykberg and Banks 2015).

The proliferation phase begins as the inflammatory phase subsides accompanied by apoptosis of immune cells. The production of growth factors and activation of dermal and epidermal cells in this phase lead to tissue granulation, angiogenesis, and epithelialization. Endothelial progenitor cells, which are important for physiological wound healing, are mobilized by VEGF, matrix metalloproteinase 9 (MMP9), and nitric oxide. Formation of extracellular matrix (ECM) rich tissue occurs in response to insulin-like growth factor (IGF) and stromal cell-derived factor (SDF) (Fromm-Dornieden and Koenen 2013; Frykberg and Banks 2015).

The last phase takes place once the wound has closed and may last 1–2 years. During this phase, matrix remodeling into organized collagen bundles (Falanga 2005; Schultz et al. 2003) and/or scar formation through cellular migration, proliferation, and angiogenic induction is initiated by TGF-A, matrix metalloproteinases (MMPs), and tumor necrosis factor (TNF) (Demidova-Rice et al. 2012).

11.1.2 Pathophysiology of Chronic Wounds

Chronic wounds are the result of an interruption in the progression of one or more of the four phases of the normal cellular and biochemical events toward the skin's integrity. Thus, they represent a failure to reach complete reepithelialization in the proper temporal sequence of tissue repair (Lazarus et al. 1994).

Ninety percent of all chronic wounds are comprised of venous ulcers, pressure ulcers, and diabetic ulcers, which are the most common types (Cherubino et al. 2011). Both local and systemic factors can be involved in chronic wounds etiopathogenesis. Ischemia, arterial/venous insufficiency, local toxins, trauma, and radiation are very important local infection factors. Aging, chronic diseases, alcoholism, smoking, drugs, nutritional deficiencies, chronic kidney disease, and neuropathies seem to be the most important systemic factors (Robertson et al. 2009). Although there are differences in etiology at molecular level, chronic wounds share certain common features, which include the prolonged inflammatory phase, the lack of appropriate metabolism, and clearance of toxic substances from the wound resulting in very high levels of pro-inflammatory cytokines, proteases, ROS, and senescent cells, the existence of persistent infection, the formation of drug-resistant microbial biofilms, the inability of dermal and/or epidermal cells to respond to regenerative stimuli (Fromm-Dornieden and Koenen 2013), and a deficiency in stem cells, which are often also dysfunctional.

The incessant and chronic inflammatory state which characterizes chronic wounds is the basis for ECM degradation due to the loss of important wound healing products, such as PDGF and hepatocyte growth factor (HGF), which are, respectively, broken down by ROS or MMPs and elastases secreted by neutrophils (Amato et al. 2015). In particular, platelet-derived factors such as transforming growth factor-beta (TGF-β) or ECM fragment molecules and microorganisms stimulate the constant influx of immune cells, due to repeated tissue injury; the pro-inflammatory cytokine cascade is therefore amplified and continues for an extended period, leading to high levels of proteases. Proteases are tightly regulated by their inhibitors in acute wounds, while in chronic wounds, protease levels exceed those of their respective inhibitors. Therefore, in chronic wound fluid, pro-inflammatory cytokines such as tumor necrosis factor-alpha (TNF-α), interleukin-1 (IL-1), as well as MMPs and neutrophil elastase are enhanced, leading to the destruction of ECM and a degradation of growth factors and their receptors. Although the production of growth factors is often increased in chronic wound fluid by comparison with acute wound fluid, their quantity and bioavailability are decreased (Demidova-Rice et al. 2012). Moreover, proteolytic destruction of ECM both precludes the wound from progressing into the proliferative phase and attracts more inflammatory cells, amplifying the inflammation cycle (McCarty and Percival 2013). Immune cells produce ROS, providing defense against microorganisms when in low concentrations. In chronic wounds, however, the predominant hypoxic and inflammatory environment boosts ROS production, damaging ECM proteins and causing damage to cells. This sequence of events brings about an enhanced stimulation of proteases and inflammatory cytokines (Schreml et al. 2010). Furthermore, chronic wounds are characterized by senescent cell populations with impaired proliferative and secretory capacities, which means that they do not respond to normal wound healing signals (Schultz et al. 2003). Accumulated data also show that chronic wounds comprise senescent keratinocytes, endothelial cells, fibroblasts, and macrophages (Bourguignon 2014; Cook et al. 2000; Telgenhoff and Shroot 2005; Wall et al. 2008). Oxidative stress is thought to be the cause of the senescent phenotype of cells in chronic wounds. This leads to DNA damage-related cell cycle arrest, or to abnormal metabolic changes in diabetic patients, which in turn bring about defects in intracellular biochemical pathways such as the GSK-3β/Fyn/Nrf2 pathway (Telgenhoff and Shroot 2005; Wall et al. 2008; Bitar 2012).

11.2 Regenerative Medicine in Wound Healing

The goal of the novel field of regenerative medicine is to re-establish function and structure in damaged tissue through the use of three tools: cell-based therapy, biomaterials (or scaffolds), and scaffolds seeded with cells (Zollino et al. 2016).

11.2.1 Cell-Based Therapy

Cell-based therapy represents a set of strategies which use live cells with therapeutic reasons. It includes stem cells, which are undifferentiated cells with the capability to autorenew and differentiate into progenitor or precursor cells of one or several specific cell types (Watt and Hogan 2000; Weissman 2000). The most commonly used stem cells in regenerative medicine are adult stem cells. In particular, mesenchymal stem cells are a cluster of stem cells originating at the mesodermal germinal layer (Caplan 1991; Friedenstein 1980) and can be isolated from various tissue sources in adults (Jin et al. 2013). The use of these cells does not procure any ethical concerns, in contrast with using embryonic cells, and they are relatively easy to obtain. Bone marrow-derived stem cells have been the focus of regenerative medicine research strategies for many years. However, current research interest is focusing on the development of ADSCs, which can be isolated directly from liposuction during plastic surgery procedures.

There have already been several clinical reports of the successful application of ADSCs in wound healing (Kim et al. 2011). Wound healing benefits because the ADSCs are multipotent, with the ability to differentiate into other specialized cells, secrete, or suppress the growth hormones and cytokines necessary in the environment. They are also capable of increasing in number while displaying a stable phenotype.

Both acute and chronic wounds can be treated using cell therapy. ADSCs can improve wound healing, reduce scar contracture, and minimize donor site morbidity in the treatment of acute wounds. Conversely, in chronic wound treatment, the wound bed is the environment where maximum healing can be achieved through the transplant of cells with excellent wound healing capacities (You and Han 2014).

We emphasize that wound bed preparation needs to be meticulous before implantation, in accordance with the tissue, infection, moisture, environment protocol (TIME protocol) (Ayello et al. 2004). The TIME protocol includes the basic principles of wound care that are critical for managing chronic wounds.

11.2.1.1 Scaffolds

The use of scaffolds and cellular matrices is essential to differentiate mesenchymal stem cells into the cells required and use them to realize a three-dimensional tissue for use in reconstructive medicine.

An essential strategy for tissue engineering is the choice and the construction of a good-quality scaffold. Preferably, the scaffold should be a structural and functional platform able to imitate the native extracellular matrix and support the morphogenesis of multiple tissues. Building on the literature, tissue-engineering scaffolds should (1) be biodegradable; (2) not trigger inflammatory responses; (3) have surface properties which enhance the attachment, proliferation, and differentiation of cells; (4) mimic skin in vitro; (5) have the relevant mechanical properties; and (6) be well suited for manufacture into different forms (Zollino et al. 2016; Croisier and Jerome 2013).

11.2.1.2 Scaffolds Seeded with Cells

Engineered tissue regeneration uses a biocompatible scaffold which replaces, repairs, or regenerates damaged tissue in combination with living cells and/or bioactive molecules. This kind of scaffold should be biocompatible, porous, and permeable to support cell adhesion and proliferation (Zollino et al. 2016).

Suitable scaffolds available for adipose tissue engineering include type I collagen sponge, non-woven polyglycolic acid (PGA), and hyaluronic acid gel. Hyaluronic acid (HA) is a naturally occurring non-sulfated glycosaminoglycan which is present in connective tissue, the synovial fluid of articular joints and the vitreous humor of the eye. It is one component of the natural extracellular matrix and is important in tissue hydration, cell differentiation, and tissue reparation.

HA is defined as a "suitable scaffold for adipose tissue engineering." It is highly biocompatible, does not cause adverse reactions, and is reabsorbed by the host tissues (Messina et al. 2005). In particular, in vivo experiments have confirmed the ideal tissue repair and restoration of full-thickness wounds for the treatment of ulcers in a placebo-controlled study (Zavan et al. 2009). Moreover, HA and its derivatives are actively angiogenic: preliminary results in in vivo models demonstrate whole vena cava regeneration inside hyaluronic acid-based prosthesis, introducing innovative perspectives in microvascular surgery applications (Messina et al. 2005; Pandis et al. 2010).

In a recent study, Altman et al. (2009) showed that a new scaffold consisting of silk fibroin-chitosan when combined with ADSCs supports the engraftment of stem cells and their differentiation into epithelial and fibro vascular components, increasing the repair and healing capacity of damaged tissues.

The microenvironment for wound regeneration depends mainly on interaction between stem cell progenitors and their niche (Wong et al. 2012); consequently, any tissue-engineered reconstruction should provide an appropriate microenvironment for the cells to proliferate and differentiate.

11.2.2 Scaffolds and Nanotechnology

Stem cells differentiation can also be induced by physical factors and modulation of extracellular matrix nanostructures. Indeed, since the majority of signaling molecules interact with stem cells at the nanoscale level, scaffolds with surface nanostructures have potential applications for stem cells in the field of tissue engineering and regenerative medicine. The literature offers several different methods to induce such differentiation through the use of high-quality nanoparticles of varying chemical composition. However, before introducing nanoparticles into clinical practice, the real biological effects of their use should be carefully assessed: releasing active peptides may conceivably cause interference with some biological functions and cellular processes (Zollino et al. 2016; Tocco et al. 2012).

11.3 Adipose-Derived Stem Cells (ADSCs) Properties

Adipose tissue is one of the largest tissues in the body, representing an important energy and endocrine reservoir (Yoshimura et al. 2009). It is mainly composed of adipocytes arranged in lobules, accounting for more than 90 % of the tissue volume, and of pericytes, fibroblasts, macrophages, vascular endothelial cells, and an extracellular matrix (Suga et al. 2008).

Zuck et al. (2001) first identified ADSCs in 2001 as a population of fibroblast cells with the ability to differentiate into myogenic, adipogenic, osteogenic, and chondrogenic cells through specific induction factors. Since this time, easier isolation without main surgical procedures and donor site morbidity, along with better availability, has sparked great clinical and research interest in ADSCs. In particular, ADSCs have attracted a lot of attention as an alternative to bone marrow stem cells. ADSCs are multipotent stem cells, with characteristics similar to bone marrow-mesenchymal stem cells. These are demonstrated by their expression of identical cell surface markers, their comparable gene expression profiles and their similar differentiation potentialities (Baglioni et al. 2009). For this reason, ADSCs have the capacity to form fat, muscle, bone, and cartilage, under appropriate stimuli. They are multipotent and produce a range of growth factors (Pu et al. 2008) such as bFGF, keratinocyte growth factor (KGF), TGF-β, HGF, and VEGF (Kim et al. 2009).

Human adipose tissue offers several advantages as a stem cell font. These include the ease with which ADSCs can be isolated and expanded, their abundance within the adipose tissue, and the frequency ranges from 1:100 to 1:1500 cells, far exceeding the prevalence of mesenchymal stem cells in bone marrow (De Ugarte et al. 2003). There is only 1 per 100,000 nucleated cells of mesenchymal stem cells in the bone marrow, and their quantity decreases with age (Sensebe and Bourin 2009).

Many groups have tried to optimize the isolation and expansion of ADSCs (Pu et al. 2008; Bunnell et al. 2008; Gnanasegaran et al. 2014; Markarian et al. 2014; Zhu et al. 2013; Gronthos and Zannettino 2011; Rose et al. 2006; Conde-Green et al. 2010a, b; Pulsfort et al. 2011; Reshak et al. 2013; Arana et al. 2013). ADSCs are commonly extracted from adipose tissue in a stepwise procedure. However, the most commonly used ADSCs isolation technique is that described by Coleman.

11.3.1 Coleman's Technique

This method was introduced by Coleman in 1995, and it was first used for facial remodeling, in breast reconstruction, and breast augmentation (Bucky and Kanchwala 2007; Moseley et al. 2006). It provides a multiple stepwise procedure including aspiration, centrifugation, and the subsequent re-injection of autologous fat. The fat donor sites are the abdomen, waist adipose deposit, flanks, and inner sides of the thigh and knees (Zollino et al. 2016).

First, Kleine's solution made up of 250 mL normal saline, 20 mL of 1 % carbocaine, 1 mL adrenaline, and 2 ml bicarbonate is injected into the fat donor area (Hunstad 1995). After diffusion of the solution, the harvesting procedure is undertaken, using a two-hole blunt cannula (Byron Medical, Tucson, AZ, USA) fitted directly to a 10 mL Luer Lock syringe (BD Syringe Luer Lock tip; Becton Dickinson, Franklin Lakes, NJ, USA), which helps to lessen the pressure caused during the harvesting procedure and preserves the fat (Mojallal et al. 2008). Following the fat harvesting, the lipoaspirate is processed via centrifugation at 3000 rpm (rotor size: 16 cm; g force: 580) for 3 min in 10 ml syringes. This separates the fat into three layers. The upper layer (supranatant) and lower layer are discarded, as they are composed of oil from destroyed fat and blood, respectively, leaving the middle layer, which contains a high concentration of stem cells (Clauser et al. 2014; Gardin et al. 2012), stromal cells, vascular endothelial cells, and mural cells, termed the "stromal vascular fraction." Once the middle layer has been extracted, the micro-fat graft is transferred into 1 ml syringes prior to injection into the tissue to be grafted (Fig. 11.1 shows the steps described above to harvest the adipose tissue and the stem cells using Coleman's technique). A multilayer technique is used to implant the aliquots of fat, with very small amounts of fatty tissue released into the recipient area in order to optimize the successful implantation of the graft. A blunt Coleman microcannula is used to deposit the micro-fat graft via a number of subdermal and hypodermal tunnels through numerous tissue planes. This technique of minimizing

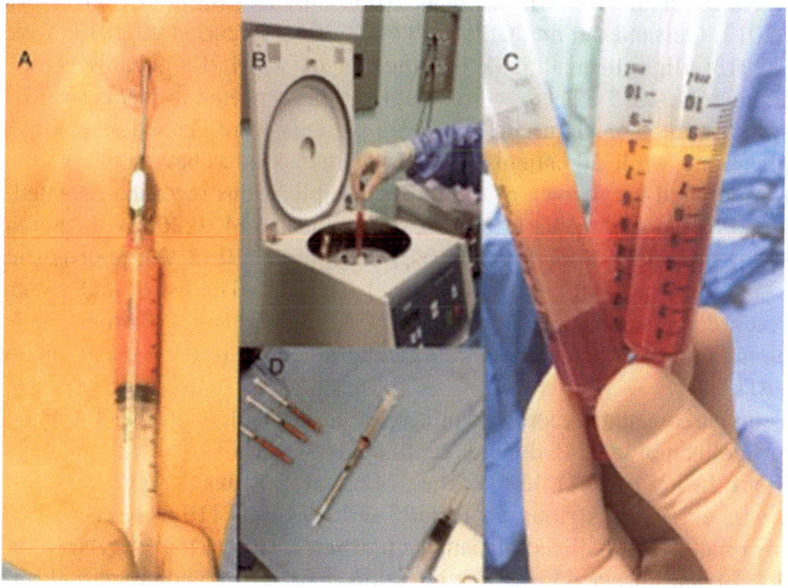

Fig. 11.1 Coleman's technique for harvesting ADSCs: (**a**) Harvesting from the periumbilical area using tumescent local anesthesia; (**b**) Centrifugation of aspirated adipose tissue; (**c**) Separation of adipose tissue components through centrifugation; (**d**) Following fat and blood elimination, the ADSCs are aspirated using a Luer Lock syringe

the amount of micro-fat graft released each time the cannula is introduced, increases the area between the grafted fat and the tissue receiving it. In this way, there is a reduction in fat damage and adipocyte necrosis, with an improvement in graft vascularization and three-dimensional fat distribution. The recently grafted fat has a readily available blood supply, which facilitates its survival and reduces the possibility of fat necrosis or calcification (Zollino et al. 2016; Coleman 2004).

11.3.1.1 Safety of Coleman's Technique for ADSCs Harvesting

The focus of current stem cells harvesting techniques is on bone marrow, umbilical cord blood, and peripheral blood. The method for harvesting and fat grafting initially introduced by Coleman (1995) yields a higher number of stem cells when related with these techniques. The stem cells can be used intraoperatively with minor organizational and legal limitations that might otherwise prove expensive for clinical use (Zollino et al. 2016).

ADSCs are also competitive with regard to complications. Zollino et al. (2016) undertook a review of the scientific literature to evaluate complications reported in studies on the fat harvesting and grafting process employing Coleman's procedure for reconstructive surgery purposes. A total of 5089 patients who underwent plastic surgery and maxillofacial surgical procedures including facial restoration and post-mastectomy reconstruction were selected (Burnouf et al. 2005; Hardy et al. 2007; Dollfus et al. 2009; Clauser et al. 2011; Guijarro-Martinez et al. 2011; Claro et al. 2012; Guisantes et al. 2012; Seth et al. 2012; Arcuri et al. 2013; Weichman et al. 2013; Biglioli et al. 2014; Endara et al. 2014; Kaoutzanis et al. 2016; Piombino et al. 2015). There were complications reported in 169 cases (3.3 % of the overall number). About 141 (2.77 %) patients were classed as having minor complications, while 28 patients (0.55 %) were classed as having major complications. A detailed breakdown of the minor complications recorded revealed nodularity and/or induration, 93 (1.83 %); dysaesthesia, 14 (0.26 %); hematoma, 12 (0.23 %); superficial infection, 11 (0.21 %); pain, 7 (0.13 %); poor cosmesis, 3 (0.06 %); and abnormal breast secretion, 1 (0.02 %). The list of major complications recorded revealed deep infection, 22 (0.43 %); sepsis, 3 (0.06 %); abdominal hematoma requiring percutaneous surgical drainage, 2 (0.04 %); and pneumothorax 1 (0.02 %).

In comparison, data recorded on the recovery and safety profiles following bone marrow collection in 9245 donors identified 345 cases (3.7 %) with potential complications. In 125 cases (1.35 % of the 9245 total) post-harvesting, these complications were classed as serious. Of the 125 serious cases, 116 were classed as being directly related to the collection, namely, mechanical injury to tissue, bone, or nerve, 69 (0.7 %); anesthesia, 45 (0.5 %); infection, 1 (0.01 %); and grand mal seizure, 1 (0.01 %). Among the 116 patients with serious complications, 67 (0.7 % of the 9245 total) experienced prolonged recovery times due to mechanical injury to tissue from needle aspirations and required interventions ranging from limited physician involvement and/or physical therapy to surgical intervention and ongoing disability

(1–10 years). Among the remaining 49 patients (0.5 % of the 9245 total) with severe reactions, most issues were due to severe acute reactions related to the anesthesia (complicated post-spinal headaches, cardiac arrhythmia, and pulmonary edema) (Miller et al. 2008; Pulsipher et al. 2014; Holtan and Weisdorf 2014).

The use of Coleman's procedure for ADSCs harvesting therefore seems advantageous when compared with other techniques for stem cells harvesting. It seems a safer option, involving less discomfort and a reduced risk of complications for the patients.

11.3.2 ADSCs in Wound Healing

As conventional treatment strategies for chronic wounds increasingly reach their limits and often fail, advanced healing therapies are being used to rectify the irregular and dysfunctional cellular pathways which are present in chronic wounds. These treatment strategies include skin substitutes, growth factor-based therapies, biological dressings, and synthetic acellular matrices (Rizzi et al. 2010).

Increasing vasculogenesis and angiogenesis is vital to the investigation of pioneering wound healing strategies (Seifter et al. 1981). Both fibroblast growth factor (FGF) and VEGF are potent angiogenic factors. Increased granulation, decreased contraction, and increased angiogenesis have been demonstrated through the application of FGF to wounds (Akasaka et al. 2007). In addition, KGF and FGF have been clarified in order to activate fibroblast and keratinocyte proliferation and migration and collagen synthesis and induce angiogenesis (Slavin 1996; Greenhalgh 1996).

Vascularization plays a crucial part in wound healing and is thus a significant parameter for new therapies (Hendrickx et al. 2010). For this reason, innovative and alternative treatment options including stem cell-based therapies have been investigated over the last decade (Cherubino et al. 2011). ADSCs offer prodigious potential through their capacity to release angiogenic factors, displaying increased angiogenesis in wound healing whenever injected or delivered via a scaffold (Branski et al. 2009). Increased vascular tissue has been appreciated in models using endothelial cells, and fibroblasts have yielded increased vascular tissue. Moreover, ADSCs have displayed improved healing in both burn models and full-thickness wound models (Cherubino et al. 2011).

ADSCs have also shown a positive impact on wound healing in clinical and preclinical studies, in addition to their angiogenic potential. Recent ADSCs applications in vitro and in vivo have demonstrated that they are attracted to the wound site and affect regeneration processes by means of paracrine mechanisms in addition to fusion and differentiation, for instance, into keratinocytes or dermal fibroblasts (Nambu et al. 2009; Ebrahimian et al. 2009; Cho et al. 2006).

Ongoing clinical trials are using ADSCs for regenerative medical and tissue engineering applications (Gimble et al. 2010, 2011). Rigotti et al. (2007) used ADSCs to treat patients with severe symptoms or irreversible function damage due to side effects from radiation treatment (LENT-SOMA scale grades 3 and 4). Irradiated areas were treated through the application of purified autologous lipoaspirates. In the majority of patients treated, both ultrastructural tissue regeneration with neovessel formation, and significant clinical improvement were observed.

Promising applications in wounds and ulcers healing have been reported, although thus far these are small studies with a total of only 98 patients (Raposio et al. 2016; Cervelli et al. 2010, 2011; Marino et al. 2013; Bura et al. 2014; Lee et al. 2012). Bartsich and Morrison (2012) have discussed the long-term treatment of chronic sickle cell ulcers and the possible use of a skin graft and fat grafting to achieve healing. Current treatment using skin grafting and local wound care is often unsuccessful over the long-term, as wounds that have healed break down again (Table 11.1). Treatment involving enduring modification of the wound bed with enrollment of a new cell population and subsequent fat grafting was satisfactorily applied in Bartsich and Morrison's study. Raposio et al. (2016) presented their experience in regenerative surgery of chronic skin ulcers, evaluating the effects related to the use of adipose-derived stem cells (ASCs) added to platelet-rich plasma (PRP), to obtain an enhanced PRP (e-PRP). e-PRP significantly enhanced wound closure rates when compared to standard wound care through a faster recovery and without causing any serious complications. In 2010 Cervelli et al. (2010) have shown the capability of combining autologous adipose grafts and PRP injected intralesionally or perilesionally to regenerate tissue and induce epithelialization with wound closure in patients with a loss of substance on the lower limb. They reported a significant decrease in healing time (57 % of patients achieved complete healing within 3 months) and a noticeable improvement in quality of life, along with a diminution in cost due to the decreased amount of medication required. Their subsequent research (Cervelli et al. 2011) suggests a new therapeutic plan: the use of enhanced stromal vascular fraction (e-SVF). As reported, e-SVF and PRP mixed with fat grafting applied to the bed of ulcers are two treatments that showed improvement in the healing process in post-traumatic extremity ulcers. Results showed that wounds treated with e-SVF healed better than wounds treated with hyaluronic acid alone. After 9.7 weeks, the patients treated with e-SVF experienced a 97.9 ± 1.5 % reepithelialization rate compared to 87.8 ± 4.4 % of the first control group (hyaluronic acid), while patients treated with PRP and lipostructure after 9.7 weeks experienced a 97.8 % ±1.5 % reepithelialization compared to 89.1 ± 3.8 % of the second control group (PRP). Marino et al. (2013) have used purified adipose-derived stem and regenerative cells (ADRCs) obtained from autologous fat for the treatment of chronic lower limb ulcers of arteriopathic patients. In all cases, injections into the edges of the ulcer produced a decrease in both the diameter and depth of the ulcer. There was healing with complete reepithelization in over half of the cases. In the first phase I trial, Bura et al. (2014) have proved that adipose-derived stromal cells (ASCs) transplantation greatly increases tissue revascularization, with ulcer evolution and wound healing improvement in non-revascularizable critical limb ischemia patients. Lee et al. (2012) have demonstrated that a safe alternative method to achieve therapeutic angiogenesis in patients with critical limb ischemia who are refractory to other treatment modalities could be the use of multiple intramuscular adipose

Table 11.1 Summary of the use of cell therapy in ulcer healing

Wound type	Cell type	Delivery system	Outcome	Reference
Chronic venous, diabetic, and ischemic ulcers ($n = 16$ patients)	e-PRP	Injection into skin edge and at the bottom of the lesion	Complete healing in 71% of patients	Raposio et al. (2016)
Non-revascularizable critical limb ischemia with/without ulcers ($n = 7$ patients)	ASCs	Intramuscular injection	Increased transcutaneous oxygen pressure, improvement of ulcer evolution and wound healing	Bura et al. (2014)
Chronic ulcers of the lower limbs of artheriopatic patients ($n = 10$ patients)	ADRCs	Injection at the edges of the ulcers	Complete healing in 60% of patients	Marino et al. (2013)
Critical limb ischemia with/without non-healing ulcers ($n = 15$ patients)	ADMSCs	Intramuscular injection	Clinical improvement occurred in 66.7% of patients (minor amputation in 5 patients)	Lee et al. (2012)
Post-traumatic lower extremity ulcers ($n = 20$ patients)	e-SVF in 10 patients, AT+PRP in 10 patients	Injection in small tunnels in the perilesional area	Complete healing in 97.9%-97.8% of patients	Cervelli et al. (2011)
Vascular, diabetes-correlated, or post-traumatic diseases with ulcers or loss of substance of the lower limbs ($n = 30$ patients)	AT+PRP	Injection of AT + PRP covered with 3-dimensional polymerized HA-medicated biological dressing	Complete healing in 57% of patients	Cervelli et al. (2010)

Key: *ASCs* (Raposio 2016) adipose-derived stem cells, *PRP* platelet-rich plasma, *e-PRP (enhanced PRP)* ASCs + PRP, *ASCs* (Bura 2014) adipose-derived stroma cells, *ADRCs* purified adipose-derived stem and regenerative cells, *ADMSCs* adipose tissue-derived mesenchymal stem cells, *e-SVF* enhanced stromal vascular fraction, *AT* adipose tissue, *HA* hyaluronic acid tissue-derived mesenchymal stem cells (ADMSC) injections. The Authors have observed clinical improvement in 66.7 % of cases. Their patients have displayed a significant decrease on the pain rating scale and improved claudication walking distance. Moreover, various studies on mouse, rat, and rabbit models have provided encouraging evidence of ulcer evolution and wound healing improvement (Nambu et al. 2007, 2009; Ebrahimian et al. 2009; Mineda et al. 2015; Kuo et al. 2016; Kinoshita et al. 2015; Shen et al. 2013; Nie et al. 2011; Hong et al. 2013; Steinberg et al. 2012; Zografou et al. 2013; Blanton et al. 2009; Kim et al. 2012; Lee et al. 2011; Lim and Yoo 2010; Lin et al. 2013; Maharlooei et al. 2011; Rodriguez et al. 2015; Amos et al. 2010). The positive results of ADSCs on wound healing have been demonstrated using animal models with chronic diseases or artificially induced impaired wound healing. An increase in collagen intensity, capillary density, VEGF, and TGF-β3 expression was displayed as a consequence of ADSCs transfer. Animals with ADSCs transplantation on wounds demonstrated appreciably increased survival, angiogenesis, and epithelialization rates (Zografou et al. 2013). The ADSCs accelerated wound healing of radiation ulcers in a modified rat model, where they were co-localized with endothelial cell markers in ulcerated tissues. Treatment using ADSCs achieved smaller wound sizes and was linked to development of new blood vessels (Huang et al. 2013). Finally, Hong et al. (2013) showed in a rabbit ear in vivo model that topically delivered rabbit ADSCs are engrafted and proliferate in wounds, where they exhibited an activated fibroblast phenotype. Furthermore, ADSCs led to augmented endothelial cell and macrophage enrollment. In contrast to bone marrow-derived stem cells and fibroblasts, they increased granulation tissue formation.

11.3.3 Limitations of Using ADSCs

Donor specificity is a recognized phenomenon. Recent studies have investigated ADSCs's function within the context of donor age and gender. Donor specificity of human ADSCs has been reported by Shu et al. (2012), who observed links between donor age and cell differentiation as well as antiapoptotic ability. Another study with human ADSCs demonstrated that equal amounts of ADSCs could be isolated, regardless of donor age. However, infant-derived cells have shown different morphology and enhanced angiogenic and osteogenic capabilities (Wu et al. 2013). Donor age specificity has also been investigated by Guercio et al. (2013) and has revealed a higher proliferation capacity of ADSCs in younger dogs compared with older animals.

There is scant evidence thus far regarding the effect of gender on ADSCs potential. Fossett and Khan (2012) surmised that females have a notably higher yield of mesenchymal stem cells than males and that estrogens play an excitatory role in controlling levels of cytokines and growth factor production. Furthermore, the gender of ADSCs donors influenced the proliferation, differentiation, paracrine, and antiapoptotic capacities of human ADSCs (Shu et al. 2012). As well as age and gender, body mass index, chronic disease, a Western lifestyle, and many other features cause donor-specific variations. With progenitor cells, it has been seen that cells harvested from patients with chronic diseases have reduced regenerative potential (Badiavas et al. 2007; Rodriguez-Menocal et al. 2012).

Another influencing factor in the applicability of ADSCs in chronic wound treatment is the composition of wound fluid, which differs markedly between acute and chronic wounds. In an in vitro wound model, wound fluid has been shown to influence ADSCs function inversely. While acute wound fluid has a strong chemotactic impact and stimulates ADSCs proliferation, chronic wound fluid has an inhibiting effect on ADSCs migration and proliferation. Chronic wound fluid strongly induces expression of bFGF, VEGF, and MMP9 (Koenen et al. 2015). Furthermore, the risk of inducing cancer by transplantation of ADSCs has not yet been fully excluded (Kim et al. 2009).

11.4 Conclusion

"Stem cell therapy" has emerged as a propitious tool in cases where wounds do not respond adequately to conventional treatment after 4 weeks. It offers a more immediate prospect for treatment through combining an autologous cell source with a well-established surgical intervention in a single procedure. Mesenchymal stem cells from adipose tissue and bone marrow alike have proved to be efficacious in several clinical approaches. However, ADSCs have been safer and less invasive to harvest.

At present, there are neither standardized protocols for the clinical application of ADSCs nor a consensus on the number of cells necessary for different therapeutic options. Therefore, standardized protocols and larger randomized controlled trials are indispensable to ensure the safety and effectiveness of ADSCs application. Moreover, donor specificity should be investigated in detail, and the underlying reasons for varying efficacy need to be clarified. However, within the scope of personalized medicine, the application of autologous ADSCs in chronic wound treatment should be considered on an individual basis.

References

Akasaka Y, Ono I, Tominaga A, Ishikawa Y, Ito K, Suzuki T, Imaizumi R, Ishiguro S, Jimbow K, Ishii T (2007) Basic fibroblast growth factor in an artificial dermis promotes apoptosis and inhibits expression of alpha-smooth muscle actin, leading to reduction of wound contraction. Wound Repair Regen 15(3):378–389. doi:10.1111/j.1524-475X.2007.00240.x

Altman AM, Yan Y, Matthias N, Bai X, Rios C, Mathur AB, Song YH, Alt EU (2009) IFATS collection: human adipose-derived stem cells seeded on a silk fibroin-chitosan scaffold enhance wound repair in a murine soft tissue injury model. Stem Cells 27(1):250–258. doi:10.1634/stemcells.2008-0178

Amato B, Coretti G, Compagna R, Amato M, Buffone G, Gigliotti D, Grande R, Serra R, de Franciscis S (2015) Role of matrix metalloproteinases in non-healing venous ulcers. Int Wound J 12(6):641–645. doi:10.1111/iwj.12181

Amos PJ, Kapur SK, Stapor PC, Shang H, Bekiranov S, Khurgel M, Rodeheaver GT, Peirce SM, Katz AJ (2010) Human adipose-derived stromal cells accelerate diabetic wound healing: impact of cell formulation and delivery. Tissue Eng Part A 16(5):1595–1606. doi:10.1089/ten.TEA.2009.0616

Arana M, Mazo M, Aranda P, Pelacho B, Prosper F (2013) Adipose tissue-derived mesenchymal stem cells: isolation, expansion, and characterization. Methods Mol Biol 1036:47–61. doi:10.1007/978-1-62703-511-8_4

Arcuri F, Brucoli M, Baragiotta N, Stellin L, Giarda M, Benech A (2013) The role of fat grafting in the treatment of posttraumatic maxillofacial deformities. Craniomaxillofac Trauma Reconstr 6(2):121–126. doi:10.1055/s-0033-1333877

Attinger CE, Janis JE, Steinberg J, Schwartz J, Al-Attar A, Couch K (2006) Clinical approach to wounds: debridement and wound bed preparation including the use of dressings and wound-healing adjuvants. Plast Reconstr Surg 117(7 Suppl):72S–109S. doi:10.1097/01.prs.0000225470.42514.8f

Ayello EA, Dowsett C, Schultz GS, Sibbald RG, Falanga V, Harding K, Romanelli M, Stacey M, TE L, Vanscheidt W (2004) TIME heals all wounds. Nursing 34(4):36–41. quiz, 41-32

Badiavas EV, Ford D, Liu P, Kouttab N, Morgan J, Richards A, Maizel A (2007) Long-term bone marrow culture and its clinical potential in chronic wound healing. Wound Repair Regen 15(6):856–865. doi:10.1111/j.1524-475X.2007.00305.x

Baglioni S, Francalanci M, Squecco R, Lombardi A, Cantini G, Angeli R, Gelmini S, Guasti D, Benvenuti S, Annunziato F, Bani D, Liotta F, Francini F, Perigli G, Serio M, Luconi M (2009) Characterization of human adult stem-cell populations isolated from visceral and subcutaneous adipose tissue. FASEB J 23(10):3494–3505. doi:10.1096/fj.08-126946

Bartsich S, Morrison N (2012) Composite fat and skin grafting for the management of chronic sickle cell ulcers. Wounds 24(3):51–54

Biglioli F, Allevi F, Battista VM, Colombo V, Pedrazzoli M, Rabbiosi D (2014) Lipofilling of the atrophied lip in facial palsy patients. Minerva Stomatol 63(3):69–75

Bitar MS (2012) The GSK-3beta/Fyn/Nrf2 pathway in fibroblasts and wounds of type 2 diabetes: on the road to an evidence-based therapy of non-healing wounds. Adipocytes 1(3):161–163. doi:10.4161/adip.20235

Blanton MW, Hadad I, Johnstone BH, Mund JA, Rogers PI, Eppley BL, March KL (2009) Adipose stromal cells and platelet-rich plasma therapies synergistically increase revascularization during wound healing. Plast Reconstr Surg 123(2 Suppl):56S–64S. doi:10.1097/PRS.0b013e318191be2d

Bourguignon LY (2014) Matrix hyaluronan-activated CD44 signaling promotes keratinocyte activities and improves abnormal epidermal functions. Am J Pathol 184(7):1912–1919. doi:10.1016/j.ajpath.2014.03.010

Branski LK, Gauglitz GG, Herndon DN, Jeschke MG (2009) A review of gene and stem cell therapy in cutaneous wound healing. Burns 35(2):171–180. doi:10.1016/j.burns.2008.03.009

Bucky LP, Kanchwala SK (2007) The role of autologous fat and alternative fillers in the aging face. Plast Reconstr Surg 120(6 Suppl):89S–97S. doi:10.1097/01.prs.0000248866.57638.40

Bunnell BA, Flaat M, Gagliardi C, Patel B, Ripoll C (2008) Adipose-derived stem cells: isolation, expansion and differentiation. Methods 45(2):115–120. doi:10.1016/j.ymeth.2008.03.006

Bura A, Planat-Benard V, Bourin P, Silvestre JS, Gross F, Grolleau JL, Saint-Lebese B, Peyrafitte JA, Fleury S, Gadelorge M, Taurand M, Dupuis-Coronas S, Leobon B, Casteilla L (2014) Phase I trial: the use of autologous cultured adipose-derived stroma/stem cells to treat patients with non-revascularizable critical limb ischemia. Cytotherapy 16(2):245–257. doi:10.1016/j.jcyt.2013.11.011

Burnouf M, Buffet M, Schwarzinger M, Roman P, Bui P, Prevot M, Deleuze J, Morini JP, Franck N, Gorin I, Dupin N (2005) Evaluation of Coleman lipostructure for treatment of facial lipoatrophy in patients with human immunodeficiency virus and parameters associated with the efficiency of this technique. Arch Dermatol 141(10):1220–1224. doi:10.1001/archderm.141.10.1220

Caplan AI (1991) Mesenchymal stem cells. J Orthop Res 9(5):641–650. doi:10.1002/jor.1100090504

Cervelli V, De Angelis B, Lucarini L, Spallone D, Balzani A, Palla L, Gentile P, Cerulli P (2010) Tissue regeneration in loss of substance on the lower limbs through use of platelet-rich plasma, stem cells from adipose tissue, and hyaluronic acid. Adv Skin Wound Care 23(6):262–272. doi:10.1097/01.ASW.0000363551.82058.36

Cervelli V, Gentile P, De Angelis B, Calabrese C, Di Stefani A, Scioli MG, Curcio BC, Felici M, Orlandi A (2011) Application of enhanced stromal vascular fraction and fat grafting mixed with PRP in post-traumatic lower extremity ulcers. Stem Cell Res 6(2):103–111. doi:10.1016/j.scr.2010.11.003

Cherubino M, Rubin JP, Miljkovic N, Kelmendi-Doko A, Marra KG (2011) Adipose-derived stem cells for wound healing applications. Ann Plast Surg 66(2):210–215. doi:10.1097/SAP.0b013e3181e6d06c

Cho HH, Kyoung KM, Seo MJ, Kim YJ, Bae YC, Jung JS (2006) Overexpression of CXCR4 increases migration and proliferation of human adipose tissue stromal cells. Stem Cells Dev 15(6):853–864. doi:10.1089/scd.2006.15.853

Claro F Jr, Figueiredo JC, Zampar AG, Pinto-Neto AM (2012) Applicability and safety of autologous fat for reconstruction of the breast. Br J Surg 99(6):768–780. doi:10.1002/bjs.8722

Clauser LC, Tieghi R, Galie M, Carinci F (2011) Structural fat grafting: facial volumetric restoration in complex reconstructive surgery. J Craniofac Surg 22(5):1695–1701. doi:10.1097/SCS.0b013e31822e5d5e

Clauser L, Ferroni L, Gardin C, Tieghi R, Galie M, Elia G, Piattelli A, Pinton P, Bressan E, Zavan B (2014) Selective augmentation of stem cell populations in structural fat grafts for maxillofacial surgery. PLoS One 9(11):e110796. doi:10.1371/journal.pone.0110796

Coelho MB, Cabral JM, Karp JM (2012) Intraoperative stem cell therapy. Annu Rev Biomed Eng 14:325–349. doi:10.1146/annurev-bioeng-071811-150041

Coleman SR (1995) Long-term survival of fat transplants: controlled demonstrations. Aesthet Plast Surg 19(5):421–425

Coleman SR (2004) Structural fat grafting. Quality Medical Publishing, St Louis

Conde-Green A, de Amorim NF, Pitanguy I (2010a) Influence of decantation, washing and centrifugation on adipocyte and mesenchymal stem cell content of aspirated adipose tissue: a comparative study. J Plast Reconstr Aesthet Surg 63(8):1375–1381. doi:10.1016/j.bjps.2009.07.018

Conde-Green A, Baptista LS, de Amorin NF, de Oliveira ED, da Silva KR, Pedrosa Cda S, Borojevic R, Pitanguy I (2010b) Effects of centrifugation on cell composition and viability of aspirated adipose tissue processed for transplantation. Aesthet Surg J 30(2):249–255. doi:10.1177/1090820X10369512

Cook H, Davies KJ, Harding KG, Thomas DW (2000) Defective extracellular matrix reorganization by chronic wound fibroblasts is associated with alterations in TIMP-1, TIMP-2, and MMP-2 activity. J Invest Dermatol 115(2):225–233. doi:10.1046/j.1523-1747.2000.00044.x

Croisier F, Jerome C (2013) Chitosan-based biomaterials for tissue engineering. Eur Polym J 49:13

De Ugarte DA, Morizono K, Elbarbary A, Alfonso Z, Zuk PA, Zhu M, Dragoo JL, Ashjian P, Thomas B, Benhaim P, Chen I, Fraser J, Hedrick MH (2003) Comparison of multi-lineage cells from human adipose tissue and bone marrow. Cells Tissues Organs 174(3):101–109. doi:71150

Demidova-Rice TN, Salomatina EV, Yaroslavsky AN, Herman IM, Hamblin MR (2007) Low-level light stimulates excisional wound healing in mice. Lasers Surg Med 39(9):706–715. doi:10.1002/lsm.20549

Demidova-Rice TN, Hamblin MR, Herman IM (2012) Acute and impaired wound healing: pathophysiology and current methods for drug delivery, part 1: normal and chronic wounds: biology, causes, and approaches to care. Adv Skin Wound Care 25(7):304–314. doi:10.1097/01.ASW.0000416006.55218.d0

Dollfus C, Blanche S, Trocme N, Funck-Brentano I, Bonnet F, Levan P (2009) Correction of facial lipoatrophy using autologous fat transplants in HIV-infected adolescents. HIV Med 10(5):263–268. doi:10.1111/j.1468-1293.2008.00682.x

Ebrahimian TG, Pouzoulet F, Squiban C, Buard V, Andre M, Cousin B, Gourmelon P, Benderitter M, Casteilla L, Tamarat R (2009) Cell therapy based on adipose tissue-derived stromal cells promotes physiological and pathological wound healing. Arterioscler Thromb Vasc Biol 29(4):503–510. doi:10.1161/ATVBAHA.108.178962

Eisenbud D, Huang NF, Luke S, Silberklang M (2004) Skin substitutes and wound healing: current status and challenges. Wounds 16:16

Endara MR, Allred LJ, Han KD, Baker SB (2014) Applications of fat grafting in facial aesthetic skeletal surgery. Aesthet Surg J 34(3):363–373. doi:10.1177/1090820X14525964

Falanga V (2005) Wound healing and its impairment in the diabetic foot. Lancet 366(9498):1736–1743. doi:10.1016/S0140-6736(05)67700-8

Fossett E, Khan WS (2012) Optimising human mesenchymal stem cell numbers for clinical application: a literature review. Stem Cells Int 2012:465259. doi:10.1155/2012/465259

Friedenstein AJ (1980) Stromal mechanisms of bone marrow: cloning in vitro and retransplantation in vivo. Haematol Blood Transfus 25:19–29

Fromm-Dornieden C, Koenen P (2013) Adipose-derived stem cells in wound healing: recent results in vitro and in vivo. OA Mol Cell Biol 1(1):1

Frykberg RG, Banks J (2015) Challenges in the treatment of chronic wounds. Adv Wound Care (New Rochelle) 4(9):560–582. doi:10.1089/wound.2015.0635

Gardin C, Bressan E, Ferroni L, Nalesso E, Vindigni V, Stellini E, Pinton P, Sivolella S, Zavan B (2012) In vitro concurrent endothelial and osteogenic commitment of adipose-derived stem cells and their genomical analyses through comparative genomic hybridization array: novel strategies to increase the successful engraftment of tissue-engineered bone grafts. Stem Cells Dev 21(5):767–777. doi:10.1089/scd.2011.0147

Gimble JM, Guilak F, Bunnell BA (2010) Clinical and preclinical translation of cell-based therapies using adipose tissue-derived cells. Stem Cell Res Ther 1(2):19. doi:10.1186/scrt19

Gimble JM, Bunnell BA, Chiu ES, Guilak F (2011) Concise review: adipose-derived stromal vascular fraction cells and stem cells: let's not get lost in translation. Stem Cells 29(5):749–754. doi:10.1002/stem.629

Gnanasegaran N, Govindasamy V, Musa S, Kasim NH (2014) Different isolation methods alter the gene expression profiling of adipose derived stem cells. Int J Med Sci 11(4):391–403. doi:10.7150/ijms.7697

Greenhalgh DG (1996) The role of growth factors in wound healing. J Trauma 41(1):159–167

Gronthos S, Zannettino AC (2011) Methods for the purification and characterization of human adipose-derived stem cells. Methods Mol Biol 702:109–120. doi:10.1007/978-1-61737-960-4_9

Guercio A, Di Bella S, Casella S, Di Marco P, Russo C, Piccione G (2013) Canine mesenchymal stem cells (MSCs): characterization in relation to donor age and adipose tissue-harvesting site. Cell Biol Int 37(8):789–798. doi:10.1002/cbin.10090

Guijarro-Martinez R, Miragall Alba L, Marques Mateo M, Puche Torres M, Pascual Gil JV (2011) Autologous fat transfer to the cranio-maxillofacial region: updates and controversies. J Craniomaxillofac Surg 39(5):359–363. doi:10.1016/j.jcms.2010.07.004

Guisantes E, Fontdevila J, Rodriguez G (2012) Autologous fat grafting for correction of unaesthetic scars. Ann Plast Surg 69(5):550–554. doi:10.1097/SAP.0b013e31821ee386

Hardy TG, Joshi N, Kelly MH (2007) Orbital volume augmentation with autologous micro-fat grafts. Ophthal Plast Reconstr Surg 23(6):445–449. doi:10.1097/IOP.0b013e31815928f8

Hendrickx B, Verdonck K, Van den Berge S, Dickens S, Eriksson E, Vranckx JJ, Luttun A (2010) Integration of blood outgrowth endothelial cells in dermal fibroblast sheets promotes full thickness wound healing. Stem Cells 28(7):1165–1177. doi:10.1002/stem.445

Holtan SG, Weisdorf DJ (2014) The donor's dilemma. Blood 123(23):3538–3539. doi:10.1182/blood-2014-04-571661

Hong SJ, Jia SX, Xie P, Xu W, Leung KP, Mustoe TA, Galiano RD (2013) Topically delivered adipose derived stem cells show an activated-fibroblast phenotype and enhance granulation tissue formation in skin wounds. PLoS One 8(1):e55640. doi:10.1371/journal.pone.0055640

Huang SP, Huang CH, Shyu JF, Lee HS, Chen SG, Chan JY, Huang SM (2013) Promotion of wound healing using adipose-derived stem cells in radiation ulcer of a rat model. J Biomed Sci 20:51. doi:10.1186/1423-0127-20-51

Hunstad JP (1995) Tumescent and syringe liposculpture: a logical partnership. Aesthet Plast Surg 19(4):321–333

Jin HJ, Bae YK, Kim M, Kwon SJ, Jeon HB, Choi SJ, Kim SW, Yang YS, Oh W, Chang JW (2013) Comparative analysis of human mesenchymal stem cells from bone marrow, adipose tissue, and umbilical cord blood as sources of cell therapy. Int J Mol Sci 14(9):17986–18001. doi:10.3390/ijms140917986

Kaoutzanis C, Xin M, Ballard TN, Welch KB, Momoh AO, Kozlow JH, Brown DL, Cederna PS, Wilkins EG (2016) Autologous fat grafting after breast reconstruction in postmastectomy patients: complications, biopsy rates, and locoregional cancer recurrence rates. Ann Plast Surg 76(3):270–275. doi:10.1097/SAP.0000000000000561

Kim WS, Park BS, Sung JH (2009) The wound-healing and antioxidant effects of adipose-derived stem cells. Expert Opin Biol Ther 9(7):879–887. doi:10.1517/14712590903039684

Kim M, Kim I, Lee SK, Bang SI, Lim SY (2011) Clinical trial of autologous differentiated adipocytes from stem cells derived from human adipose tissue. Dermatol Surg 37(6):750–759. doi:10.1111/j.1524-4725.2011.01765.x

Kim H, Choi K, Kweon OK, Kim WH (2012) Enhanced wound healing effect of canine adipose-derived mesenchymal stem cells with low-level laser therapy in athymic mice. J Dermatol Sci 68(3):149–156. doi:10.1016/j.jdermsci.2012.09.013

Kinoshita K, Kuno S, Ishimine H, Aoi N, Mineda K, Kato H, Doi K, Kanayama K, Feng J, Mashiko T, Kurisaki A, Yoshimura K (2015) Therapeutic potential of adipose-derived SSEA-3-positive muse cells for treating diabetic skin ulcers. Stem Cells Transl Med 4(2):146–155. doi:10.5966/sctm.2014-0181

Koenen P, Spanholtz TA, Maegele M, Sturmer E, Brockamp T, Neugebauer E, Thamm OC (2015) Acute and chronic wound fluids inversely influence adipose-derived stem cell function: molecular insights into impaired wound healing. Int Wound J 12(1):10–16. doi:10.1111/iwj.12039

Kuo YR, Wang CT, Cheng JT, Kao GS, Chiang YC, Wang CJ (2016) Adipose-derived stem cells accelerate diabetic wound healing through the induction of autocrine and paracrine effects. Cell Transplant 25(1):71–81. doi:10.3727/096368915X687921

Lauritano D, Palmieri A, Vinci R, Azzi L, Taglabue A, Carinci F (2014) Adipose derived stem cells: basic science fundaments and clinical application. An update. Minerva Stomatol 63(7–8):273–281

Lazarus GS, Cooper DM, Knighton DR, Percoraro RE, Rodeheaver G, Robson MC (1994) Definitions and guidelines for assessment of wounds and evaluation of healing. Wound Repair Regen 2(3):165–170. doi:10.1046/j.1524-475X.1994.20305.x

Lee SH, Lee JH, Cho KH (2011) Effects of human adipose-derived stem cells on cutaneous wound healing in nude mice. Ann Dermatol 23(2):150–155. doi:10.5021/ad.2011.23.2.150

Lee HC, An SG, Lee HW, Park JS, Cha KS, Hong TJ, Park JH, Lee SY, Kim SP, Kim YD, Chung SW, Bae YC, Shin YB, Kim JI, Jung JS (2012) Safety and effect of adipose tissue-derived stem cell implantation in patients with critical limb ischemia: a pilot study. Circ J 76(7):1750–1760

Lim JS, Yoo G (2010) Effects of adipose-derived stromal cells and of their extract on wound healing in a mouse model. J Korean Med Sci 25(5):746–751. doi:10.3346/jkms.2010.25.5.746

Lin YC, Grahovac T, Oh SJ, Ieraci M, Rubin JP, Marra KG (2013) Evaluation of a multi-layer adipose-derived stem cell sheet in a full-thickness wound healing model. Acta Biomater 9(2):5243–5250. doi:10.1016/j.actbio.2012.09.028

Maharlooei MK, Bagheri M, Solhjou Z, Jahromi BM, Akrami M, Rohani L, Monabati A, Noorafshan A, Omrani GR (2011) Adipose tissue derived mesenchymal stem cell (AD-MSC)

promotes skin wound healing in diabetic rats. Diabetes Res Clin Pract 93(2):228–234. doi:10.1016/j.diabres.2011.04.018

Marino G, Moraci M, Armenia E, Orabona C, Sergio R, De Sena G, Capuozzo V, Barbarisi M, Rosso F, Giordano G, Iovino F, Barbarisi A (2013) Therapy with autologous adipose-derived regenerative cells for the care of chronic ulcer of lower limbs in patients with peripheral arterial disease. J Surg Res 185(1):36–44. doi:10.1016/j.jss.2013.05.024

Markarian CF, Frey GZ, Silveira MD, Chem EM, Milani AR, Ely PB, Horn AP, Nardi NB, Camassola M (2014) Isolation of adipose-derived stem cells: a comparison among different methods. Biotechnol Lett 36(4):693–702. doi:10.1007/s10529-013-1425-x

McCarty SM, Percival SL (2013) Proteases and delayed wound healing. Adv Wound Care (New Rochelle) 2(8):438–447. doi:10.1089/wound.2012.0370

Messina A, Bortolotto SK, Cassell OC, Kelly J, Abberton KM, Morrison WA (2005) Generation of a vascularized organoid using skeletal muscle as the inductive source. FASEB J 19(11):1570–1572. doi:10.1096/fj.04-3241fje

Miller JP, Perry EH, Price TH, Bolan CD Jr, Karanes C, Boyd TM, Chitphakdithai P, King RJ (2008) Recovery and safety profiles of marrow and PBSC donors: experience of the National Marrow Donor Program. Biol Blood Marrow Transplant 14(9 Suppl):29–36. doi:10.1016/j.bbmt.2008.05.018

Mineda K, Feng J, Ishimine H, Takada H, Doi K, Kuno S, Kinoshita K, Kanayama K, Kato H, Mashiko T, Hashimoto I, Nakanishi H, Kurisaki A, Yoshimura K (2015) Therapeutic potential of human adipose-derived stem/stromal cell microspheroids prepared by three-dimensional culture in non-cross-linked hyaluronic acid gel. Stem Cells Transl Med 4(12):1511–1522. doi:10.5966/sctm.2015-0037

Mizuno H, Nambu M (2011) Adipose-derived stem cells for skin regeneration. Methods Mol Biol 702:453–459. doi:10.1007/978-1-61737-960-4_33

Mojallal A, Auxenfans C, Lequeux C, Braye F, Damour O (2008) Influence of negative pressure when harvesting adipose tissue on cell yield of the stromal-vascular fraction. Biomed Mater Eng 18(4–5):193–197

Moseley TA, Zhu M, Hedrick MH (2006) Adipose-derived stem and progenitor cells as fillers in plastic and reconstructive surgery. Plast Reconstr Surg 118(3 Suppl):121S–128S. doi:10.1097/01.prs.0000234609.74811.2e

Nambu M, Ishihara M, Nakamura S, Mizuno H, Yanagibayashi S, Kanatani Y, Hattori H, Takase B, Ishizuka T, Kishimoto S, Amano Y, Yamamoto N, Azuma R, Kiyosawa T (2007) Enhanced healing of mitomycin C-treated wounds in rats using inbred adipose tissue-derived stromal cells within an atelocollagen matrix. Wound Repair Regen 15(4):505–510. doi:10.1111/j.1524-475X.2007.00258.x

Nambu M, Kishimoto S, Nakamura S, Mizuno H, Yanagibayashi S, Yamamoto N, Azuma R, Nakamura S, Kiyosawa T, Ishihara M, Kanatani Y (2009) Accelerated wound healing in healing-impaired db/db mice by autologous adipose tissue-derived stromal cells combined with atelocollagen matrix. Ann Plast Surg 62(3):317–321. doi:10.1097/SAP.0b013e31817f01b6

Nie C, Yang D, Xu J, Si Z, Jin X, Zhang J (2011) Locally administered adipose-derived stem cells accelerate wound healing through differentiation and vasculogenesis. Cell Transplant 20(2):205–216. doi:10.3727/096368910X520065

Pandis L, Zavan B, Abatangelo G, Lepidi S, Cortivo R, Vindigni V (2010) Hyaluronan-based scaffold for in vivo regeneration of the rat vena cava: preliminary results in an animal model. J Biomed Mater Res A 93(4):1289–1296. doi:10.1002/jbm.a.32626

Phillips TJ (2001) Current approaches to venous ulcers and compression. Dermatol Surg 27(7):611–621

Piombino P, Marenzi G, Dell'Aversana Orabona G, Califano L, Sammartino G (2015) Autologous fat grafting in facial volumetric restoration. J Craniofac Surg 26(3):756–759. doi:10.1097/SCS.0000000000001663

Pu LL, Coleman SR, Cui X, Ferguson RE Jr, Vasconez HC (2008) Autologous fat grafts harvested and refined by the Coleman technique: a comparative study. Plast Reconstr Surg 122(3):932–937. doi:10.1097/PRS.0b013e3181811ff0

Pulsfort AK, Wolter TP, Pallua N (2011) The effect of centrifugal forces on viability of adipocytes in centrifuged lipoaspirates. Ann Plast Surg 66(3):292–295. doi:10.1097/SAP.0b013e3181c7140e

Pulsipher MA, Chitphakdithai P, Logan BR, Navarro WH, Levine JE, Miller JP, Shaw BE, O'Donnell PV, Majhail NS, Confer DL (2014) Lower risk for serious adverse events and no increased risk for cancer after PBSC vs BM donation. Blood 123(23):3655–3663. doi:10.1182/blood-2013-12-542464

Raposio E, Bertozzi N, Bonomini S, Bernuzzi G, Formentini A, Grignaffini E, Pio Grieco M (2016) Adipose-derived stem cells added to platelet-rich plasma for chronic skin ulcer therapy. Wounds 28(4):6

Reiber GE, Boyko EJ, Smith DG (1995) Lower extremity foot ulcers and amputations in diabetes. Diabetes in America, 2nd edn. U. S. Government Printing Office, Washington, DC. First published

Reshak AH, Shahimin MM, Buang F (2013) Comparative study on human and bovine AT-SC isolation methods. Prog Biophys Mol Biol 113(2):295–298. doi:10.1016/j.pbiomolbio.2013.09.001

Rigotti G, Marchi A, Galie M, Baroni G, Benati D, Krampera M, Pasini A, Sbarbati A (2007) Clinical treatment of radiotherapy tissue damage by lipoaspirate transplant: a healing process mediated by adipose-derived adult stem cells. Plast Reconstr Surg 119(5):1409–1422 . doi:10.1097/01.prs.0000256047.47909.71discussion 1423–1404

Rizzi SC, Upton Z, Bott K, Dargaville TR (2010) Recent advances in dermal wound healing: biomedical device approaches. Expert Rev Med Devices 7(1):143–154. doi:10.1586/erd.09.57

Robertson L, Lee AJ, Gallagher K, Carmichael SJ, Evans CJ, McKinstry BH, Fraser SC, Allan PL, Weller D, Ruckley CV, Fowkes FG (2009) Risk factors for chronic ulceration in patients with varicose veins: a case control study. J Vasc Surg 49(6):1490–1498. doi:10.1016/j.jvs.2009.02.237

Rodriguez J, Boucher F, Lequeux C, Josset-Lamaugarny A, Rouyer O, Ardisson O, Rutschi H, Sigaudo-Roussel D, Damour O, Mojallal A (2015) Intradermal injection of human adipose-derived stem cells accelerates skin wound healing in nude mice. Stem Cell Res Ther 6:241. doi:10.1186/s13287-015-0238-3

Rodriguez-Menocal L, Salgado M, Ford D, Van Badiavas E (2012) Stimulation of skin and wound fibroblast migration by mesenchymal stem cells derived from normal donors and chronic wound patients. Stem Cells Transl Med 1(3):221–229. doi:10.5966/sctm.2011-0029

Rose JG Jr, Lucarelli MJ, Lemke BN, Dortzbach RK, Boxrud CA, Obagi S, Patel S (2006) Histologic comparison of autologous fat processing methods. Ophthal Plast Reconstr Surg 22(3):195–200. doi:10.1097/01.iop.0000217710.09941.10

Schreml S, Szeimies RM, Prantl L, Karrer S, Landthaler M, Babilas P (2010) Oxygen in acute and chronic wound healing. Br J Dermatol 163(2):257–268. doi:10.1111/j.1365-2133.2010.09804.x

Schultz GS, Sibbald RG, Falanga V, Ayello EA, Dowsett C, Harding K, Romanelli M, Stacey MC, Teot L, Vanscheidt W (2003) Wound bed preparation: a systematic approach to wound management. Wound Repair Regen 11(Suppl 1):S1–28

Seifter E, Rettura G, Padawer J, Stratford F, Kambosos D, Levenson SM (1981) Impaired wound healing in streptozotocin diabetes. Prevention by supplemental vitamin A. Ann Surg 194(1):42–50

Sensebe L, Bourin P (2009) Mesenchymal stem cells for therapeutic purposes. Transplantation 87(9 Suppl):S49–S53. doi:10.1097/TP.0b013e3181a28635

Seth AK, Hirsch EM, Kim JY, Fine NA (2012) Long-term outcomes following fat grafting in prosthetic breast reconstruction: a comparative analysis. Plast Reconstr Surg 130(5):984–990. doi:10.1097/PRS.0b013e318267d34d

Shen T, Pan ZG, Zhou X, Hong CY (2013) Accelerated healing of diabetic wound using artificial dermis constructed with adipose stem cells and poly (L-glutamic acid)/chitosan scaffold. Chin Med J 126(8):1498–1503

Shu W, Shu YT, Dai CY, Zhen QZ (2012) Comparing the biological characteristics of adipose tissue-derived stem cells of different persons. J Cell Biochem 113(6):2020–2026. doi:10.1002/jcb.24070

Slavin J (1996) The role of cytokines in wound healing. J Pathol 178(1):5–10. doi:10.1002/(SICI)1096-9896(199601)178:1<5::AID-PATH443>3.0.CO;2-W

Steinberg JP, Hong SJ, Geringer MR, Galiano RD, Mustoe TA (2012) Equivalent effects of topically-delivered adipose-derived stem cells and dermal fibroblasts in the ischemic rabbit ear model for chronic wounds. Aesthet Surg J 32(4):504–519. doi:10.1177/1090820X12442679

Stojadinovic A, Carlson JW, Schultz GS, Davis TA, Elster EA (2008) Topical advances in wound care. Gynecol Oncol 111(2 Suppl):S70–S80. doi:10.1016/j.ygyno.2008.07.042

Suga H, Matsumoto D, Inoue K, Shigeura T, Eto H, Aoi N, Kato H, Abe H, Yoshimura K (2008) Numerical measurement of viable and nonviable adipocytes and other cellular components in aspirated fat tissue. Plast Reconstr Surg 122(1):103–114. doi:10.1097/PRS.0b013e31817742ed

Telgenhoff D, Shroot B (2005) Cellular senescence mechanisms in chronic wound healing. Cell Death Differ 12(7):695–698. doi:10.1038/sj.cdd.4401632

Tocco I, Zavan B, Bassetto F, Vindigni V. Nanotechnology-based therapies for skin wound regeneration. J Nanomater 2012:714134 1–11. doi:10.1155/2012/714134

Uysal CA, Tobita M, Hyakusoku H, Mizuno H (2014) The effect of bone-marrow-derived stem cells and adipose-derived stem cells on wound contraction and epithelization. Adv Wound Care (New Rochelle) 3(6):405–413. doi:10.1089/wound.2014.0539

Wall IB, Moseley R, Baird DM, Kipling D, Giles P, Laffafian I, Price PE, Thomas DW, Stephens P (2008) Fibroblast dysfunction is a key factor in the non-healing of chronic venous leg ulcers. J Invest Dermatol 128(10):2526–2540. doi:10.1038/jid.2008.114

Watt FM, Hogan BL (2000) Out of Eden: stem cells and their niches. Science 287(5457):1427–1430

Weichman KE, Broer PN, Tanna N, Wilson SC, Allan A, Levine JP, Ahn C, Choi M, Karp NS, Allen R (2013) The role of autologous fat grafting in secondary microsurgical breast reconstruction. Ann Plast Surg 71(1):24–30. doi:10.1097/SAP.0b013e3182920ad0

Weissman IL (2000) Stem cells: units of development, units of regeneration, and units in evolution. Cell 100(1):157–168

Wong VW, Levi B, Rajadas J, Longaker MT, Gurtner GC (2012) Stem cell niches for skin regeneration. Int J Biomater 2012:926059. doi:10.1155/2012/926059

Woo K, Ayello EA, Sibbald RG (2007) The edge effect: current therapeutic options to advance the wound edge. Adv Skin Wound Care 20(2):99–117. quiz 118–119

Wu W, Niklason L, Steinbacher DM (2013) The effect of age on human adipose-derived stem cells. Plast Reconstr Surg 131(1):27–37. doi:10.1097/PRS.0b013e3182729cfc

Yolanda MM, Maria AV, Amaia FG, Marcos PB, Silvia PL, Dolores E, Jesùs OH (2014) Adult stem cell therapy in chronic wound healing. J Stem Cell Res Ther 4:1

Yoshimura K, Suga H, Eto H (2009) Adipose-derived stem/progenitor cells: roles in adipose tissue remodeling and potential use for soft tissue augmentation. Regen Med 4(2):265–273. doi:10.2217/17460751.4.2.265

You HJ, Han SK (2014) Cell therapy for wound healing. J Korean Med Sci 29(3):311–319. doi:10.3346/jkms.2014.29.3.311

Zavan B, Vindigni V, Vezzu K, Zorzato G, Luni C, Abatangelo G, Elvassore N, Cortivo R (2009) Hyaluronan based porous nano-particles enriched with growth factors for the treatment of ulcers: a placebo-controlled study. J Mater Sci Mater Med 20(1):235–247. doi:10.1007/s10856-008-3566-3

Zhu M, Heydarkhan-Hagvall S, Hedrick M, Benhaim P, Zuk P (2013) Manual isolation of adipose-derived stem cells from human lipoaspirates. J Vis Exp 79:e50585. doi:10.3791/50585

Zografou A, Papadopoulos O, Tsigris C, Kavantzas N, Michalopoulos E, Chatzistamatiou T, Papassavas A, Stavropoulou-Gioka C, Dontas I, Perrea D (2013) Autologous transplantation of adipose-derived stem cells enhances skin graft survival and wound healing in diabetic rats. Ann Plast Surg 71(2):225–232. doi:10.1097/SAP.0b013e31826af01a

Zollino I, Zuolo M, Gianesini S, Pedriali M, Sibilla MG, Tessari M, Carinci F, Occhionorelli S, Zamboni P (2016) Autologous adipose-derived stem cells: basic science, technique, and rationale for application in ulcer and wound healing. Phlebology. doi:10.1177/0268355516641546
Zuk PA, Zhu M, Mizuno H, Huang J, Futrell JW, Katz AJ, Benhaim P, Lorenz HP, Hedrick MH (2001) Multilineage cells from human adipose tissue: implications for cell-based therapies. Tissue Eng 7(2):211–228. doi:10.1089/107632701300062859

Chapter 12
Adipose Tissue-Derived Stem Cells in Regenerative Medicine and Plastic Surgery: Perspective from Personal Practice

Dana Jianu, Oltjon Cobani, and Stefan Jianu

Abbreviations

ADSCs	Adipose-derived stem cells
AFT	Autologous fat transfer
BM	Bone marrow
BMSCs	Bone marrow stem cells
EM	Electromagnetic
HLLT	High-level laser treatment
HSCs	Hematopoietic stem cells
LAL	Laser-assisted lipolysis
LLLT	Low level laser treatment
MSCs	Mesenchymal stem cells
PRP	Platelet-rich plasma
PS	Plastic surgery
RM	Regenerative medicine
RPS	Regenerative plastic surgery
SCNs	Stem cells niches
SVF	Stromal vascular fraction
TM	Translational medicine

D. Jianu, MD, PhD (✉) • O. Cobani • S. Jianu
ProEstetica Hospital, Research Center in Regenerative Medicine and Surgery of Romanian Academy of Medical Sciences, 38-40, Tudor Stefan Street, Bucharest, Romania
e-mail: djianu02@gmail.com

P.V. Pham (ed.), *Pancreas, Kidney and Skin Regeneration*, Stem Cells in Clinical Applications, DOI 10.1007/978-3-319-55687-1_12

12.1 Background

For years after the discovery of liposuction, fat graft, and "lipostructures," fat tissues have been used as fillers in plastic surgery. In fact, plastic surgery and regenerative medicine have focused on the use of adipose-derived stem cells (ADSCs), in addition to other regenerative cells and agents, which has advanced research from the laboratory to the clinical field (Coleman and Mazzola 2009).

Clinical observations have led to the hypothesis that adipose tissue is more than simply a filler. Adipose tissue may contour the irregularities of fillers and enhance the quality and function of the recipient soft tissues, including skin. At the same time, research worldwide since the 1990s has been increasingly dedicated to investigations of adipose-derived fillers in regenerative medicine and the rationale and potential benefit for their use (Coleman 1997).

From both basic research and clinical findings, it was revealed that fat tissue has a very complex cellular composition. Adipose tissue is one of the richest sources of mesenchymal multipotent stem cells in the whole body (Dinescu et al. 2015). The improvements observed in the structure and quality of grafted soft tissues are the consequences of the activity of a complex population of cells injected as a "fat graft." These injected cells include mesenchymal stem cells (MSCs), which are surrounded by a glycoproteic scaffold, composed of tissue factors, preadipocytes, and mature adipocytes. The MSCs form "stem cell niches" (SCNs).

Grafted adipose tissue acts not only like a mattress or filler but also a regenerating agent with angiogenic, anti-fibrotic, analgesic, and anti-inflammatory effects. The physiopathology of these affects is not completely elucidated. The angiogenic, anti-fibrotic, and anti-inflammatory effects seem to be partly mediated by release of growth factors (Rigotti and colab 2009). As a result, the long-term clinical outcomes consist in a natural appearance and palpation of the treated areas. A reduction of activity of T lymphocytes was observed during the occurrence of an anti-inflammatory effect on immune cells (Rigotti and colab 2009). However, the underlying mechanisms for the other effects, such as analgesic effect, are still not completely understood (Rigotti and colab 2009).

Indeed, one of the most important factors in the field of regenerative plastic surgery is ADSCs and the SCNs they create (Scadden 2006). Both concepts will be discussed herein.

12.2 Definitions Used in Regenerative Medicine and Surgery

The consensus is that a stem cell is any cell that is able to self-renew indefinitely, without turning into a cancer cell or losing its genomic organization (Ghanem 2011). There are many sources of MSCs, including adipose tissue, bone marrow, periodontal ligament, trabecular bone, synovial membrane, periosteum muscle, embryonic tissue (Wharton's jelly – the gelatinous substance from umbilical cord), the nervous system, and skin (Kokai et al. 2005).

Adipose-derived stem cells (ADSCs) are a relatively new subtype of MSCs which exhibit multiple advantages. Firstly, they can be obtained from adults by less invasive methods, such as liposuction, which is often desired by patients to correct fat deposits. As such, they can be obtained in larger quantities than any other MSCs. Secondly, they have a multi-lineage differentiation capacity similar to that of bone marrow (BM)-derived MSCs. Thirdly, they can be easily grown in standard tissue culture conditions (Tholpady et al. 2005).

Adipose tissue represents an abundant source of autologous cells for regenerative medicine applications (Jianu et al. 2014). Cells derived from discarded human fat have many properties that make them ideal for tissue engineering and applications related to immunocompatibility, multipotency, and gene therapy. Moreover, they are abundant, are capable of self-renewal, and present minimal ethical conflicts (Pa et al. 2001).

Currently, bone marrow is the most known and preferable source of adult stem cells. As explained, nowadays fatty tissue appears to be the most convenient source of stem cells. As human beings age, the proportion of BMSCs decreases 200-fold, from one in 10,000 (at birth) to one in 2,000,000 (at age 80). Globally and chronologically, the population and individuals gain weight, making ADSCs rise in number to 500 more cells per volume compared to bone marrow (Ghanem 2011). When compared to BMSCs, ADSCs are proving to be superior; they have minimal donor site morbidity, easier procurement, greater availability, rapid growth rate, and lower risk of contamination during ex vivo expansion (Yoshimura et al. 2011).

Adult tissues have specific SCNs, which serve as a source of replacement cells during normal cell turnover and tissue regeneration following injury. SCNs represent the anatomical foundation of regenerative mechanisms and are comprised of stem cell themselves and a surrounding system of structures capable of regulating its progression toward differentiation, maturity, and development of tissues (Scadden 2006).

The epidermis, hair, HSCs – hematopoietic stem cells – and gastrointestinal tract represent examples of tissues with niches that supply stem cells during normal cellular turnover. The exact location of these stem cell niches is poorly understood, but there is growing evidence suggesting a close relationship with pericytes (Orbay et al. 2012).

Recently, a new subspecialty of plastic surgery (PS) was born: regenerative plastic surgery (RPS). This new category is an extension of regenerative surgery and is a part of regenerative medicine (RM), according to the National Institutes of Health. RM represents an interdisciplinary field of research and clinical applications which focuses on the restoration, replacement, and/or regeneration of cells, tissues, or organs in order to restore normal function (Daar and Greenwood 2007).

The main goal of RM is to regenerate (human) cells, tissue, or organs in vivo/ex vivo by utilizing their own intrinsic regeneration potential. In order to achieve its purpose, RM uses cells and biomaterials which are specifically created. Therapies based on RM are already being used in many clinical trials related to most surgical specialties (Yoshimura et al. 2008).

Moreover, RPS applies the principles of cellular and tissue regeneration in plastic surgery applications. Plastic surgery is, by definition, close to regenerative medicine due to similar goals: restoration and reconstruction of shape and function of the human body surface and hands. Another rationale why PS is close to RM could be that PS has soft tissues, including adipose tissue, as the active cell types.

The link between RPS and research in the field of stem cells is a very good example of translational medicine (TM). TM represents a discipline which transforms the scientific findings from clinical trials, lab studies, or populational studies into clinical applications meant to reduce disease incidence, morbidity, or mortality (Rubio et al. 2010). TM also represents a field of study in biomedical research and public health, with the aim to rapidly transform scientific results into diagnostic tools, therapies, or procedures by using a multidisciplinary, highly collaborative, "bench-to-bedside" approach. Thus, TM is bidirectional; the clinical side (bedside) impacts research (bench side), inspiring optimizations, new directions, and future strategies (Marincola 2003).

12.3 Personal Experience and Contributions in Regenerative Plastic Surgery

12.3.1 Introduction

Our personal experience with autologous fat graft or autologous fat transfer (AFT) has reached several milestones. Our experience began in the 1990s with face corrections and rejuvenation. From year 2000 onward, we moved into body re-contouring and composite breast reconstruction postmastectomy, using fat graft and prostheses. Between 2000 and 2010, fat grafting began to be considered by the community of plastic surgeons, including our group, as a regenerative procedure due to the inclusion of progenitor cells. As a result, a gradual implementation in our current practice was the increasing use of techniques and products of regenerative plastic surgery to address cervicofacial, breast, hands, and scars.

Giorgio Fischer, Pierre Fournier, and Gerard Yves Illouz are considered the "fathers" of lipoaspiration and lipo-sculpture techniques which have revolutionized the PS field. They opened the doors to the idea of fat management with scientific strategy. Moreover, the contributions of W. Futrell, Coleman, Rigotti, Chajchir, Biggs, Khouri, Yoshimura, and Rubin – just to mention a few – who were prominent plastic surgeons as well as researchers, led to an enormous progress in our understanding of adipose tissue transplant.

There was an increased interest in ADSCs in treating patients with various conditions, from pathological to aesthetic and rejuvenation purposes. Also, through the findings or technological innovations of the above surgeons, other prominent biomedical experts and practitioners (e.g., E. Alt, Zuk, M. Costache, N. Pallua,

L. Badimon, L. Simion, Labusca, V. Cervelli, M. Simionescu, D. Simionescu, and V. Paunescu) began to contribute in a major way to enrich the plethora of knowledge pertaining to the applications of regenerative sciences. There were an increasing number of patients in need of regenerative plastic surgery seeking less invasive but visible corrections using their own cells to treat breast, face, neck, limb, and body anomalies.

Chronologically, our personal experience with autologous fat transplant (AFT) in our practice in the 1990s centered on "lipo-filling" or "fat grafts." Cells were harvested with needle (14–18G) or microcannulas when used for facial or scar corrections or with 3.5 mm cannulas when used for body contour corrections. Although, clinically, we observed volume corrections, we also noted an important improvement of skin texture. We were not aware at that time about the regenerative properties of AFT.

12.3.2 Areas of Interest

Nowadays, in our practice, we are currently offering treatments for different areas affected by congenital, posttraumatic, or aging conditions.

The regenerative procedures that our team performs are:

- Autologous fat transplant (AFT)
- ADSC treatment (as AFT enrichment or stand-alone procedure)
- Platelet-rich plasma (PRP)
- Laser treatments: fractional CO2 superficial ablative resurfacing and laser-assisted lipolysis (LAL)

They are applied for treating:

- Face ("AdipoLaser reJuvenation" personal, original technique)
- Breast regeneration (hipotrophic or postmastectomy)
- Flaccid: abdomen, inner thighs, inner arms
- Hands (rejuvenation or Dupuytren diseases)
- Scars (these applications will be specifically mentioned in other chapters)

12.3.3 The Combination of Regenerative Techniques and Factors: An Innovative Approach

12.3.3.1 Autologous Fat Transfer (AFT) and Fat Management

We have refined and enriched our techniques and equipment. Since 2008, in the Plastic Surgery Department of ProEstetica Medical Center, over 300 patients have undergone single (initially) and combined regenerative closed surgery, based on the following methods introduced chronologically:

1. Autologous fat transplant (micrografts) since the 1990s and nanografts since 2013
2. Lipoaspiration (fine lipoaspiration cannulas, since the 1990s, have been mostly used for reshaping by defatting, consequent skin retraction, or to mobilize the cutaneous and subcutaneous tissues as a preparation for laser lipolysis and superficial nanofat grafting)
3. Laser-assisted lipolysis (LAL) with optical fiber (diode laser λ =980 nm) since 2008; fractional CO2 laser (λ = 10,600 nm) for resurfacing and fat graft stimulation since 2008
4. Platelet-rich plasma (PRP) treatment (injections and spreading) since 2012
5. Additional ADSC standardized enzymatic digestion from lipoaspirate since 2013

It is important to highlight here that all abovementioned techniques are performed together during the same procedure (Jianu et al. 2012). The simultaneous combination of these five regenerative techniques (regenerative factors) is considered innovative. They represent steps of the same surgical session, with the main goal of the complex procedure being their potentiated clinical outcomes as a result of the synergetic interaction of the regenerative side of each factor. Therefore, the postoperative healing and further biological status of the targeted tissues are improved. When used as simultaneous combinations, we can succeed in optimizing results and decreasing the side effects and complications of each technique, compared to when used as a single procedure.

Autologous fat transplant has proven to be an important regenerative clinical application due to its composition of progenitor (stem) cells and other regenerative factors. From clinical experience, we have observed the following:

A. ADSCs act with better results when combined with fatty tissue (enriched AFTs) (Cervelli et al. 2009).
B. The fat graft has a better survival rate and less complications when it is enriched with ADSCs (e.g., in breast correction).
C. Additional laser treatments, especially fractional CO2 laser application, optimize the regenerative activity of ADSCs and autologous fat graft. Taking clinical observations into consideration, we have learned even more through research about optimizations of fat and ADSCs when laser exposure is performed.
D. Growth factors and cytokines, contained in PRP and administrated simultaneously with ADSCs and AFT, can enhance AFT (Van Pham et al. 2013).

The outcomes we obtained through regenerative techniques are reflected at the macro- and microlevel. At the macro-level, we are performing a real sculpture of treated areas with redistribution and rebalance of volumes and curves. At the same time, we are changing the skin texture by making it tighter and thicker with a brighter, more even structured surface.

At microlevel, we produce tissular and cellular regeneration due to:

1. The imported population of fat cells: ADSCs, preadipocytes, adipocytes, stem cell niches, SVF – stromal vascular fraction

2. Supplementary ADSCs
3. Growth factors and cytokines from PRP
4. The laser effects: HLLT (high-level laser treatment) and LLLT (low-level laser therapy)

The equipment we use are as follows:

1. For lipoaspiration and for fat harvesting: Khouri cannulas with 12, 9, and 6 holes and needles 14G and 16G. (Miami Breast Center, 580 Crandon Blvd Suite 102, Key Biscayne, FL 33149)
2. For fat graft: Fischer curved cannulas, ϕ 1, 2 mm, Khouri cannulas, Coleman cannulas
3. For laser lipolysis = diode laser 980 nm (MedArt Diode Laser 980 nm, ASAH MEDICO A/S, Valseholmen 11–13, Denmark)
4. For fractional laser: FRX CO2 laser (λ = 10,600 MedArt 610 FRx, ASAH MEDICO A/S, Valseholmen 11–13, Denmark)
5. For ADSCs – InGeneron equipment (enzymatic digestion)
6. For platelet-rich plasma: PRP – Glofinn (Fig. 12.1)

The most important regenerative factors, in our experience, are fat tissue and stem cells. Autologous ADSCs exist as single entities or are contained in AFT. The optimizing factors of regenerative cell activity are the growth factors from PRP and light energy delivered by laser, in synergic action, during the same procedure.

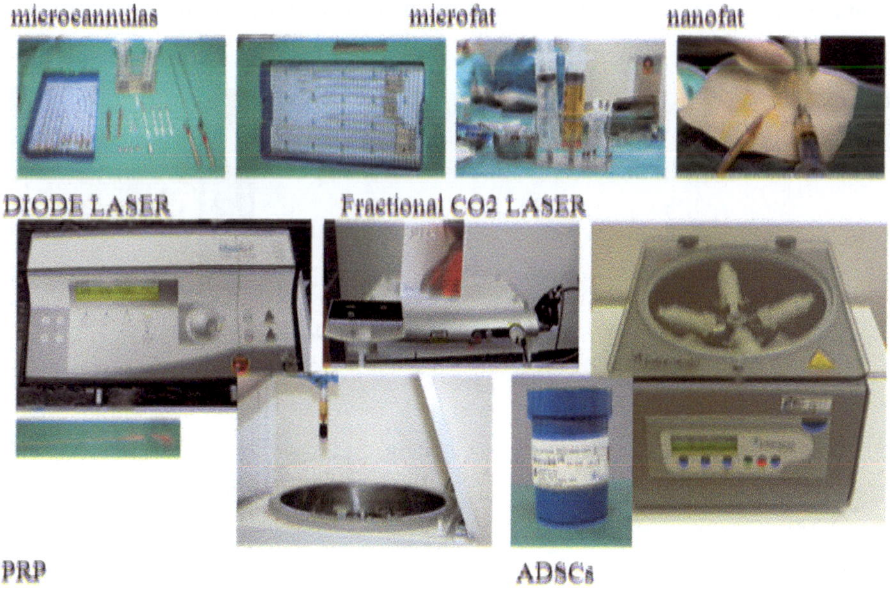

Fig. 12.1 The equipment we used in this process

12.3.3.2 Adipose-Derived Stem Cells (ADSCs) in Our Practice of Plastic Regenerative Surgery

ADSCS are fresh autologous cells isolated usually from the superficial, abdominal infraumbilical fat or other areas like flanks or external/internal part of thighs. Of note, 120 cc of fat is aspirated with a Khouri cannula with 12 or 9 holes in 60 cc Luer-lock syringes. The isolating procedure continues with processing the harvested fat with InGeneron's proprietary Matrase™ reagent (collagenase I and II and a neutral protease) followed by a separation phase, filtering, and washing (Song and Prant 2011) The last phase involves administering the regenerative cells (stem cells and progenitor cells) to the tissue of need (Fig. 12.2).

This equipment has the following functions: heating for optimal reagent activity, agitation, and centrifugation. It exists as a single automated system and has an invertible swinging bucket rotor which can centrifuge cells, Matrix, or PRP (Song et al. 2011). The obtained ADSCs are used in recombination with fat graft (from AFT) for breast grafting in breast regenerative surgical treatments and with "nanofat" in face, neck, hand, or body injections. The number of ADSCs is 800,000–1,000,000 per gram of fat suspension (Figs. 12.3 and 12.4).

12.3.3.3 Platelet-Rich Plasma (PRP) in Our Practice and Basic Theory

PRP is a blood (autologous) product with high concentration of platelets (PLT) in a limited volume of plasma. Platelet-poor plasma (PPP) is a blood (autologous) product with low concentration of platelets (PLT) in a limited volume of plasma. PRP contains growth factors and bioactive proteins that regulate differentiated cells and

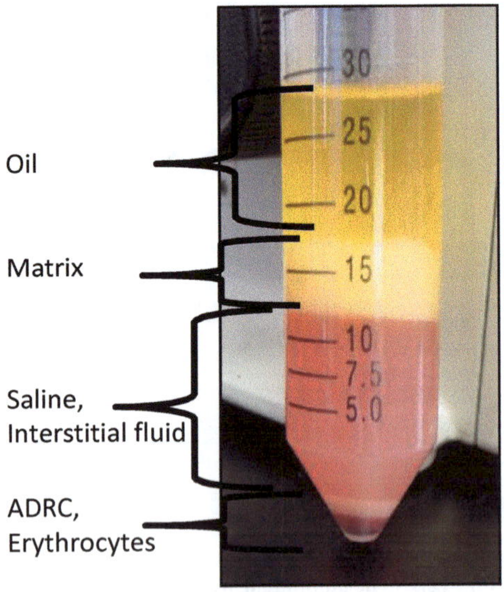

Fig. 12.2 The tissue processing unit belongs to InGeneron products

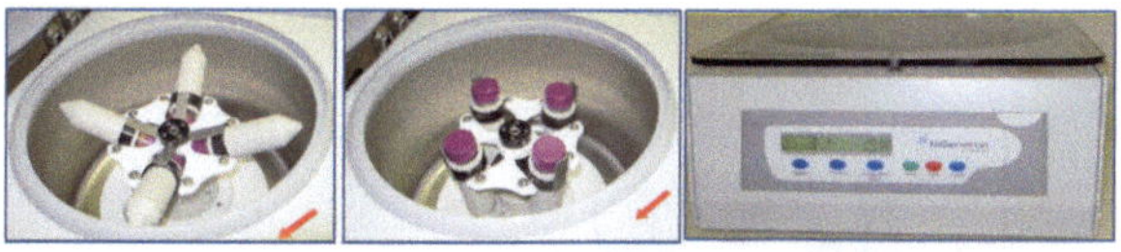

Fig. 12.3 InGeneron – tissue processing unit

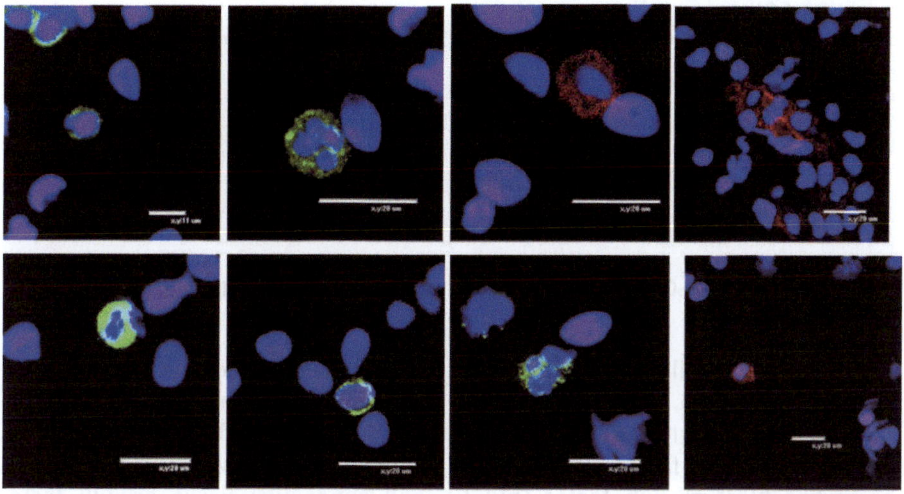

Fig. 12.4 InGeneron – small cells positive for stem cell markers among freshly isolated cells from human adipose tissue (Courtesy of Prof. Alt Eckhard)

modulate cell growth and activity. They are vital for tissue regeneration and the fabrication of engineered tissues (Fig. 12.5).

PRP was first identified in the early 1990s through the use of plasmapheresis and PRP sequestration. Platelets are known to perform multiple functions during injury and tissue repair and initiate the body's response to a normal sequence of events that provide clotting and healing of the damaged tissue (Cervelli et al. 2009). The four phases of wound healing:

1. Hemostasis (clot formation)
2. Inflammation (cleanup and recruitment)
3. Proliferation (regeneration)
4. Tissue remodeling

Cytokines are released from both white cells and platelets and attract neutrophils and fibroblasts. Growth factors are proteins released from degranulated platelets (among other numerous substances). They are contained in alpha granules of platelets, macrophages, and endothelium.

Growth factors are needed to start the proliferative phase or tissular regeneration. They are also responsible for the early migration of the cells to the injury site and are the trigger of mitosis of the cells once at the site. Within the clot, growth factors send out signals to trigger cell division. The newly created blood

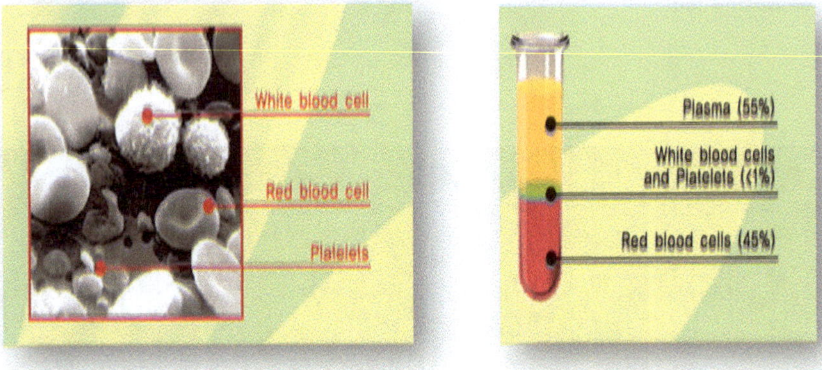

Fig. 12.5 GloTech, GLO PRP. Constituent parts of the blood and separated blood

vessels and blood flow bring necessary nutrients and oxygen for optimal healing. In this process, the following growth factors play an important role: transforming growth factor-beta (TGF-β), platelet-derived growth factor (PDGFa-b), epidermal growth factor (EGF), vascular endothelial growth factor (VEGF), basic fibroblast growth factor (bFGF), and connective tissue growth factor (CTGF) (Marx 2004).

The specific platelet-isolated growth factors have specific functions:

(a) PDGF initiates connective tissue healing, bone generation, and repair, increases mitogenesis of fibroblast, stimulates angiogenesis in the wound bed, and activates macrophages.
(b) TGF-beta promotes cell mitosis and differentiation for connective tissue and bone and acts on stem cells, osteoblast precursors, and fibroblasts.
(c) VEGF stimulates angiogenesis and related vascular permeability enhancing activities specific for endothelial cells and chemoattractant for osteoblasts.
(d) EGF induces epithelial development and promotes angiogenesis.

Research indicates that acceleration of the wound healing process requires a product with:

(a) Minimal red blood cells (RBCs)
(b) Platelet concentration four to five times above baseline
Most individuals have a platelet count near the range of 250,000/cubic milliliter and a PRP count of 1,000,000/cubic milliliter; this has become the therapeutic level. Increased number (concentration) of degranulating platelets will produce increased level of growth factors in PRP (PDGF, TBF-beta, EGF, etc.) (Eppley et al. 2004). The products we are using in our clinic meet the therapeutic requirements.

12.3.3.4 Laser as a Regenerative Factor

Basic Theory

From studying and using lasers in our current practice, we were led to new findings about their photobioactivation and optimizing effects on ADSCs and fat grafts. Laser is an acronym for "light amplification by stimulated emission of radiation." Light is electromagnetic (EM) radiation in the visible wavelength from 380 to 750 nm (i.e., just one part of a broader EM spectrum). The laser radiation length is between 100 and 10,600 nm.

Albert Einstein developed the theoretical concept of light travelling in waves of particles (photons) and of "stimulated emission."

- Speed of light $c = \lambda \bullet \nu$

  C = vacuum ~300 mil.m/s
  ν = frequency of EM wave
  Λ = wavelength

- E photon = $h \bullet \nu$ – the photon has no mass but carries energy and momentum (wave-particle duality)
- Momentum: p = E photon = $h\nu$

Lasers are a special type of light with the characteristics of monochromaticity, directionality, and coherence. The optimization effect of laser over ADSCs and fat graft is explained by two laws related to the low-level laser therapy (LLLT) or theorethical bases.

One law is "*THE SECOND LAW OF PHOTO BIOLOGY: EINSTEIN – STARK LAW OF PHOTO CHEMICAL EQUIVALENCE*," which states that "for every photon absorbed – if a photochemical reaction is the result – an activated particle, such as an atom molecule of the radical, is formed (primary reaction)." The primary reaction is followed by a secondary reaction with liberation of *PHOTOPRODUCTS* (*LATTER DISSIPATION PATHWAYS*). This is the basis of PHOTO*ACTIVATION IN BIOLOGIC TISSUE*.

The second law is "*THE LAW OF ARNDT – SCHULTZ*," which states that "in a biologic system, weak stimuli will elicit strong reactions, medium stimuli will cause moderate reactions, moderately strong stimuli will slightly inhibit the system and very strong stimuli will completely retard it."

Any surgical ablative laser is producing energy absorbed by the tissues with two types of simultaneous reactions:

1. Destructive:

   (a) *PHOTOTHERMIC* (*HEAT*): high-level laser treatment (HLLT), Ohshiro
   (b) *NONPHOTOTHERMIC*

2. Nondestructive:
    (c) *PHOTOTHERMIC*
    (d) *NONPHOTOTHERMIC, PHOTOBIOACTIVATION*

LLLT means that a low energy is absorbed by membrane and cellular organelles with synthesis of materials by photochemical, photodynamic, photo-enzymatic, and photo-immune reactions. This complex process is a regenerative one (Ohshiro 1991).

At the tissue level, neovascularization is produced. Clinically, we observe less post-operatory edema and a faster recovery time of the wound (Kauvar and Warycha 2011). The result is very explicit (Fig. 12.6).

The LLLT effect triggers a successive series of reactions: increased phagogytic reactions, increased number of fibroblasts, increased healing on site and at distant sites (other sites with triggered release and activation), greater neovascular proliferation, and greater photobioactivation of nerve endings and other cells with electrodense nuclei. These are all part of the regenerative processes (Ohshiro 2011).

Ablative lasers (fractional CO2 laser and diode laser) have a concomitant photo-destructive effect on the HLLT and LLLT components of the beam periphery. The LLLT effect possibly prolongs the life and improves the take of the fat grafts, with the well-documented HLLT effect inducing swift collagenesis and better remodeling of the dermal matrix (Jianu et al. 2012).

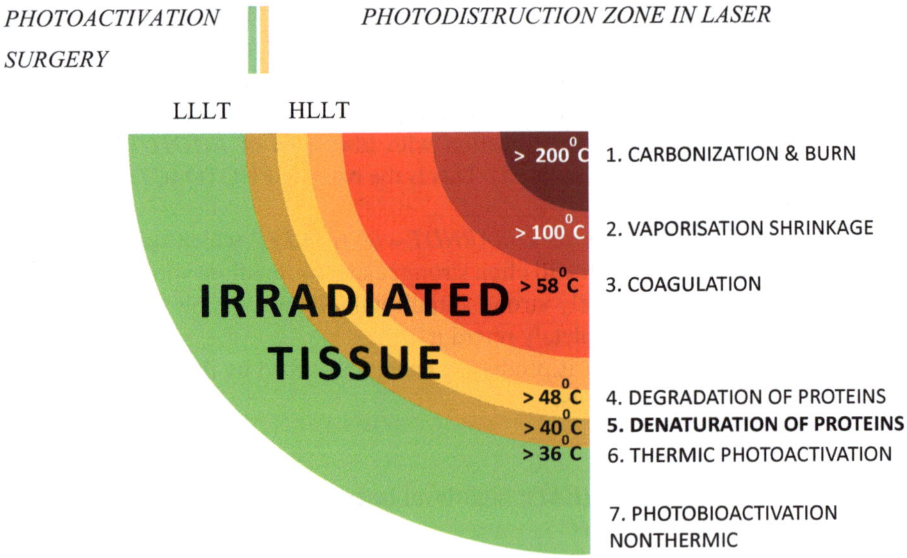

Fig. 12.6 Bioreactions and associated temperature intervals of the laser impact (Oshiro et al. 2000)

Translational Medicine in Our Practice: Research for Laser as a Regenerative Factor

For a better understanding of the mechanisms involved in the stimulatory effects of LLLT, we developed a collaborative experimental study (with the Institute of Cellular Biology and Pathology; N. Simionescu) aimed at investigating the effects of CO2 fractional laser on the regenerative capacities of ADSCs.

Human ADSCs isolated from lipoaspirates were expanded in culture for four passages and then exposed, or not (control), to a single irradiation of a 10,600-nm fractional CO2 laser with the following parameters: spot density of 9×9 spots/cm^2, 4 ms per shot, at 5, 8, and 9 W energy power. We searched for the viability and proliferation of ADSCs after LLL treatment by MTT assay. The effect of laser treatment on protein expression of intracellular and secreted VEGF, MMP2, and MMP9 were analyzed by Western blot and ELISA assays, respectively. The molecules involved in the redox signaling (NF-kB, SOD2, PGC1-a) were assessed by Western blot.

We found that compared to controls, ADSCs irradiated with CO2 laser at 9 W exhibited increased viability (~twofold) and enhanced proliferation rate (~sixfold). In addition, the CO2 laser induced in ADSCs led to an increased level of VEGF, MMP-2, and MMP-9 secreted in the culture media (about two-, eight-, and twofold, respectively). Also, augmented protein levels of p65 subunits of NF-kB complex involved in redox signaling and increased expression of mitochondrial proteins SOD2 and PGC-1α have been found. Moreover, CO2 laser treatment corrected mitochondrial dysfunction induced in ADSC by rotenone (an inhibitor of the complex I of the mitochondrial electron transport chain).

These data revealed that low-level CO2 laser applied to human adipose stromal cells stimulates mitochondria, which, by activation of redox signaling pathways, increase cell proliferation and enhance secretion of VEGF, MMP2, and MMP9. We can safely assume that the increased VEGF may favor angiogenesis, and the enhanced secretion of MMP2 and MMP9 could assist in tissue remodeling. These results explain, in part, the mechanisms implicated in the increased regenerative capacity of ADSCs under effects of low-level laser therapy effect of CO2 laser (preliminary results have been previously presented)[1] (Constantin et al. 2017).

Histological evidence by averaged biopsies taken from patients enrolled in a pilot trial in ProEstetica Hospital (IRB Resolution 7/2011) shows an increased tissular activity after CO2 fractional laser applied on the fat grafted skin. Larger hypodermic septs (i.e., fat niches or matrix) in the laser-stimulated area as well as increased fibroblast and young collagen fibers, and increased preadipocytes compared with nonstimulated fat grafted skin, are evident (see Fig. 12.7a, b) (Jianu et al. 2012).

[1] This work was funded by the Romanian Academy. Alina Constantin acknowledges the financial support of the European Social Found within the Sectorial Operational Program Human Resources Development 2007-2013 (ID: POSDRU/159/1.5/S/133391).

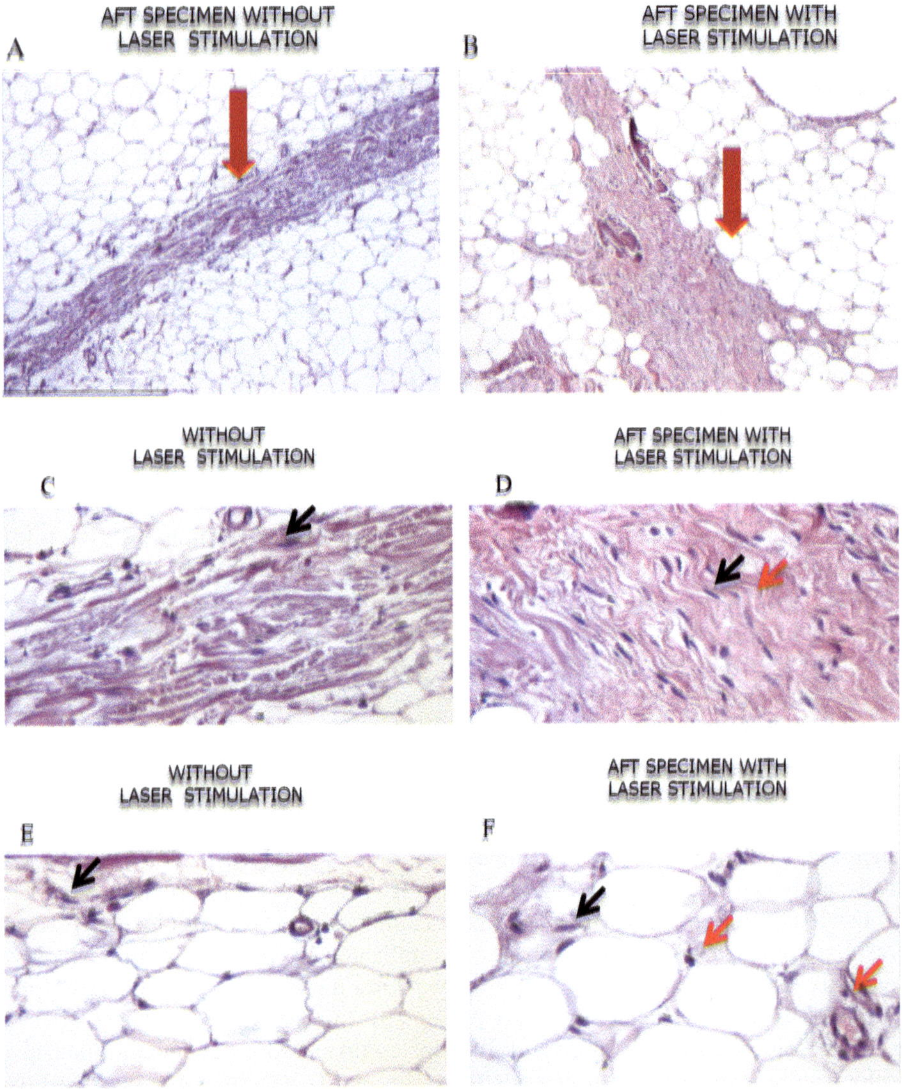

Fig. 12.7 (**a, b**) The septs ("fat niches MATRIX") are larger in the CO2 laser-stimulated area (H&E, 100×). (**b, c**) A richer cellularity in the stimulated area, with numerous fibroblasts (*black arrow*) surrounded with collagen fibers (*red*). (**e, f**) H&E staining of specimens collected 3 months after fat grafting (AFT) from PATIENT P.S., 42 y.o. Preadipocytes (*red arrow*) and fibroblasts (*black arrow*) are more evident in the stimulated tissue (**a, b**: 100× magnification; **c, d**: 100× magnification; **e, f**: 400× magnification)

12.4 Conclusions

Clinical applications of regenerative plastic surgery in our practice are oriented toward reconstruction, correction, and rejuvenation of the soft tissues of different anatomic areas, mainly the face, neck, hands, limbs, and abdomen, affected by congenital, oncological, posttraumatic, or aging conditions. The regenerative treatment consists in a closed surgery with visible outcomes explained by the synergic action of multiple methods (three to five) performed in a single surgical session, under general anesthesia. The three main methods in combination are autologous microfat transplant combined with fractional CO2 laser superficial ablative resurfacing and PRP injected intradermal (of this, PPP is spread over the lasered skin). The additional regenerative methods are ADSC enrichment of AFT for better intake and laser-assisted lipolysis that treats the flaccidity of skin and underlying tissues. The equipment is based on microcannulas and other devices for fat manipulation both in closed and opened systems, fractional CO2 laser (10,600 nm λ) and diode laser (860 nm λ). The outcomes bring regeneration of normal and pathological skin and underlying soft tissues and the decrease of side effects and complications in number and intensity of each technique compared to when used as a single procedure. The regenerative effects are visible starting 2 months postoperative.

References

Cervelli V, Gentile P, Scioli MG, Grimaldi M, Casciani CU, Spagnioli LG, Orlandi A (2009) Application of platelet-rich plasma in plastic surgery: clinic and in vitro evaluation. Tissue Eng Plast C Methods 15(4):625–634

Coleman SR (1997) Facial recontouring with lipostructure. Clin Plast Surg 24:347–367

Coleman SR, Mazzola RF (2009) Fat injection from filling to regeneration. Quality Medical Publishing. ISBN 978-1-57626-284-9 (hardcover) 5–11

Constantin A, Dumitrescu M, Nemecz MA, Tanko G, Popov D, Jianu D, Simionescu M (2015) Redox homeostasis regulation after in vitro exposure of human adipose stromal cells to low-level CO2 LASER. 33rd Annual Scientific Session of the Romanian Society for Cell Biology, Book of Abstracts, p 102

Constantin A, Dumitrescu M, Mihai Corotchi MC, Jianu D, Simionescu M (2017) CO2 laser increases the regenerative capacity of human adipose-derived stem cells by a mechanism involving the redox state and enhanced secretion of pro-angiogenic molecules. Lasers Med Sci. doi:10.1007/s10103-016-2093-6

Daar AS, Greenwood HL (2007) A proposed definition of regenerative medicine. J Tissue Eng Regen Med 1(3):179–184

Dinescu S, Galateanu B, Radu E, Hermenean A, Lungu A, Stancu IC, Jianu D, Tumbar T, Costache M (2015) Research Article "A 3D porous gelatin-alginate-based-IPN acts as an efficient promoter of chondrogenesis from human adipose-derived stem cells". Hindawi Publishing Corporation, Stem Cells International, Article ID 252909

Eppley BL, Woodell JE, Higgins J (2004) Platelet quantification and growth factor analysis from platelet-rich plasma, implication for wound healing. Plast Reconstr Surg 1114(6):1502–1508

Ghanem AM (2011) Stem cell theory. Body Lang 43:53–54

Jianu DM, Filipescu M, Jianu SA, Nita AC, Chirita DA (2012) The synergy between LASERS and adipose surgery in face and neck rejuvenation: a new approach from personal experience. Laser Theraphy J 21(3.) ISSN 08985901:215–222

Jianu DM, Jianu S, Filipescu M, Cobani O. Regenerative surgery of periocular area, p.249–254, Stem cells derived from fat tissue – personal experience in research and therapeutic applications, p. 301–306. Dermatology at the interface with other specialties" coord by Daciana Elena Branisteanu Edit. "Gr.T.Popa and Pharmacy Iasi 2014, ISBN 978–606–544-221-4

Kauvar ANB, Warycha MA (2011) Wrinkles and acne scars: fractional ablative lasers. In: Raulin C, Karsai S (eds) Laser and IPL technology in dermatology and aesthetic medicine. Springer, Berlin, pp 307–318

Kokai LE, Rubin JP, Marra KG (2005) The potential of adipose-derived adult stem cells as a source of neuronal progenitor cells. Plast Reconstr Surg 116(5):1453–1460

Marincola FM (2003) J Transl Med 1:1

Marx E (2004) Platelete-rich plasma, evidence to support its use. J Oral Maxillofac Surg 62(4):489–496

Ohshiro T (1991) Lower reactive level LASER therapy, practical applications. Wiley, pp 6–76

Ohshiro T (2011) New classification for single-system light treatment. Laser Ther 20(1):11–15

Orbay H, Tobita M, Mizuno H (2012) Mesenchymal stem cells isolated from adipose and other tissues: basic biological properties and clinical applications. Stem Cells Int 2012:461718

Oshiro T, Popa M, Niculae B, Savu B (2000) Laser therapy and laser surgery in dermatology, printed at SC National IMPRIM SA

Pa Z, Zhu M, Mizuno H, Huang J, Futrell JW, Katz AJ, Benhaim P, Lorenz HP, Hedrick MH (2001) Multilineage cells from human adipose tissue: implications for cell based therapies. Tissue Eng 7(2):211–228

Rigotti G et al (2009) Adipose – derived mesenchymal stem cells: past, present and future. Aesthet Plast Surg 33(3):271

Rubio DMG, Schoenbaum EE, Lee LS, Schteingart DE, Marantz PR, Anderson KE, Platt LD, Baez A, Esposito K (2010) Defining translational research: implications for training. Acad Med 85(3):470–475. doi:10.1097/ACM.0b013e3181ccd618

Scadden (2006) The stem-cell niche as an entity of action. Nature 441:1075–1079

Song Y-H, Prant L, Alt E (2011) Differentiation and plasticity of stem cells for tissue engineering. Tissue Engineering. Springer, Berlin/Heidelberg, pp 1–18

Tholpady SS, Aojanepong C, Llull R, Jeong J-H, Mason AC, Futrell JW, Ogle RC, Katz AJ (2005) The cellular plasticity of human adipocytes. Ann Plast Surg 54(6):651–656. Original Article

Van Pham P, Bui KH-T, Ngo DQ, Vu NB, Truong NH, Phan NL-C, Le DM, Duong TD, Nguyen TD, Le VT, Phan NK (2013) Activated platelet-rich plasma improves adipose-derived stem cell transplantation efficiency in injured articular cartilage. Stem Cell Res Ther 4(4):91. doi:10.1186/scrt277

Yoshimura K, Sato A, Aoi N, Kurita M, Inoue K, Suga H, Eto H, Kato H, Hirohi T, Harii K (2008) tCell-assisted lipotransfer for facial lipoatrophy: efficacy of clinical use of adipose-derived stem cells. Dermatol Surg 34(9):1178–1185

Yoshimura K, Eto H, Kato H, Doi K, Aoi N (2011) In vivo manipulation of stem cells for adipose tissue repair/reconstruction. Regen Med 6(6 Suppl):33–41. doi:10.2217/rme.11.62

Chapter 13
Stem Cell Applications in Rejuvenation

Aurora Almadori and Deepak M. Kalaskar

13.1 Ageing

13.1.1 Introduction

The progress of biomedical science and public heath interventions, together with improvements in water and food quality and the prevention of many infectious diseases, has resulted in a remarkable increasing survival, leading to demographic shift in favour of older people (Valerio and Nisoli 2015; Partridge 2014). The World Health Organization estimates that by 2050 nearly 25% of the world's population will be over 60 years of age (Auley and Mooney 2015; WHO 2011). However, increasing lifespan doesn't necessarily increase health span: although we are living longer, the extra few years that are gained are not necessarily spent in optimum health (Fries et al. 2011; Partridge 2014; Auley and Mooney 2015). For example, in the UK 37.1% of individuals older than 85 years have underlying cardiovascular disease (CVD), which hampers their quality of life (Capewell et al. 2008; Auley and Mooney 2015). Advanced age in humans is considered the largest risk factor for a range of diseases, including cancer, Alzheimer's disease (AD), and cardiovascular and atherosclerosis syndromes (Butler et al. 2008; Sikora et al. 2011). For this reason, it is imperative to

A. Almadori, MD
UCL Division of Surgery & Interventional Science,
9th Floor, Royal Free Campus, Rowland Hill Street, London NW3 2PF, UK

Department of Plastic Surgery, Royal Free Hospital
NHS Foundation Trust Hospital, London, UK
e mail: aurora.almadori@nhs.net

D.M. Kalaskar, PhD (✉)
UCL Division of Surgery & Interventional Science,
9th Floor, Royal Free Campus, Rowland Hill Street, London NW3 2PF, UK
e-mail: d.kalaskar@ucl.ac.uk

P.V. Pham (ed.), *Pancreas, Kidney and Skin Regeneration*, Stem Cells in Clinical Applications, DOI 10.1007/978-3-319-55687-1_13

gain a deep understanding of the impact of ageing on the regulation of biological systems and to find therapeutic strategies to keep people healthy for longer (Partridge 2014; Auley and Mooney 2015). Although ageing is complicated and variable, with diverse kinds of damage and pathology accumulating in a way that varies between both different body tissues and individual organisms, the process has proved to be malleable. Several factors are involved, and in particular, stem cells seem to be extremely important either as affected by ageing or as therapeutic remedy to regenerate ageing tissues and reduce inflammation level associated with ageing (Partridge 2014).

13.1.1.1 Ageing and Cellular Senescence

Mammalian ageing has been defined as a reduction in the capacity to adequately maintain tissue homeostasis or to repair tissues after injury (Jeyapalan and Sedivy 2008; Beltrami et al. 2011). From an evolutionary point of view, the process of ageing may be regarded as a consequence of the attenuation of natural selection pressures rather than a programmed adaptation (Kirkwood 2002; Kappei and Londoño-Vallejo 2008).

Cellular senescence is defined as a specialized form of growth arrest, confined to mitotic cells, induced by various stressful stimuli and characterized by several markers (Sharpless and DePinho 2007; Beltrami et al. 2011). The process of cellular senescence was first described by Hayflick in 1961 observing that normal human fibroblasts were able to enter a state of irreversible growth arrest after serial cultivation in vitro, while cancer cells did not enter this growth arrest state and proliferated indefinitely (Collado et al. 2007). Senescent cells are characterized by remarkable changes in gene expression that involve both cell cycle-related genes and genes that encode secreted proteins (Gonos et al. 1998; Shelton et al. 1999; Semov et al. 2002; Pascal et al. 2005).

13.1.1.2 Somatic Cell Senescence in Ageing and Age-Related Diseases

Senescent cells are suggested to participate both in ageing and age-related pathologies (Sikora et al. 2011), and the presence of senescence markers at sites of age-related pathologies has provided links between cellular senescence and ageing (Jeyapalan and Sedivy 2008). Recently, it has been found that cellular senescence is closely interconnected with ageing, longevity and age-related disease, either by sharing common genes and regulators or by protein–protein interactions and by common signalling pathways (Tacutu et al. 2011; Beltrami et al. 2011). Notably, cells that express senescence markers increase with age in several mammalian tissues like in skin, retina and liver (Chimenti et al. 2003; Janzen et al. 2006; Krishnamurthy et al. 2006; Molofsky et al. 2006); additionally, senescent cells are found at sites of age-related pathologies, such as osteoarthritis (Price et al. 2002), atherosclerosis (Minamino et al. 2002), chronic heart failure (Torella et al. 2004; Urbanek et al. 2005; Beltrami et al. 2011) and hyperplastic prostate (Shawi and Autexier 2008).

The association between cellular senescence and organismal ageing is highly suggestive of a causal link between these two processes, although establishing a direct causative relationship is challenging. First, accumulation of senescent cells in tissues may reach a point that compromises functionality, and, second, senescence may affect also stem cells, limiting their regenerative potential.

13.1.1.3 Stem Cell Senescence in Ageing and Age-Related Diseases

Tissue homeostasis and regenerative capacity are considered to be related to the stem cell pool present in every tissue. For this reason, ageing and age-related disease characterized by altered tissue homeostasis and impaired regenerative capacity can be viewed as a consequence of the reduction in stem cell number and function (Sharpless and DePinho 2007; Rossi et al. 2008; Beltrami 2011). On the one hand, ageing could be produced by accumulation of senescent cells; on the other hand, the accumulation of senescent cells per se could be harmless, and ageing could result primarily from the exhaustion of the regenerative potential of stem cells (Collado et al. 2007). Therefore, senescence may contribute to ageing not only by accumulation of senescent cells in tissues but also by limiting the regenerative potential of stem cell pools. These two mechanisms—accumulation of senescent cells and loss of stem cell function—probably contribute to ageing simultaneously (Collado et al. 2007).

13.1.2 Causes of Ageing

Ageing is a complex process characterized by both intrinsic (genetic) and extrinsic (environmental) responsible for cell senescence, stem cell exhaustion, altered intercellular communication and overall increased inflammation level (Beltrami et al. 2011; Partrige 2014; Sorrentino et al. 2014) (Fig. 13.1).

13.1.2.1 Intrinsic Factors

Telomere Attrition

Normal ageing is accompanied by telomere attrition or shortening (Lopez-Otin et al. 2013). Telomeres consist of tandem repeats of TAG sequences that are bound by specialized nucleoprotein complex (Palm and de Lange 2008; Lopez-Otin et al. 2013), the shelterin–telosome protein complex, that make them invisible to the DNA repair machinery (Blasco 2007; Beltrami et al. 2011). Due to the intrinsic inability of the replication machinery to copy the ends of linear molecules, telomeres become progressively shorter with every round of cell division (Blasco 2007). When telomeres reach a critically short length, activate the p53 tumour suppressor protein resulting in telomere-initiated senescence or apoptosis (de Lange 2005; von Zglinicki and Martin-Ruiz 2005).

Telomerase is a ribonucleoprotein with DNA polymerase activity that elongates telomeres (Greider and Blackburn 1985), but its level of activity in most adult tissues is not sufficient to compensate for the progressive telomere attrition that occurs

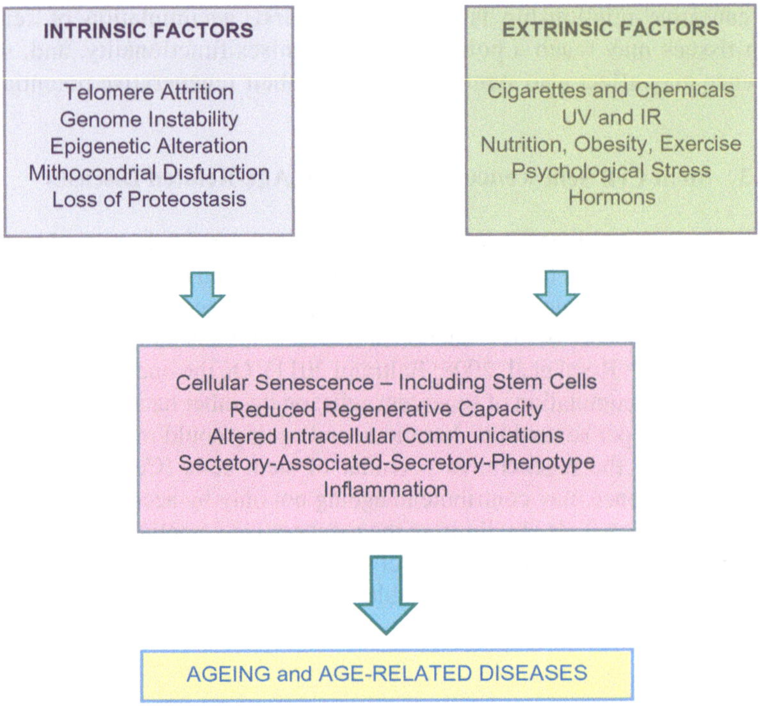

Fig. 13.1 The diagram illustrates the mechanisms responsible for ageing and age-related diseases. Both intrinsic and extrinsic factors induce cellular senescence and stem cell senescence, with stem cell exhaustion and reduction of the regenerative capacity. Furthermore, senescent cells present alterations in intercellular communication and increased secretion of pro-inflammatory factors

with ageing (Collins and Mitchell 2002; Collado et al. 2007). Most somatic cells do not express telomerase, and this leads to the progressive and cumulative loss of telomere-protective sequences from chromosome ends (Lopez-Otin et al. 2013).

Recently, it has been shown that short telomeres can up-regulate the expression of interferon-stimulated gene 15 (ISG15) independent of DNA damage signalling, possibly contributing to the chronic inflammation associated with human ageing (Lou et al. 2009; Beltrami et al. 2011).

Recent studies demonstrate that telomere length and possibly lifespan can be affected by environmental factors, such as psychological stress, obesity and cigarette smoking (Canela et al. 2007; Epel et al. 2004; Valdes et al. 2005). Epel et al. (2004) studied the implications on psychological stress and telomere length and telomerase activity of peripheral blood mononuclear cells by examining 58 healthy premenopausal women who cared for either healthy or chronically ill children. They concluded that both perceived stress and chronicity of stress are significantly associated with higher oxidative stress, lower telomerase activity and shorter telomere length (Canela et al. 2007). Because obesity and smoking are notable risk factors for age-related diseases, Valdes et al. (2005) investigated telomere lengths of

1,122 white women of different age groups using DNA from white blood cells. They concluded that telomere length within the cohort decreased at an average rate of 27 base pairs per year. Furthermore, the telomeres of obese women and smokers were significantly shorter than those of the cohort who were lean and did not smoke (Shawi and Autexier 2008).

Genomic Instability: DNA Damage

One common denominator of ageing is the accumulation of genetic damage throughout life (Moskalev et al. 2012). Moreover, numerous premature ageing diseases, such as Werner syndrome and Bloom syndrome, are the consequence of increased DNA damage accumulation (Burtner and Kennedy 2010). The integrity and stability of DNA are continuously challenged by environmental factors (see later) as well as by endogenous threats inducing DNA replication errors, spontaneous hydrolytic reactions and reactive oxygen species (ROS) (Hoeijmakers 2009). The genetic lesions arising from extrinsic or intrinsic damage are highly diverse and include point mutations, translocations, chromosomal gains and losses and gene disruption. To minimize these lesions, organisms have evolved a complex network of DNA repair mechanisms that are collectively capable of dealing with most of the damage inflicted to nuclear DNA (Lord and Ashworth 2012; Lopez-Otin et al. 2013).

The response of a cell to DNA damage is strictly dependent on the cellular background, the extent of DNA damage and the efficacy of DNA repair mechanisms (Bernhard et al. 2007). Especially severe DNA damage or irreparable damage induces cellular senescence or apoptosis (Bitomsky and Hofmann 2009; Beltrami et al. 2011).

Mitochondrial Dysfunction and mtDNA Damage

The relation between mitochondrial dysfunction and ageing has been extensively investigated, and it is now well established that mitochondria have a profound impact on the ageing process and their dysfunction can accelerate ageing in mammals (Kujoth et al. 2005; Trifunovic et al. 2004; Vermulst et al. 2008; Lopez-Otin et al. 2013).

The mitochondrion is considered to be crucial in the development of ageing, since it is involved not only in generating the majority of energy in the form of ATP via the respiratory chain and the oxidative phosphorylation system but also in the production of reactive oxygen species (Lenaz 2001). The importance of ROS and mitochondria in organism ageing led to the postulation of the 'free radical theory of aging' (Harman 1956) and the 'mitochondrial theory of aging' (Wei 1992). Ageing is characterized both by a decline in the mitochondrial respiratory function and by an increased production of reactive oxygen species (ROS) (Wei 1992; Sohal et al. 1994; Richter 1995; Barja 2004). This is coupled with an increase in the oxidative damage to mitochondrial DNA (mtDNA), accumulation of mutations (Michikawa et al. 1999; Wang et al. 2001c; Khaidakov et al. 2003) and altered expression of genes involved in intermediary metabolism of mitochondrial respiration (Ma et al. 2009). Accumulation of mutation and oxidative damage to mtDNA may result in respiratory chain dysfunction (Passos et al. 2007b) triggering a vicious cycle characterized by further production of ROS and induction of mtDNA mutations (Ma et al. 2009). Therefore, mitochondrial compromise could contribute to organ dysfunction through decreased ATP generation and changes in mitochondrial metabolism. Human disease and transgenic murine models support the causal role of mtDNA mutations in ageing; individuals with mitochondrial genetic diseases (Wallace 2005) and mice that generate frequent mutations in mitochondrial DNA (Trifunovic et al. 2004; Kujoth et al. 2005; Trifunovic et al. 2005) display phenotypes that resemble premature ageing (Beltrami et al. 2011).

Epigenetic Alteration

There are multiple lines of evidence suggesting that ageing is accompanied by epigenetic changes and that epigenetic perturbations can provoke progeroid syndromes (Lopez-Otin et al. 2013) and contribute to the development and progression of several age-related disease (Hamm and Costa 2015).

The term epigenetics was described in the 1940s and defines the interactions between the genome and the environment that leads to specific phenotypes. Epigenetics encompasses alterations in gene expression without changes in the DNA coding sequence (Hamm and Costa 2015), including alterations in DNA methylation patterns, post-translational modification of histones and chromatin remodelling (Lopez-Otin et al. 2013). Epigenetic changes are reversible (as opposed to genetic alterations which are not reversible); therefore, epigenetic-based therapeutics offers an exciting opportunity to reverse disease-associated epigenetic abnormalities (Hamm and Costa 2015).

Loss of Proteostasis

Ageing and some ageing-related diseases are linked to impaired protein homeostasis or proteostasis (Powers et al. 2009). All cells take advantage of quality control mechanisms to preserve the stability and functionality of their proteomes. Proteostasis involves mechanisms for the stabilization of correctly folded proteins, most prominently the heat-shock family of proteins, and mechanisms for the degradation of proteins by the proteasome or the lysosome (Hartl et al. 2011; Koga et al. 2011; Mizushima et al. 2008). There are several regulators of age-related proteotoxicity, and function in a coordinated fashion to restore the structure of misfolded polypeptides or to remove and degrade them completely, thus preventing the accumulation of damaged components and assuring the continuous renewal of intracellular proteins. Accordingly, many studies have demonstrated that proteostasis is altered with ageing. Additionally, chronic expression of unfolded, misfolded or aggregated proteins contributes to the development of some age-related pathology, such as Alzheimer's disease, Parkinson's disease and cataracts (Powers et al. 2009; Lopez-Otin et al. 2013).

13.1.2.2 Extrinsic Factors

The concept that environmental exposures can promote ageing is not new. For example, Tom Perls noted that the rate of physiological ageing is determined largely (50–75%) by non-genetic factors (Perls and Puca 2002). Moreover, George Martin reasoned that there must be environmental agents that accelerate the rate of molecular ageing and coined the term 'gerontogen' for such age-promoting toxicants (Martin 1987; Sorrentino et al. 2014).

According to this theory, differential exposure to largely unknown gerontogens explains much of the non-genetic variation in the rates of human physiological ageing (Sorrentino et al. 2014).

Chemical Factors: Arsenic, Benzenes, Cigarettes Smocking and Chemotherapy

Arsenic is a common toxicant found in ground water that has been linked to age-related phenotypes, including skin and bladder cancer, type 2 diabetes, neurodegenerative disease and atherosclerosis (Kapaj et al. 2006; Migliore and Coppede 2009). In both rodents and humans, arsenic has been reported to decrease aspects of DNA repair (Banerjee et al. 2008; Andrew et al. 2006), suggesting a plausible link to senescence and physiologic ageing. This effect might be due to indirect effects on DNA damage repair or direct, as yet poorly understood pro-ageing effects of arsenic that are DNA damage independent.

Benzene exposure occurs from cigarette smoke, car exhaust and industrial emissions (Wallace 1989) and has been linked to age-related cancers such as non-Hodgkin's lymphoma and leukaemia (Smith et al. 2007). Benzene exposure may lead to leukocyte telomere shortening (Hoxta et al. 2009; Zhang et al. 2013a, b), suggesting that chronic exposure could promote haematopoietic ageing through the accumulation of senescent cells in vivo, and therefore benzene may be an environmental gerontogen.

Cigarette smoke contains over 4,000 potential toxicants, and a significant smoking history decreases a human lifespan on average by 7 years (Behrnard 2007). Moreover, exposure to both primary and secondary smoke is associated with human diseases, including atherosclerosis, cancer and emphysema (Ito and Barnes 2009). Smoking has also been shown to decrease capillary blood flow, contributing to the higher rate of facial wrinkling in smokers (Grady and Ernster 1992). Research suggests that mutagens in cigarette smoke, including formaldehyde, carbon monoxide and nicotine, directly promote DNA damage (DeMarini 2004; Sorrentino et al.

2014). Many other incidental exposures such as lead, nitrosamines and pesticides have been related to chronic and age-related disease.

Other class of gerontogens are chemotherapy drugs, such as cytotoxic chemotherapy and HIV therapy. These therapeutic treatments present the induction of DNA damage as side effects. For example, chemotherapy can causes lasting non-haematologic toxicities in cancer survivors that can manifest as an acceleration of an ageing phenotype, including frailty and cognitive decline (Ness et al. 2013). The ability to measure the age-promoting effects of chemotherapy will be important in routine oncologic care, allowing clinicians to balance the long-term gerontogenic effects with the intended therapeutic effects of such agents (Sorrentino et al. 2014).

Physical Factors: UV Light and Ionizing Radiation

UV light leads to age-related pathologies of the skin. There is strong evidence indicating that UV light accelerates photo ageing and skin cancer (Kligman 1989), in part through direct DNA damage (Setlow and Carrier 1966).

Exposure to IR occurs in medical, occupational and industrial settings and produces single- and double-strand DNA breaks that seem to promote cellular and organismal ageing. Excess IR exposure causes a number of cancers but is also associated with other age-related pathologies as well as a shortened lifespan (Upton 1957).

Both types of radiation seem to promote molecular ageing, causing chronic changes in exposed tissues in the expression of multiple senescence biomarkers, including expression of SASP factors (Rodier et al. 2009; Sorrentino et al. 2014).

Nutritional Factors: Obesity, Caloric Restriction and Physical Exercise

Obesity has been suggested to lead to age-related diseases such as cancer, cardiovascular disease and diabetes (Flegal et al. 2007). In many studies caloric restriction has been shown to slow ageing resulting in a longer lifespan (Masoro 2005) and to enhance stem cell function (Cerletti M 2012). Therefore, limited caloric intake is generally considered to be anti-ageing, whereas excess calories might be considered to be pro-ageing. Some work using telomere length as a biomarker for ageing showed that telomere length has been inversely correlated with BMI and hip circumference (Cassidy et al. 2010) showing the effect of high-fat diet and obesity on physiological age.

Furthermore, obesity is a condition associated with systemic elevated inflammatory markers, and adipose tissue itself may represent a major source of immune cell-derived inflammatory factors called adipokines (Baker et al. 2011). Systemic alterations are also seen in cellular markers of inflammation: in obese population leucocyte numbers are increase, as well as soluble factors including TNF, IL-1, IL-6 and IL-8 (Sharpless et al. 2007). It is particularly striking that obesity even in adolescents is also associated with increased immune deregulation and the appearance of a senescent T cell profile (Carroll et al. 1992). Thus, obesity-linked inflammation and earlier development of features of immunosenescence could be a major contributor to the increasing burden of chronic disease and precox ageing (see Inflammageing) (Marino et al. 1997; Pawelec et al. 2014).

Several studies evaluated the effect of exercise on ageing in humans, and results showed that exercise results in a younger physiologic age as measured by ageing biomarkers such as longer telomere length (Ludlow et al. 2008) and reduced SA-cytokines (Drela et al. 2004) among exercisers (Sorrentino et al. 2014).

Hormonal Factors: Psychological Stress and Psycho-Neuro-Endocrine-Immune Axis

Stress is a state of threatened homeostasis by intrinsic or extrinsic factors, the organism's response to which is essential for every day function and survival (Pacák and Palkovits 2001). However, severe or prolonged stress can exacerbate pathophysiology in chronic and age-related diseases: recent studies suggest that perceived psychological stress leads to decreased telomerase activity and lower telomere length (Epel et al. 2004) (Sorrentino et al. 2014) and that chronic psychological stress may lead to age-related pathologies such as cardiovascular disease, decreased immunity and neural degeneration (Liu and Mori 1999) including Alzheimer's disease (AD) (Carroll et al. 2012). Since the first studies on immunomodulation of the endocrine system in the 1970s (Besedovsky et al. 1975), evidence demonstrates that the immune system influences the endocrine system and vice versa (Blalock 1994; Straub 2000). Immune and endocrine system are strictly related: not only cells of both systems have receptors to cytokines, neuropeptides and neurotransmitters but also endocrine mediators modulate the immune system, and vice versa immune structures mediators may affect the endocrine system. The strict correlation between central nervous system, endocrine glands and immune cells increased scientific community's interest in the multidisciplinary field of research of PNEI (psycho-neuro-endocrine-immune) axis (Fig. 13.2). For example, recent finding showed that

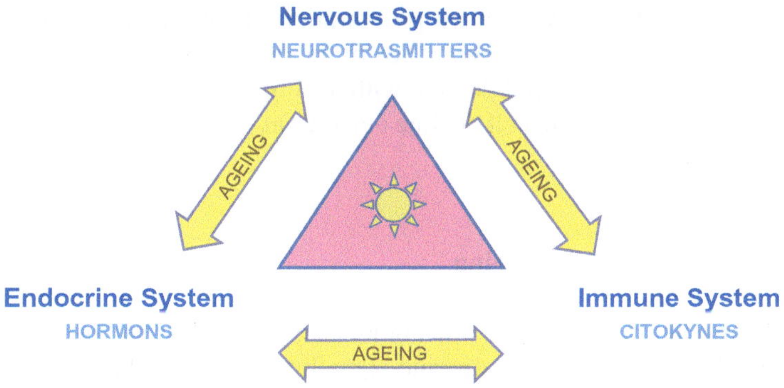

Fig. 13.2 PNEI can explain the intercorrelation of psychological stress, CNS, hormones and pro-inflammatory cytokines involved in ageing. These systems regulate each other in a finely orchestrated way by releasing chemical signals that bind specific receptors shared by neural cells, endocrine glands and immune cells

stress responses activate NF-κB in the hypothalamus and induce a signalling pathway that results in reduced production of gonadotropin-releasing hormone (GnRH) by neurons (Zhang et al. 2013a, b). The GnRH decline can contribute to numerous ageing-related changes such as bone fragility, muscle weakness, skin atrophy and reduced neurogenesis. These findings suggest that the hypothalamus may modulate systemic ageing by integrating NF-kB-driven inflammatory responses with GnRH-mediated neuroendocrine effects (see PNEI) (Lopez-Otin et al. 2013).

Other studies demonstrated that hormones released by endocrine system modulate immune functions, such as inhibition of cytokines by cortisol (Mendelson and Glasgow 1966), estrogens (Ralston et al. 1990), testosterone (Kanda et al. 1996) and dehydroepiandrosterone (DHEA) (Straub et al. 1998). On the other hand, the hypothalamic–pituitary–thyroid axis can be inhibited by IL-1, tumour necrosis factor (TNF) and IL6, and the hypothalamic–pituitary–adrenal axis may influence immune function with glucocorticoids suppressing the maturation, differentiation and proliferation of immune cells (Medeiros and Maitelli 2007). The hypothalamic–pituitary axis can also modulate the immune function. Gonadotropin-releasing hormone (GnRH) is also involved in the process of developing and modulating the immune system (Ho et al. 1995; Gameo and Romao 2010).

13.1.3 Stem Cell Decline During Ageing

13.1.3.1 Stem Cell Exhaustion

Stem cells are important to keep homeostasis of tissues by replenishing cells lost due to damage (Molofsky et al. 2006). Only a small number of stem cells, however, have to differentiate into new cell types, while the remaining has to self-renew their population. With age, a reduction of stem cells is observed, which may affect the maintenance of tissue function (Janich et al. 2011; Fonseca Costa and Ripperger 2015). The decline in the regenerative potential of tissues is one of the most obvious characteristics of ageing. A deficient proliferation of stem and progenitor cells is detrimental for the long-term maintenance of the organism. Telomere shortening is an important cause of stem cell decline with ageing in multiple tissues (Flores and Blasco 2010); however, stem cell decline emerges as the integrative consequence of multiple types of damage.

13.1.3.2 Stem Cell Niche Ageing

Stem cells respond to and depend upon other cells for support and decision making, either via cell–cell contact or paracrine signalling. This cellular and non-cellular environment influencing stem cells is referred to as the niche (Rando et al. 2014).

Stem cell niches are specific microenvironments that anatomically harbour stem cells, govern their survival and self-renewal ability and protect them from exhaustion (Lymperi et al. 2010). Main features of stem cell niches include a typical spatial localization, the anchorage of stem cells to supporting cells, the presence of a typical extracellular matrix and the integration of stimulatory as well as inhibitory signals (Moore and Lemischka 2006; Beltrami et al. 2011).

Stem cell ageing is driven not only by stem cell intrinsic factors but also by the ageing niche. Several experimental works demonstrated that niche ageing contributes to the age-related declines in stem cell characteristics and function (Geiger et al. 2013; Rando 2014; Pan et al. 2007; Beltrami 2011).

13.1.4 *Inflammageing*

Researchers have been exploring the complex relationship between inflammation and ageing and its associated chronic diseases. Chronic inflammation activates the process of damage and deterioration in target cells and organs, which further leads to chronic disease (Lavrovsky et al. 2000). Thus, evidence indicates that chronic inflammation with advancing age can precede several chronic diseases (Prasad 2012).

13.1.4.1 Senescence-Associated Secretory Phenotype, SASP

Cellular senescence is a complex biological process with both cell autonomous and paracrine effects, having a significant impact on the microenvironment. Senescent cells are far from passive: traditionally viewed as a passive state, it is now well established that they are metabolically and transcriptionally active and secrete molecules continuously for prolonged periods. This phenomenon was observed for the first time by Campisi's group and was called senescent-associated secretory phenotype, SASP (Lasri and Ben-Neriah 2015).

Specifically, senescent cells are characterized by an altered secretome, whose main functions are to remodel the extracellular matrix and to modulate the immune response (Prasad et al. 2012). Therefore, senescent cells secreting matrix metalloproteinase and inflammatory cytokines alter the surrounding tissue structure. For example, senescent cells have been implicated in altering the tissue microenvironment in the skin, where the secretion of growth factors, degradative enzymes and inflammatory cytokines by the ageing dermal fibroblasts has been suggested to contribute to the characteristic aged skin morphology (Jeyapalan and Sedivy 2008). Fibroblasts, which are the stromal support for most renewable epithelial tissues, can produce degradative enzymes and inflammatory cytokines upon senescence. This secretory phenotype of senescent fibroblasts could translate into disturbed tissue structure and function and, additionally, create a favourable environment in which preneoplastic cells could proliferate (Krtolica et al. 2001; Shawi and Autexier 2008).

13.1.4.2 Alteration of Intracellular Communication

Intercellular communication allows for synchronization of the entire population of cells within a tissue. The impact of ageing on this process involves local inflammation and the concomitant communication of tissue and immune cells by cytokines and other mediators (Fonseca Costa et al. 2015). There is compelling evidence that ageing is not an exclusively cell biological phenomenon and that it is coupled to a general alteration in intercellular communication (Laplante and Sabatini 2012; Rando and Chang 2012; Russell and Kahn 2007; Zhang et al. 2013a, b).

A prominent ageing-associated alteration in intercellular communication is called 'inflammageing', a pro-inflammatory phenotype that accompanies ageing in mammals (Salminen et al. 2012; Lopez-Otin et al. 2013).

13.1.4.3 Inflammation and Ageing: 'Inflammageing'

Ageing is characterized by the presence of higher levels of systemic inflammatory markers than generally seen in the young. For this reason, the term 'inflammageing' has been introduced by Claudio Franceschi in 2000 (Franceschi et al. 2000; Pewelec et al. 2014). Inflammageing may result from multiple causes such as the accumulation of pro-inflammatory tissue damage, dysfunctional immune system and the propensity of senescent cells to secrete pro-inflammatory cytokines and by psychological stress (Salminen et al. 2012). These alterations result in an enhanced activation of pro-inflammatory pathways, finally leading to increased production of IL-1ß, tumour necrosis factor, interferon gamma and others cytokines (Green et al. 2011; Lopez-Otin et al. 2013).

Several age-related diseases are associated with inflammation, and most chronic diseases are preceded by a chronic low level of inflammation. Elevated systemic indicators of inflammation in the elderly are implicated in carcinogenesis, sarcopenia, atherosclerosis, diabetes, neurodegeneration and other chronic diseases of ageing (Pawelec et al. 2014). Ageing results in an increase of inflammatory cytokines that contribute to the progression of many degenerative diseases (McGeer and McGeer 2004). As people grow and age, inflammation starts due to several environmental and physiological factors. Chronic inflammation damages cells of the brain, heart, arterial walls and other body structures, leading to various inflammatory diseases such as heart disease, Alzheimer's disease, Parkinson's disease, rheumatoid arthritis, psoriasis and prostatitis. As a person ages, the levels of the inflammatory markers are often sharply elevated, indicating the presence of an underlying inflammatory chronic disorder (Bremmer et al. 2008; Kriete and Mayo 2009; Prasad et al. 2012).

13.2 Stem Cell Potentialities in Rejuvenation

13.2.1 Rejuvenation

13.2.1.1 Rejuvenation and Longevity

Rejuvenation represents a well-organized and tightly regulated cellular process whereby senescent and revert specific properties acquired during previous steps of maturation to restore again a younger phenotype (Half 2009). In contrast to rejuvenation, lifespan extension—also termed longevity—does not represent a retrograde development but an overall prolonged function of tissues and organs (Half 2009).

The process of rejuvenation is observed in distinct cell populations and includes a coordinated multistep network of transduction cascades with extracellular signalling and cell-to-cell communication to relay intracellular pathways. This provides a certain tissue homeostasis by a regenerative potential and renewal upon tissue-specific repair requirements in addition to residual stem cells, which can vary among different organs to extend their lifespan (Half 2009).

13.2.1.2 Rejuvenation by Regeneration

One property of ageing is supposed to be a decline in regenerative capacity, and conversely, increasing regenerative potential should represent a characteristic of rejuvenation. Whereas the regeneration of tissues and organs is performed by proliferating and differentiating precursor cells and stem cells, it seems plausible to postulate that stem cell ageing may trigger organismal ageing (Half 2009).

Stem cells would function permanently to restore tissues and organs after providing an appropriate microenvironment; however, in an aged microenvironment, their function of self-renewal is limited or abolished (Half 2009), suggesting that declining options for cell renewal and a reduced regenerative potential of certain tissues in combination with environmental factors influence the lifespan of a corresponding organism. Conversely, rejuvenation of distinct cells which enhance the regenerative potential and maintenance of tissue homeostasis contribute to a lifespan extension (Half 2009).

13.2.2 Stem Cells and Rejuvenation

Over the last years, the scientific community has accepted the hypothesis of the active involvement of stem cell senescence in ageing and age-related diseases. Cellular senescence is now considered to be an active phenotype, profoundly impacting on organ pathophysiology through the induction of inflammation, functional deterioration and aberrant growth in a paracrine fashion. Given the abovementioned impact of pathology on stem cell senescence, several investigators have started to interfere with this phenomenon in an attempt to delay organ and organism ageing. The interventions that have been experimented could be broadly divided into two categories: to 'maintain stem cells' and to 'cure with stem cells'. Stem cell maintenance is done with stem cell rejuvenation and stem cell protection (Rando and Wyss-Coray 2014).

13.2.2.1 Stem Cell Maintenance: Protection and Rejuvenation

An alternative to stem cell rejuvenation is the protection of stem cells from cellular senescence. It is reasonable to hypothesize that this goal may be reached by increasing the resistance of stem cells to stress, possibly by intervening on molecular pathways that have been associated with lifespan extension.

Interestingly, several authors have demonstrated that drugs that are utilized in the clinical practice to treat patients affected by cardiovascular pathologies such as nitric oxide NO, statins and ACE inhibitors may exert their actions preventing endothelial progenitor cell senescence (Spyridopoulos et al. 2004). Furthermore, the upregulation of DNA repair enzymes has been associated with prevention of stem cell senescence. Specifically, APE1/Ref-1 overexpression suppressed superoxide production and b-galactosidase expression in cultured MSCs. Sirtuins are another category associated with longevity: several studies demonstrated that Sirt1 may exert both a protective role on cellular senescence (Orimo et al. 2009) and a proliferative effect on muscle precursor cells (Rathbone et al. 2009). With regard to TOR, recent experimental evidence indicates the pivotal role played by this molecule in mediating stem cell senescence and organism ageing (Blagosklonny 2010). The recently published data on the effects of rapamycin on median and maximal lifespan extension in mice corroborate this hypothesis. Chen and colleagues recently reported that pulsed administration of rapamycin may rejuvenate haematopoietic stem cells of old mice (Chen et al. 2009). Therefore, a new field of research aimed at modulating TOR activity on stem cells as a way to protect them from exhaustion is just at its beginning (Rando and Wyss-Coray 2014).

Furthermore, recent studies have shown that an increase in FGF2 signalling in the aged muscle stem cell niche results in the loss of quiescence and eventually in stem cell depletion and diminished regenerative capacity, while suppression of this signalling pathway rescues these defects (Chakkalakal et al. 2012). This opens the possibility of designing strategies aimed at inhibiting FGF2 signalling to reduce stem cell exhaustion during ageing. An important debate regarding the decline in stem cell function is the relative role of cell-intrinsic pathways compared to cell-extrinsic ones (Conboy and Rando 2012). Recent work has provided strong support for the latter. In particular, DR increases intestinal and muscle stem functions through cell-extrinsic mechanisms (Cerletti et al. 2012). Likewise, transplantation of muscle-derived stem cells from young mice to progeroid mice extends lifespan and improves degenerative changes of these animals even in tissues where donor cells are not detected, suggesting that their therapeutic benefit may derive from systemic effects caused by secreted factors (Lavasani et al. 2012). Furthermore, parabiosis experiments have demonstrated that the decline in neural and muscle stem cell function in old mice can be reversed by systemic factors from young mice (Conboy et al. 2005; Villeda et al. 2011). Pharmacological interventions are also being explored to improve stem cell function. In particular, mTORC1 inhibition with rapamycin, which can postpone ageing by improving proteostasis (see section on Loss of Proteostasis), may also improve stem cell function in the epidermis, in the haematopoietic system and in the intestine (Castilho et al. 2009). This illustrates the difficulty of disentangling the mechanistic basis for the anti-ageing activity of rapamycin and underscores the interconnectedness between the different hallmarks of ageing. Stem cell exhaustion unfolds as the integrative consequence of multiple types of ageing-associated damages and likely constitutes one of the ultimate culprits of tissue and organismal ageing. Recent promising studies suggest that stem cell rejuvenation may reverse the ageing phenotype at the organismal level (Lopez-Otin et al. 2013).

13.2.2.2 Stem Cell Therapy

Stem cell potentiality gives therapeutic opportunities via transplantation of stem cells from diverse sources for virtually every age-related disease (Rando and Wyss-Coray 2014).

However, stem cell therapy has to consider that the role of the stem cell within specific tissue types is highly variable. Some tissues, such as cells of the blood and epithelia of the skin and intestines, engage in constant turnover throughout life, attesting to the essential role of stem and progenitor cells in those tissues for the maintenance of organismal viability. Other tissues, such as the vasculature and skeletal muscle, exhibit far less turnover in the absence of injury or disease but have remarkable regenerative potential when the tissue is damaged. In those tissues, the stem cells are more like reserve cells. Central nervous system is highly stable with little regenerative capacity, properties that contribute to the devastating consequences of strokes and spinal cord injury, but has restricted areas of active stem cell function throughout life, giving hope for the potential for stem cell therapeutics (Rando and Wyss-Coray 2014).

Once placed in injured tissues, stem cells may exert their therapeutic effect via different mechanism, including cell differentiation and paracrine secretion of tropic and immunomodulatory factors.

Stem Cell Paracrine Action Versus Differentiation

While the original hypothesis underlying stem cell therapy was based on functional recovery as a consequence of stem cell differentiation, it is now clear that mechanisms other than cell differentiation are involved. In particular, trophic factors secreted by stem cells have been shown to mediate functional improvements in several preclinical studies.

In several studies, functional improvements with little evidence of stem cell differentiation in the target tissue demonstrated a paracrine action of stem cells. For example, the therapeutic effects of ASCs on liver injury have been demonstrated to be due to several growth factors that were secreted by ADSC at high levels.

Another study demonstrated that ASCs were able to promote neuroregeneration by secretion of CXCL5 cytokine which has neurotrophic properties.

Stem Cell Immunomodulation

The above-mentioned paracrine actions are mostly concerned with their trophic effects. However, in a broader sense, paracrine actions also include the immunomodulatory effect. In fact, many studies show that immunomodulation is a key mechanism through which stem cells exert therapeutic efficacy.

In particular, stem cell immunomodulatory effects are related to the immunosuppressive phenotype. These include therapeutic effects on multiple sclerosis, rheumatoid arthritis, colitis, allergic rhinitis and asthma. In addition, MSC immunomodulatory property has also been exploited as an effective method to attenuate graft-versus-host reaction, and several studies have also demonstrated the feasibility of using MSCs for the induction of tolerance in xenotransplantation.

Property of stem cells is extremely intriguing: immunological studies have shown that chronic inflammation predisposes individuals to reduced organ function, age-related diseases (Franceschi et al. 2000) and various types of cancer (Mantovani et al. 2008); therefore, the ability to suppress inflammation might lead to reduction of inflammageing and age-related diseases.

13.2.3 Sources of Stem Cells

13.2.3.1 Embryonic Stem Cells

Human embryonic stem cells (ESC) were first isolated in 1998 (Cedar and Minger 2008). Human ESC are characterized by their ability to form all the germ layers of the embryo and therefore are called pluripotent, in contrast to adult stem cells found in the various compartments of the body which are thought to be restricted to forming cells within their particular compartment and are thus termed multipotent. ESCs are thought to be able to self-renew indefinitely as well as maintaining pluripotency while somatic stem cells undergo replicative senescence (Amit et al. 2000). These two characteristics, potency to form many cell types and self-renewal, have made ESC of interest in various fields of biology and medicine, for example, in regenerative medicine looking to replace degenerated tissues in diseases such as myocardial infarction, Alzheimer's and Parkinson's and in cancer, as cancer may be caused by mutations in stem cells (Cedar and Minger 2008).

ESCs could be the ideal source for stem cell therapy because of their ability to self-renew indefinitely and to differentiate into cells from all three embryonic germ layers. However, their clinical application is limited because they represent an allogenic resource and thus have the potential to evoke an immune response, for ethical reason because the derivation of ECS cells requires the destruction of embryos and for the high risk of developing teratomas (Hipp and Atala 2008). Therefore, other stem cell sources are preferred to overcome these limitations (Hipp and Atala 2008).

13.2.3.2 Adult Stem Cells

Adult stem cells have the ability to renew themselves as well as the ability to differentiate into various cell types (Hipp and Atala 2008). However, adult stem cells tend to be tissue specific and can differentiate into cell types associated with the organ system in which they reside. Currently, it is known that stem cells reside in many tissues, such as adipose tissue, bone marrow, brain, liver, skin, skeletal muscle, gastrointestinal tract, pancreas, eyes, blood and dental pulp. Of these, the most studied are the mesenchymal stem cell, easy to obtain and able to differentiate into numerous tissue types (Hipp and Atala 2008).

Mesenchymal Stem Cells: ASCs and BMSCs

MSCs have been successfully isolated from various tissues including bone marrow, peripheral blood, amnion, adipose tissues and many others, These cells can differentiate into adipocytes, osteoblasts, chondrocytes, neurons, skeletal muscle cells, endothelial cells and vascular smooth muscle cells (Smith et al. 2010). MSCs have been considered ideal for cell therapy due to their simple isolation techniques, easy expandability, low immunogenicity and pluripotency (Liu et al. 2013).

The main sources of MSC is represented by adipose tissue (ASCs), followed by bone marrow (BMSCs). The therapeutic use of ASCs or BM-MSCs requires large quantities of cells for infusion or co-transplantation. A large quantity of ASCs can be easily obtained by liposuction or lipectomy. In contrast, BMSCs are present in lower number, and the extraction procedure is associated with higher morbidity. Furthermore, cultured ASCs seem to display an increased in vitro proliferative potential compared with BMSCs and could generate a clinically effective cell dose more rapidly than the same number of BMSCs (De Ugarte et al. 2003).

One of the most intriguing characteristic of these cells is their ability to induce immunosuppression. Several studies reported on the in vitro and in vivo immunosuppressive properties of BMSCs and ASCs (Bartholomew et al. 2002). These properties strengthen the clinical relevance of these cells in allogenic transplantation by reducing the incidence and severity of graft-versus-host disease (GVHD). Furthermore, these cells were shown escape to the immune system because they do not express major histocompatibility complex (MHC) class II or co-stimulatory molecule B7, and consequently, they do not induce allospecific T cell proliferative responses (Plock et al. 2013).

13.2.3.3 iPSCs

Recently, it has been discovered that somatic cells could be reset to a pluripotent state through somatic cell nuclear transfer. The first experiment was performed by Takahashi and Yamanaka, demonstrating that adult somatic cells could be restored to pluripotency through the exogenous expression of four transcription factors: Oct4, Sox2, Klf4 and c-Myc. These induced pluripotent stem cells (iPSCs) expressed markers exclusive to embryonic stem cells (ESCs), mimed their morphology and growth properties and could differentiate into all three germ layers (Takahashi and Yamanaka 2006; Rohani et al. 2014). Human iPSCs have been generated from various differentiated cells of healthy donors as well as patients suffering from genetic disorders including progeroid ones, such as dyskeratosis congenita and Hutchinson–Gilford progeria syndrome. iPSCs were obtained from in vitro senescing cells as well as very old patients, including centenarians (Lapasset et al. 2011). iPSCs generated from senescent and centenarian fibroblasts had reset telomere size, gene expression profiles, oxidative stress and mitochondrial metabolism. In other studies, it was shown that reprogramming restores telomere elongation in dyskeratosis congenita cells despite genetic lesions affecting telomerase (Agarwal et al. 2010). It seems that this strategy can be beneficial as it leads to cell rejuvenation.

iPS cells derived by cellular dedifferentiation are virtually indistinguishable from embryonic stem (ES) cells (Maherali et al. 2007) and thereby could potentially replace ES cells for various clinical applications, circumventing crucial ethical concerns regarding destroying embryos. Notably, iPS cells also present the benefit of being patient-specific autologous cells that should avoid immune rejection if used for cell therapy in regenerative medicine (Selvaraj et al. 2010). The ‘cellular U-turn’ approach by which patient-specific somatic cells can be dedifferentiated into iPS cells and subsequently redifferentiated into target cells opens the door to personalized medicine (Selvaraj et al. 2010). However, so far, there is a serious limitation in using this sort of cells as a therapeutic tool. The fact is that upon transplantation, iPSCs, like ESCs, have the propensity to form teratomas (Takahashi and Yamanaka 2006; Sikora 2011).

However, there are currently conflicting data suggesting that iPSCs may harbour a higher number of genetic and epigenetic abnormalities (Pera 2011) and that cells derived from iPSCs may be subject to premature senescence (Rohani et al. 2014). For example, Feng et al. differentiated human iPSCs into multiple cell types and results show that, unlike cells derived from ESCs, somatic cells derived from iPSCs exhibited early senescence and possessed dramatic defects in expansion capability (Feng et al. 2010; Rohani et al. 2014).

In conclusion, the discovery that somatic cells can be induced into a pluripotent state by the expression of reprogramming factors has enormous potential for therapeutics and human disease modelling. With regard to age-related disease and rejuvenation, the ability to reprogramme a cell to a youthful state without affecting the differentiation programme may be an effective strategy for rejuvenating an aged organism (Rando and Chang 2012): the reprogramming process resets an aged, somatic cell to a more youthful state, elongating telomeres, rearranging the mitochondrial network, reducing oxidative stress, restoring pluripotency and making numerous other alterations (Rohani et al. 2014). It is quite clear that the reprogramming reverses many aspects of ageing, and even iPSCs derived from senescent and centenarian cells exhibit a more youthful signature, displaying elongated telomeres and gene expression profiles comparable to ESCs (Lapasset et al. 2011; Rohani L et al. 2014).

13.2.3.4 Optimal Stem Cell Source

In conclusion, several sources of stem cells are available, and all of those presents advantages and disadvantages (Table 13.1). Adult stem cells are autologous and easily available. Furthermore, they present restricted differentiation potential that can be advantageous because they have a lower tendency to form tumours and mixed phenotypes. ES cells have the ability to grow indefinitely and differentiate into cells of all three germ layers; however, ethical concerns regarding the use of embryos, the formation of teratomas and the potential of ES cells to evoke an immune response currently dampen enthusiasm in their clinical potential (Rohani et al. 2014). The recently introduced iPSCs can counteract some of the disadvantages of the other

Table 13.1 The table illustrates advantages (Pros) and disadvantages (Cons) of each stem cell type

Cell type	Pros	Cons
ESCs	Indefinite grow Differentiation towards all tissues	Ethical reason Risk of teratomas Induction of immune response
ASCs	Autologous Easily available No immune reaction Limited differentiation	Limited differentiation
BMSCs	Autologous No immune reaction Limited differentiation	Morbidity of harvesting Limited differentiation
iPSCs	Autologous Easily available No immune reaction Differentiation towards all tissues	Precox senescence Risk of teratomas

stem cells, and the ability to reprogramme a cell to a youthful state without affecting the differentiation programme may be an effective strategy for rejuvenating an aged organism. However, data are controversial, and different studies have been published regarding the extent to which reprogramming rejuvenates aged somatic cells and whether iPSCs exhibit ageing signatures (Rohani et al. 2014).

13.3 Clinical Applications

13.3.1 Stem Cells and Cardiac Rejuvenation

With the ageing of population worldwide, cardiovascular diseases remain a major cause of morbidity and mortality. The heart is a self-renewing organ regulated by a compartment of multipotent cardiac stem cells (CSCs) capable of regenerating myocytes and coronary vessels. Cardiovascular ageing is a physiological process gradually leading to structural degeneration and functional loss of all the cardiac and vascular components. Ageing affects the growth and differentiation potential of CSCs interfering not only with their ability to sustain physiological cell turnover but also with their capacity to adapt to increases in pressure and volume loads (Anversa et al. 2005). Loss of self-renewing capacity, forced entry in an irreversible quiescent state and increased asymmetric and symmetric division towards lineage commitment may result in a critical reduction of the CSC pool and myocardial ageing (Anversa et al. 2005). Therefore, ageing effects on myocytes and coronary vasculature may be due to CSC ageing. However, an important factor in the pathophysiology of ageing is represented also by the extracellular matrix (ECM) that not only structurally supports the heart but also regulates the biological signalling important for cellular function and tissue homeostasis. Age-associated alterations of cardiac ECM are able to profoundly affect the function of the conduction system (Spadaccio et al. 2015). During ageing, ECM presents increased collagen content, together with the enhanced interstitial and reactive fibrosis. This is due to an imbalance in the metalloproteinase (MMPs) and their specific tissue inhibitors (TIMPs) leading to reduced ECM degradative ability and profibrotic shift occurring in aged ventricles (Bonnema et al. 2007).

Stem cell therapies were initially hypothesized to directly replace lost or non-viable myocardium with newly generated myocytes from transplanted cells. However, novel regenerative paradigm showed that stem cell therapy is able to create a healing microenvironment by paracrine signalling, reducing inflammation, modulating the ECM and reducing fibrosis.

The stem cell source most used for heart repair is the bone marrow and adipose tissue. Several studies showed that ASCs in pig and rodent models of acute myocardial infarction have a cardio protective influence, preserving ejection fraction and reducing end systolic volumes (Alt et al. 2010; Valina et al. 2007). In chronic heart disease, different studies on either pig, rodent or rabbit models of subacute myocardial infarction investigating the effect of ASCs directly delivered into the peri-infarction region of the myocardium showed an overall improvement in myocardial pump function and remodelling (Miyahara et al. 2006; Schenke-Layland et al. 2009). Stem cells can be delivered in the hearth via four routes: peripheral intravenous injection, intracoronary catheter-based delivery, catheter-based endocardial transplantation and epicardial injection during cardiothoracic surgery (Bartunek et al. 2009).

In conclusion, stem cell therapy demonstrated significant capacity to improve myocardial function contributing both via direct differentiation and indirect regenerative paracrine signalling in the myocardial microenvironment. Thus, stem cell, in particular ASCs, presents high potentiality to generate therapeutic treatment of cardiovascular disease.

13.3.2 Stem Cells and Neurological Tissue Rejuvenation

Among organ systems, rejuvenation of the central nervous system (CNS) is the most complex and challenging due to its structural and functional complexity and its restricted capacity for repair. Thus, the prospect of CNS rejuvenation has seemed extremely challenging. However, advances in stem cell science are beginning to challenge this assumption opening new perspectives (Bouchard et al. 2015).

Critical for CNS regeneration throughout life is the maintenance of endogenous tissue-specific adult stem cells, which have the ability to self-renew and produce new cells in adult tissue (Conboy and Rando 2005; Rando 2006). With age comes functional failure of adult stem cells, serving as one potential contributor to the regenerative decline of both peripheral tissues and the CNS during ageing (Conboy and Rando 2005; Bouchard et al. 2015).

Interestingly, the adult CNS was traditionally thought to be devoid of significant regenerative capacity. However, we now know that specific brain regions including the sub-ventricular zone (SVZ) lining the lateral ventricle and the subgranular zone (SGZ) of the dentate gyrus in the hippocampus harbour adult neural stem cells capable of generating neurons (neurogenesis) (Lee et al. 2012a, b; Bouchard et al. 2015). Neurogenic stem cells are principally concentrated in two spatially and functionally distinct zones in the human brain: the sub-ventricular zone (SVZ), lining the walls of the lateral ventricles, and the subgranular zone (SGZ) of the hippocampal dentate gyri (Fig. 13.1). The cellular, architectural and signalling milieus of these zones, or niches, are specialized to support stem cell function (Marr et al. 2010), in contrast to the relatively inhospitable microenvironment of the remainder of the brain. A third population of progenitor cells, known as oligodendrocyte progenitor cells (OPCs), are diffusely distributed in the brain and spinal cord. OPCs are multipotent, giving rise chiefly to myelinating oligodendrocytes, and also to Schwann cells, astrocytes and possibly neurons, but the question of whether they constitute bone fide stem cells is a subject of ongoing debate (Bouchard et al. 2015). In the spinal cord, adult neural stem cells have been reported to be quiescent and located in the ependymal layer or in the subependymal zone of the central canal (Johansson et al. 1999; Weiss et al. 1996). A fourth group of putative neural progenitor cells are reportedly scattered throughout the CNS in regions classically considered to be non-neurogenic (Arsenijevic et al. 2001). Although it appears that these cells may have neurogenic potential when cultured in vitro, it is not clear whether this capacity is realized in vivo, and it is uncertain how these cells differ, if at all, from OPCs (Bouchard et al. 2015). Furthermore, these neural progenitor cells are deeply localized in brain or spinal cord and, thus, are not easily accessible for autogenic cell therapeutics (Bouchard et al. 2015).

Consistent with peripheral tissues, the levels of adult neurogenesis also decline with age (Bondolfi et al. 2004). While adult neurogenesis has been shown to regulate cognitive functions in young adults (Clelland et al. 2009), data still remain inconclusive and predominantly correlative with regard to decreased neurogenesis and cognitive dysfunction in the elderly (Lee et al. 2012). Notwithstanding, this does not preclude the possibility that increasing adult neurogenesis in the aged brain may facilitate cognitive improvements in the elderly (Lee et al. 2012). Correspondingly, strategies that enhance regeneration in both peripheral tissues and the CNS may prove efficacious at counteracting the deleterious effects of ageing on tissue function as they arise with age (Bouchard et al. 2015).

Allogeneic foetal midbrain tissues have been used for the treatment of Parkinson's disease (Freed et al. 2001) or Huntington's disease (Bachoud-Lévi et al. 2006), and these foetal tissues have been found to alleviate the symptoms of such patients. However, ethical questions regarding the use of tissue derived from aborted human foetuses as well as issues of tissue availability, limited safety and quality control have historically been raised. Moreover, transplantation of allogeneic foetal tissue or neural stem cells carries the risk of disease transmission to the recipient and immune rejection of the donor cells (Illouz and Sterodimas 2011).

Oligodendrocyte progenitor cells and progenitors of motor neurons can be induced from mice and human ES cells, and these cells participate in the functional repair of spinal cord injuries (Deshpande et al. 2006; Keirstead et al. 2005). Nevertheless, ethical issues involving the use of ES cells, as well as problems related to tumorigenesis, will need to be overcome for future application in regenerative medicine (Illouz and Sterodimas 2011).

Induced pluripotent stem (iPS) cells are alternative cell sources for treating neural disease, and several papers have reported that dopaminergic neurons generated from iPS cells alleviated some deficits in an animal model of Parkinson's disease (Illouz and Sterodimas 2011; Soldner et al. 2009).

In recent years, numerous papers have reported that mesenchymal stem cells (MSCs) from bone marrow can be used to treat neurological disorders or diseases (Kassis et al. 2008). Some reports have stated that murine MSCs differentiate to astrocytes, giving rise to neuronal phenotypes after transplantation into mouse brain (Mezey et al. 2000). Other studies have attempted to differentiate MSCs to neural cells. Although several studies resulted in cells that exhibited neural cell phenotypes after induction in vitro, direct evidence is lacking as to whether these cells can functionally behave like neuronal or glial cells in vivo (Illouz and Sterodimas 2011).

Recent investigations of neurogenesis-promoting functions of MSCs in central nervous system (CNS) injury models have focused mainly on other characteristics of MSCs such as their secretion of neurotrophic factors including glial cell-derived neurotrophic factor (GDNF), brain-derived neurotrophic factor (BDNF), nerve growth factor (NGF), angiopoietin-1 and vascular endothelial growth factor (Yilmaz et al. 2010; Zhang et al. 2002). Additionally, MSCs have also been shown to have an immunomodulating function in vivo (Gordon et al. 2008; Cassis et al. 2008). Through the actions of these secreted trophic factors and this immunomodulating function, MSCs stimulate angiogenesis and neural stem cell migration to the injury site, enhancing neural cell survival and differentiation, eventually improving neural function in CNS injury or disease (Illouz and Sterodimas 2011).

13.3.3 Stem Cells and Rejuvenation in Urology

Chronic kidney disease is a major health problem worldwide with an overall mortality rate of 50–80% (Chhabra and Brayman 2009). Current treatments are met with an acute shortage of compatible organs and limited adaptability of dialysis techniques. Therefore, alternative treatments such as stem cell-based therapy are currently being investigated. However, due to its anatomic complexity, the kidney has proven to be a very difficult organ for such applications (Yokoo et al. 2008).

For renal diseases, transplantation of stem cells into the kidney of an ischemia–reperfusion mouse model showed that ASCs differentiated towards renal tubular epithelium at an early stage of renal injuries. Moreover, the differentiated donor cells appeared to contribute to promoting host cell proliferation as well. Thus, both cell differentiation and paracrine actions appeared to play a role in ADSC's ability to repair injured kidney. For urethral diseases, studies have demonstrated the angiogenic potential of bone marrow-derived cells to increase the survival of urothelial cells seeded on decellularized or synthetic urethral grafts. Using ASCs for the purpose of bladder augmentation has been demonstrated in recent studies with encouraging results. Transplantation of ASCs into the bladder or through intravenous injection has also shown promises in treating hyperlipidaemia-associated overactive bladder. A clinical trial on the use of ASCs to treat postprostatectomy has also obtained favourable outcomes. Intracavernous injection of ADSC has also demonstrated efficacy in treating various types of erectile dysfunction, including hyperlipidaemia-associated, diabetic and cavernous nerve injury (Illouz and Sterodimas 2011).

Despite these encouraging advances, the application of stem cells for treating urological diseases is still in its infancy, and the most of the studies have been done only in preclinical settings.

13.3.4 Stem Cells and Skin Rejuvenation

In recent years, there has been an increased interest in the rejuvenating properties of ASCs and autologous fat grafting, in particular in aged skin of face and hands (Coleman 2006; Mojallal et al. 2009). Several animal studies described an improvement in skin texture, elasticity and aesthetic aspect, including scar tissue, radiation-induced fibrosis, sequelae of burns and scleroderma after treatment with ASCs and fat grafting (Kim et al. 2011).

ASCs had been used in association of adipose tissue as scaffold or in combination with PRP platelet-rich plasma (Serra-Mestre et al. 2014; Gentile et al. 2012). However, the biological mechanism responsible for the rejuvenating and regenerative effect is not yet clearly understood. Several authors hypothesize that ASCs have a beneficial effect in skin rejuvenation because of its paracrine actions, inducing angiogenesis, antioxidant effects and immunomodulation (Lee 2012). In particular, the new formation of micro-vessels seems to play a key role in the modification observed after treatment of aged skin. A recent study evaluating the effect of expanded ASCs and fat with its stromal vascular fraction demonstrates histological and ultrastructural improvements of aged facial skin, presenting reorganization of the reticular dermis with reduced collagen fibres, reabsorption of the elastic component in the reticular dermis and new formation of elastic fibres in the papillary dermis (Charles-de-Sà et al. 2015). No significant differences have been reported between the two groups, and the authors suggested that fat grafting plus stromal vascular fraction should be preferred to expanded ASCs because it is easier to perform; it presents lower costs and avoids regulatory issues applied to cell culturing for implantation in human (Charles-de-Sà et al. 2015).

References

Agarwal S et al (2010) Telomere elongation in induced pluripotent stem cells from dyskeratosis congenita patients. Nature 464:292–296

Alt E et al (2010) Effect of freshly isolated autologous tissue resident stromal cells on cardiac function and perfusion following acute myocardial infarction. Int J Cardiol 144(1):26–35

Amit M et al (2000) Clonally derived human embryonic stem cell lines maintain pluripotency and proliferative potential for prolonged periods of culture. Dev Biol 227:271–278

Andrew AS et al (2006) Arsenic exposure is associated with decreased DNA repair in vitro and in individuals exposed to drinking water arsenic. Environ Health Perspect 114:1193–1198

Anversa P et al (2005) Myocardial aging a stem cell problem. Basic Res Cardiol 100:482–493

Arsenijevic Y et al (2001) Isolation of multipotent neural precursors residing in the cortex of the adult human brain. Exp Neurol 170(1):48–62

Auley MT, Mooney K (2015) Lipid metabolism and aging. Comput Struct Biotechnol J 13:38–46

Bachoud-Lévi AC et al (2006) Effect of fetal neural transplants in patients with Huntington's disease 6 years after surgery: a long-term follow-up study. Lancet Neurol 5:303–309

Baker DJ et al (2011) Clearance of p16Ink4a-positive senescent cells delays ageing-associated disorders. Nature 479:232–236

Banerjee M et al (2008) DNA repair deficiency leads to susceptibility to develop arsenic-induced premalignant skin lesions. Int J Cancer 123:283–287
Barja G (2004) Free radicals and aging. Trends Neurosci 27:595–600
Bartholomew et al (2002) Mesenchymal stem cells suppress lymphocyte proliferation in vitro and prolong skin graft survival in vivo. Exp Hematol 30(1):42–48
Bartunek J et al (2009) Delivery of biologics in cardio- vascular regenerative medicine. Clin Pharmacol Ther 85(5):548–552
Beltrami AP et al (2011) At the stem of youth and health. Pharmacol Ther 129:3–20
Bernhard D et al (2007) Cigarette smoke—an aging accelerator? Exp Gerontol 42:160–165
Besedovsky H et al (1975) Changes in blood hormone levels during the immune response. Proc Soc Exp Biol Med 150(2):466–470
Bitomsky N, Hofmann TG (2009) Apoptosis and autophagy: regulation of apoptosis by DNA damage signalling - roles of p53, p73 and HIPK2. FEBS J 276(21):6074–6083
Blagosklonny MV (2010) Calorie restriction: decelerating mTOR-driven aging from cells to organisms (including humans). Cell Cycle 9(4):683–688. Epub 2010 Mar 2
Blalock JE (1994) The syntax of immune-neuroendocrine communication. Immunol Today 15(11):504–511
Blasco MA (2007) Telomere length, stem cells and aging. Nat Chem Biol 3:640–649
Bondolfi L, Ermini F, Long JM, Ingram DK, Jucker M (2004) Impact of age and caloric restriction on neurogenesis in the dentate gyrus of C57BL/6 mice. Neurobiol Aging 25:333–340
Bonnema DD et al (2007) Effects of age on plasma matrix metalloproteinases (MMPs) and tissue inhibitor of metalloproteinases (TIMPs). J Card Fail 13:530–540
Bouchard J et al (2015) Ageing and brain rejuvenation as systemic events. J Neurochem 132:5–19
Bremmer MA et al (2008) Inflammatory markers in late-life depression: results from a population-based study. J Affect Disord 106:249–255
Burtner CR, Kennedy BK (2010) Progeria syndromes and ageing: what is the connection? Nat Rev Mol Cell Biol 11(8):567–578
Butler RN et al (2008) New model of health promotion and disease prevention for the 21st century. BMJ 337:a399
Canela A et al (2007) High-throughput telomere length quantification by FISH and its application to human population studies. Proc Natl Acad Sci U S A 104(13):5300–5305
Capewell S et al (2008) Modelling the UK burden of cardiovascular disease to 2020 a research report for the cardio & vascular coalition and the British heart foundation. British Heart: Foundation
Carroll KK et al (1992) Dolichol: function, metabolism, and accumulation in human tissues. Biochem Cell Biol 70:382–384
Carroll BJ et al (2012) Pathophysiology of hypercortisolism in depression: pituitary and adrenal responses to low glucocorticoid feedback. Acta Psychiatr Scand 125(6):478–491
Cassidy A et al. (2010) Associations between diet, lifestyle factors, and telomere length in women. Am J Clin Nutr 91(5):1273–1280
Cassis LA et al (2008) Local adipose tissue renin-angiotensin system. Curr Hypertens Rep 10(2):93–98
Castilho RM et al (2009) mTOR mediates Wnt-induced epidermal stem cell exhaustion and aging. Cell Stem Cell 5(3):279–289
Cedar S, Minger SR (2008) Human embryonic stem cells a model for human ageing. Exp Gerontol 43:1005–1008
Cerletti M et al (2012) Short-term calorie restriction enhances skeletal muscle stem cell function. Cell Stem Cell 10:515–519
Chakkalakal JV et al (2012) The aged niche disrupts muscle stem cell quiescence. Nature 490(7420):355–360
Charles-de-Sá L et al (2015) Antiaging treatment of the facial skin by fat graft and adipose-derived stem cells. PRS 135:6

Chen C et al (2009) mTOR regulation and therapeutic rejuvenation of aging hematopoietic stem cells. Sci Signal 2(98):ra75
Chhabra P, Brayman KL (2009) The use of stem cells in kidney disease. Curr Opin Organ Transplant 14:72–78
Chimenti C et al (2003) Senescence and death of primitive cells and myocytes lead to premature cardiac aging and heart failure. Circ Res 93(7):604–613
Clelland CD et al (2009) A functional role for adult hippocampal neurogenesis in spatial pattern separation. Science 325:210–213
Coleman SR (2006) Structural fat grafting: more than a permanent filler. Plast Reconstr Surg 118(Suppl):108S–120S
Collado M et al (2007) Cellular senescence in cancer and aging. Cell 130(2):223233
Collins K, Mitchell JR (2002) Telomerase in the human organism. Oncogene 21:564–579
Conboy IM, Rando TA (2005) Aging, stem cells and tissue regeneration: lessons from muscle. Cell Cycle 4:407–410
Conboy IM, Rando TA (2012) Heterochronic parabiosis for the study of the effects of aging on stem cells and their niches. Cell Cycle 11(12):2260–2267
Conboy IM et al (2005) Rejuvenation of aged progenitor cells by exposure to a young systemic environment. Nature 433(7027):760–764
De Ugarte DA et al (2003) Comparison of multi-lineage cells from human adipose tissue and bone marrow. Cells Tissues Organs 174(3):101–1092003
DeMarini DM (2004) Genotoxicity of tobacco smoke and tobacco smoke condensate: a review. Mutat Res 567:447–474
Deshpande DM et al (2006) Recovery from paralysis in adult rats using embryonic stem cells. Ann Neurol 60:32–44
Drela N et al (2004) Moderate exercise may attenuate some aspects of immunosenescence. BMC Geriatr 4:8
Epel ES et al (2004) Accelerated telomere shortening in response to life stress. Proc Natl Acad Sci U S A 101:17312–17315
Feng Q et al (2010) Hemangioblastic derivatives from human induced pluripotent stem cells exhibit limited expansion and early senescence. Stem Cells 28:704–712
Flegal KM et al (2007) Cause-specific excess deaths associated with underweight, overweight, and obesity. JAMA 298:2028–2037
Flores I, Blasco MA (2010) The role of telomeres and telomerase in stem cell aging. FEBS Lett 584(17):3826–38302010 Sep 10
Fonseca Costa SS (2015) Impact of the circadian clock on the ageing process. Front Neurol 6:43
Franceschi C et al (2000) Inflamm-aging. An evolutionary perspective on immunosenescence. Ann N Y Acad Sci 908:244–254
Freed CR et al (2001) Transplantation of embryonic dopamine neurons for severe Parkinson's disease. N Engl J Med 344:710–719
Fries JF et al (2011) Compression of morbidity 1980-2011: a focused review of paradigms and progress. J Aging Res 2011:261702
Gameiro C, Romao F (2010) Changes in the immune system during menopause and aging. Front Biosci E2:1299–1303
Geiger H et al (2013) The ageing haematopoietic stem cell compartment. Nat Rev Immunol 13:376–389
Gentile P et al (2012) Concise review: adipose-derived stromal vascular fraction cells and platelet-rich plasma: basic and clinical implications for tissue engineering therapies in regenerative surgery. Stem Cells Transl Med 1:230–236
Gonos ES et al (1998) Cloning and identification of genes that associate with mammalian replicative senescence. Exp Cell Res 240(1):66–74
Gordon D et al (2008) Human mesenchymal stem cells abrogate experimental allergic encephalomyelitis after intraperitoneal injection, and with sparse CNS infiltration. Neurosci Lett 448:71–73

Grady D, Ernster V (1992) Does cigarette smoking make you ugly and old? Am J Epidemiol 135:839–842
Green DR, Galluzzi L, Kroemer G (2011) Mitochondria and the autophagy-inflammation-cell death axis in organismal aging. Science 333:1109–1112
Greider CW, Blackburn EH (1985) Identification of a specific telomere terminal transferase activity in Tetrahymena extracts. Cell 43:405–413
Half R (2009) Rejuvenation in distinct cell populations. Exp Gerontol 44:634–638
Hamm C, Costa F (2015) Epigenomes as therapeutic targets. Pharmacol Ther. doi:10.1016/j.pharmthera.2015.03.003
Harman D (1956) Aging: a theory based on free radical and radiation chemistry. J Gerontol 11(3):298–300
Hartl FU et al (2011) Molecular chaperones in protein folding and proteostasis. Nature 475:324–332
Hipp J, Atala A (2008) Sources of stem cells. Stem Cell Rev 4:3–11
Ho HN et al (1995) In vivo CD3+CD25+ lymphocyte subpopulation is down-regulated without increased serum- soluble interleukin-2 receptor (sIL-2R) by gonadotropin releasing hormone agonist (GnRH-a). Am J Reprod Immunol 33:134–139
Hoeijmakers JH (2009) DNA damage, aging, and cancer. N Engl J Med 361(15):1475–1485
Hoxha M et al (2009) Association between leukocyte telomere shortening and exposure to traffic pollution: a cross-sectional study on traffic officers and indoor office workers. Environ Health 8:41
Illouz YG, Sterodimas A (2011) Adipose stem cells and regenerative medicine. Springer, New York
Ito K, Barnes PJ (2009) COPD as a disease of accelerated lung aging. Chest 135:173–180
Janich P et al. (2011) The circadian molecular clock creates epidermal stem cell heterogeneity. Nature 480(7376):209–214
Janzen V et al (2006) Stem-cell ageing modified by the cyclin-dependent kinase inhibitor p16INK4a. Nature 443(7110):421–426
Jeyapalan J, Sedivy JM (2008) Cellular senescence and organismal aging. Mech Ageing Dev 129:467–474
Johansson CB et al (1999) Identification of a neural stem cell in the adult mammalian central nervous system. Cell 96:25–34
Kanda N et al (1996) Testosterone inhibits immunoglobulin production by human peripheral blood mononuclear cells. Clin Exp Immunol 106:410–415
Kapaj S et al (2006) Human health effects from chronic arsenic poisoning—a review. J Environ Sci Health A Tox Hazard Subst Environ Eng 41:2399–2428
Kappei D, Londoño-Vallejo JA (2008) Telomere length inheritance and aging. Mech Ageing Dev 129:17–26
Kassis I et al (2008) Neuroprotection and immunomodulation with mesenchymal stem cells in chronic experimental autoimmune encephalomyelitis. Arch Neurol 65:753–761
Keirstead HS et al (2005) Human embryonic stem cell- derived oligodendrocyte progenitor cell transplants remyeli- nate and restore locomotion after spinal cord injury. J Neurosci 25:4694–4705
Khaidakov M et al (2003) Accumulation of point mutations in mitochondrial DNA of aging mice. Mutat Res 526(1–2):1–7
Kim JH et al (2011) Adipose-derived stem cells as a new therapeutic modality for ageing skin. Exp Dermatol 20:383–387
Kirkwood TB (2002) Evolution in ageing. Mech Ageing Dev 123:737–745
Kligman LH (1989) Photoaging. Manifestations, prevention, and treatment. Clin Geriatr Med 5:235–251
Koga H et al (2011) Protein homeostasis and aging: the importance of exquisite quality control. Ageing Res Rev 10:205–215
Kriete A, Mayo KL (2009) Atypical pathways of NF-kappaB activation and aging. Exp Gerontol 44:250–255

Krishnamurthy J et al (2006) p16INK4a induces an age-dependent decline in islet regenerative potential. Nature 443(7110):453–457
Krtolica A et al (2001) Senescent fibroblasts promote epithelial cell growth and tumorigenesis: a link between cancer and aging. Proc Natl Acad Sci U S A 98(21):12072–12077
Kujoth GC et al (2005) Mitochondrial DNA mutations, oxidative stress, and apoptosis in mammalian aging. Science 309(5733):481–484
de Lange T (2005) Shelterin: the protein complex that shapes and safeguards human telomeres. Genes Dev 19:2100–2110
Lapasset L et al (2011) Rejuvenating senescent and centenarian human cells by reprogramming through the pluripotent state. Genes Dev 25:2248–2253
Laplante M, Sabatini DM (2012) mTOR signaling in growth control and disease. Cell 149:274–293
Lasri A, Ben-Neriah B (2015) Senescence-associated inflammatory responses. Trends Immunol 36(4):217–228
Lavasani et al (2012) Muscle-derived stem/progenitor cell dysfunction limits health span and lifespan in a murine progeria model. Nat Commun 3:6082012 Jan 3
Lavrovsky Y et al (2000) Role of redox-regulated transcription factors in inflammation, aging and age-related diseases. Exp Gerontol 35:521–532
Lee SH et al (2012a) Paracrine effects of adipose-derived stem cells on keratinocytes and dermal fibroblasts. Ann Dermatol 24:136–143
Lee SW et al (2012b) New neurons in an aged brain. Behav Brain Res 227:497–507
Lenaz G (2001) The mitochondrial production of reactive oxygen species: mechanisms and implications in human pathology. IUBMB Life 52(3–5):159–164
Liu J, Mori A (1999) Stress, aging, and brain oxidative damage. Neurochem Res 24(11): 1479–1497
Liu X et al (2013) Direct comparison of the potency of human mesenchymal stem. Int J Mol Med 31:407–415
Lopez-Otin C et al (2013) The hallmarks of aging. Cell 153(6):1194–1217
Lord CJ, Ashworth A (2012) The DNA damage response and cancer therapy. Nature 481(7381):287–294
Lou Z et al (2009) Telomere length regulates ISG15 expression in human cells. Aging 1(7):608–621
Ludlow AT et al (2008) Relationship between physical activity level, telomere length, and telomerase activity. Med Sci Sports Exerc 40:1764–1771
Lymperi S et al (2010) The HSC niche concept has turned 31. Has our knowledge matured? Ann N Y Acad Sci 1192(1):12–18
Ma D et al (2009) Metabonomic studies of schizophrenia and psychotropic medications: focus on alterations in CNS energy homeostasis. Bioanalysis 1(9):1615–1626
Maherali N et al (2007) Directly reprogrammed fibroblasts show global epigenetic remodeling and widespread tissue contribution. Cell Stem Cell 1:55–70
Mantovani A et al (2008) Cancer-related inflammation. Nature 454(7203):436–4442008 Jul 24
Marino M et al (1997) Accumulation of dolichol and impaired signal transduction in aging. Aging (Milano) 9:433–434
Marr RA (2010) Insights into neurogenesis and aging: potential therapy for degenerative disease? Future Neurol 5(4):527–541
Martin GM (1987) Interactions of aging and environmental agents: the gerontological perspective. Prog Clin Biol Res 228:25–80
Masoro EJ (2005) Overview of caloric restriction and ageing. Mech Ageing Dev 126:913–922
McGeer PL, McGeer EG (2004) Inflammation and the degenerative diseases of aging. Ann N Y Acad Sci 1035:104–116
Medeiros SF, Maitelli A (2007) A.P.B. Nince: Efeitos da terapia hormonal na menopausa sobre o sistema imune. Rev Brás Ginecol Obestet 29:593–601

Mendelson J, Glasgow LA (1966) The in vitro and in vivo effects of cortisol on interferon production and action. J Immunol 96:345–352
Mezey E et al (2000) Turning blood into brain: cells bearing neuronal antigens generated in vivo from bone marrow. Science 290:1779–1782
Michikawa Y et al (1999) Aging-dependent large accumulation of point mutations in the human mtDNA control region for replication. Science 286(5440):774–779
Migliore L, Coppede F (2009) Environmental-induced oxidative stress in neurodegenerative disorders and aging. Mutat Res 674:73–84
Minamino T et al (2002) Endothelial cell senescence in human atherosclerosis: role of telomere in endothelial dysfunction. Circulation 105(13):1541–1544
Miyahara Y et al (2006) Monolayered mesenchymal stem cells repair scarred myocardium after myocardial infarction. Nat Med 12(4):459–465
Mizushima N et al (2008) Autophagy fights disease through cellular self digestion. Nature 451:1069–1075
Mojallal A et al (2009) Improvement of skin quality after fat grafting: clinical observation and an animal study. Plast Reconstr Surg 124:765–774
Molofsky AV et al (2006) Increasing p16INK4a expression decreases forebrain progenitors and neurogenesis during ageing. Nature 443(7110):448–452
Moore KA, Lemischka IR (2006) Stem cells and their niches. Science 311(5769):1880–1885
Moskalev EA et al (2012) RNA-directed epigenomic reprogramming: an emerging principle of a more targeted cancer therapy? Genes Chromosomes Cancer 51(2):105–110
Ness KK et al (2013) Physiologic frailty as a sign of accelerated aging among adult survivors of childhood cancer: a report from the st jude lifetime cohort study. J Clin Oncol 31:4496–4503
Orimo et al (2009) Protective role of SIRT1 in diabetic vascular dysfunction. Arterioscler Thromb Vasc Biol 29(6):889–8942009 Jun
Pacák K, Palkovits M (2001) Stressor specificity of central neuroendocrine responses: implications for stressrelated disorders. Endocr Rev 22(4):502–548
Palm W, de Lange T (2008) How shelter in protects mammalian telomeres. Annu Rev Genet 42:301–334
Pan L et al (2007) Stem cell aging is controlled both intrinsically and extrinsically in the Drosophila ovary. Cell Stem Cell 1(4):458–469
Partridge L (2014) Intervening in ageing to prevent the diseases of ageing. Trends Endocrinol Metab 25(11): 555–557
Pascal T et al (2005) Comparison of replicative senescence and stress-induced premature senescence combining differential display and low-density. DNA Arrays 579(17):3651–3659
Passos JF et al (2007) DNA damage in telomeres and mitochondria during cellular senescence: is there a connection? Nucleic Acids Res 35(22):7505–7513
Pawelec G et al (2014) Inflammation, ageing and chronic disease. Curr Opin Immunol 29:23–28
Pera MF (2011) Stem cells: the dark side of induced pluripotency. Nature 471:46–47
Perls T, Puca A (2002) The genetics of aging—implications for pharmacogenomics. Pharmacogenomics 3:469–484
Plock et al (2013) Perspectives on the use of mesenchymal stem cells in vascularized composite allotransplantation. Front Immunol. doi:10.3389/fimmu.2013.00175
Powers ET et al (2009) Biological and chemical approaches to diseases of proteostasis deficiency. Annu Rev Biochem 78:959–991
Prasad K (2012) Resveratrol, wine, and atherosclerosis. Int J Angiol 21(1):7–18
Prasad S et al (2012) Age-associated chronic diseases require age-old medicine: role of chronic inflammation. Prev Med 54:S29–S37
Price JS et al (2002) The role of chondrocyte senescence in osteoarthritis. Aging Cell 1(1):57–65
Ralston SH et al (1990) Estrogen inhibits release of tumor necrosis factor from peripheral blood mononuclear cells in postmenopausal women. J Bone Miner Res 5:983–988
Rando TA (2006) Stem cells, ageing and the quest for immortality. Nature 441:1080–1086

Rando TA, Chang HY (2012) Aging, rejuvenation, and epigenetic reprogramming: resetting the aging clock. Cell 148:46–57
Rando AT, Wyss-Coray T (2014) Stem cells as vehicles for youthful regeneration of aged tissues. J Gerontol A Biol Sci Med Sci 69(S1):S39–S42
Rathbone et al (2009) Sirt1 increases skeletal muscle precursor cell proliferation. Eur J Cell Biol 88(1):35–442009
Richter F (1995) Survival and morphology of isolated pancreatic acinar cells from rats with induced acute pancreatitis are not improved with anti-inflammatory drugs. Int J Pancreatol 18(2):145–152
Rodier F et al (2009) Persistent DNA damage signalling triggers senescence-associated inflammatory cytokine secretion. Nat Cell Biol 11:973–979
Rohani L et al (2014) The aging signature hallmark of induced pluripotent stem cells. Aging Cell 13(1):2–7
Rossi DJ et al (2008) Stems cells and the pathways to aging and cancer. Cell 132(4):681–696
Russell SJ, Kahn CR (2007) Endocrine regulation of ageing. Nat Rev Mol Cell Biol 8:681–691
Salminen A et al (2012) Inflammaging: disturbed interplay between autophagy and inflammasomes. Aging 4:166–175
Schenke-Layland K et al (2009) Adipose tissue-derived cells improve cardiac function following myocardial infarction. J Surg Res 153(2):217–223
Selvaraj V et al (2010) Induced pluripotent stem cells and lineage reprogramming technologies. Trends Biotechnol 28(4):214–223
Semov A et al (2002) Microarray analysis of E-box binding-related gene expression in young and replicatively senescent human fibroblasts. Anal Biochem 302(1):38–51
Serra-Mestre JM et al (2014) Platelet-rich plasma mixed-fat grafting: a reasonable Prosurvival strategy for fat grafts? Aesth Plast Surg 38:1041–1049
Setlow RB, Carrier WL (1966) Pyrimidine dimers in ultraviolet-irradiated DNA's. J Mol Biol 17:237–254
Sharpless NE, DePinho RA (2007) How stem cells age and why this makes us grow old. Nat Rev Mol Cell Biol 8(9):703–713
Shawi M, Autexier C (2008) Telomerase, senescence and ageing. Mech Ageing Dev 129:3–10
Shelton DN et al (1999) Microarray analysis of replicative senescence. Curr Biol 9(17):939–945
Sikora E (2011) Immunoregulation by interference RNA (iRNA) – mechanisms, role, perspective. Postepy Hig Med Dosw 65:482–495
Sikora E et al (2011) Impact of cellular senescence signature on ageing research. Ageing Res Rev 10:146–152
Smith MT et al (2007) Benzene exposure and risk of non-Hodgkin lymphoma. Cancer Epidemiol Biomark Prev 16:385–391
Smith AN et al (2010) Mesenchymal stem cells induce dermal fibroblast responses to injury. Exp Cell Res 316:48–54
Sohal RS et al (1994) Effect of age and caloric restriction on DNA oxidative damage in different tissues of C57BL/6 mice. Mech Ageing Dev 76(2–3):215–224
Soldner F et al (2009) Parkinson's disease patient-derived induced pluripotent stem cells free of viral reprogramming factors. Cell 136:964–977
Sorrentino A et al (2014) Defining the toxicology of aging. Trends Mol Med 20(7):375–384
Spadaccio C et al (2015) The role of extracellular matrix in age-related conduction disorders: a forgotten player? J Geriatr Cardiol 12:76–82
Spyridopoulos et al (2004) Statins enhance migratory capacity by upregulation of the telomere repeat-binding factor TRF2 in endothelial progenitor cells. Circulation 110(19):3136–3142Nov 9
Straub RH (2000) Cytokines and hormones as possible links between endocrinosenescence and immunosenescence. J Neuroimmunol 109:10–15
Straub RH et al (1998) Serum dehydroepiandrosterone (DHEA) and DHEA sulfate are negatively correlated with serum interleukin-6 (IL-6), and DHEA inhibits IL-6 secretion from mononuclear cells in man in vitro: possible link between endocrinosenescence and immunosenescence. J Clin Endocrinol Metab 83:2012–2017

Tacutu R et al (2011) Molecular links between cellular senescence, longevity and age-related diseases – a systems biology perspective. Aging (Albany NY) 3(12):1178–1191
Takahashi K, Yamanaka S (2006) Induction of pluripotent stem cells from mouse embryonic and adult fibroblast cultures by defined factors. Cell 126(4):663–676. Epub 2006 Aug 10
Torella D et al (2004) Cardiac stem cells and myocyte aging, heart failure, and insulin-like growth factor1 overexpression. Circ Res 95:514–524
Trifunovic A et al (2004) Premature ageing in mice expressing defective mitochondrial DNA polymerase. Nature 429(6990):417–423
Trifunovic A et al (2005) Somatic mtDNA mutations cause aging phenotypes without affecting reactive oxygen species production. Proc Natl Acad Sci U S A 102(50):17993–17998
Upton AC (1957) Ionizing radiation and the aging process; a review. J Gerontol 12:306–313
Urbanek K et al (2005) Myocardial regeneration by activation of multipotent cardiac stem cells in ischemic heart failure. Proc Natl Acad Sci U S A 102(24):8692–8697
Valdes AM et al (2005) Obesity, cigarette smoking, and telomere length in women. Lancet 366(9486):662–664
Valerio A, Nisoli E (2015) NO to slow ageing. Front Cell Dev Biol 3:6
Valina C et al (2007) Intracoronary administration of autologous adipose tissue-derived stem cells improves left ventricular function, perfusion, and remodelling after acute myocardial infarction. Eur Heart J 28(21):2667–2677
Vermulst M et al (2008) Quantification of random mutations in the mitochondrial genome. Methods 46(4):263–268
Villeda SA et al (2011) The ageing systemic milieu negatively regulates neurogenesis and cognitive function. Nature 477(7362):90–94Aug 31
Wallace LA (1989) Major sources of benzene exposure. Environ Health Perspect 82:165–169
Wallace DC (2005) Mitochondria and cancer: Warburg addressed. Cold Spring Harb Symp Quant Biol 70:363–374
Wang F et al (2001) Preliminarily functional analysis of a cloned novel human gene ADAM29. Sci China C Life Sci 44(4):392–399
Wei JY (1992) Age and the cardiovascular system. N Engl J Med 327(24):1735–1739
Weiss S et al (1996) Multipotent CNS stem cells are present in the adult mammalian spinal cord and ventricular neuroaxis. J Neurosci 16:7599–7609
WHO (2011) WHO, US National Institute of Aging: Global Health and Ageing.
Yilmaz G et al (2010) Induction of neuro-protective/regenerative genes in stem cells infiltrating post-ischemic brain tissue. Exp Transl Stroke Med 2:11
Yokoo T et al (2008) Kidney organogenesis and regeneration: a new era in the treatment of chronic renal failure? Clin Exp Nephrol 12:326–331
von Zglinicki T, Martin-Ruiz CM (2005) Telomeres as biomarkers for ageing and age-related diseases. Curr Mol Med 5:197–203
Zhang ZG et al (2002) Angiopoietin-1 reduces cerebral blood vessel leakage and ischemic lesion volume after focal cerebral embolic ischemia in mice. Neuroscience 113:683–687
Zhang X et al (2013a) Environmental and occupational exposure to chemicals and telomere length in human studies. Occup Environ Med 70:743–749
Zhang GL et al (2013b) Hypothalamic programming of systemic ageing involving IKK-beta, NF-kappaB and GnRH. Nature 497:211–216. [PubMed: 23636330]

Index

P.V. Pham (ed.), *Pancreas, Kidney and Skin Regeneration*, Stem Cells in Clinical Applications, DOI 10.1007/978-3-319-55687-1

MIX
Papier aus verantwortungsvollen Quellen
Paper from responsible sources
FSC® C105338

If you have any concerns about our products,
you can contact us on
ProductSafety@springernature.com

In case Publisher is established outside the EU,
the EU authorized representative is:
Springer Nature Customer Service Center GmbH
Europaplatz 3, 69115 Heidelberg, Germany

Printed by Libri Plureos GmbH
in Hamburg, Germany